Polymerization Kinetics
and Technology

Polymerization Kinetics and Technology

Norbert A. J. Platzer, *Editor*

A symposium co-sponsored by the Division of Industrial and Engineering Chemistry and the Division of Polymer Chemistry at the 163rd Meeting of the American Chemical Society, Boston, Mass., April 10–14, 1972.

ADVANCES IN CHEMISTRY SERIES **128**

AMERICAN CHEMICAL SOCIETY

WASHINGTON, D. C.   1973

ADCSAJ 128 1-288

Copyright © 1973

American Chemical Society

All Rights Reserved

Library of Congress Catalog Card 73-91733

ISBN 8412-0188-9

PRINTED IN THE UNITED STATES OF AMERICA

Advances in Chemistry Series
Robert F. Gould, *Editor*

Advisory Board

Bernard D. Blaustein

Paul N. Craig

Ellis K. Fields

Edith M. Flanigen

Egon Matijević

Thomas J. Murphy

Robert W. Parry

Aaron A. Rosen

Charles N. Satterfield

FOREWORD

ADVANCES IN CHEMISTRY SERIES was founded in 1949 by the American Chemical Society as an outlet for symposia and collections of data in special areas of topical interest that could not be accommodated in the Society's journals. It provides a medium for symposia that would otherwise be fragmented, their papers distributed among several journals or not published at all. Papers are refereed critically according to ACS editorial standards and receive the careful attention and processing characteristic of ACS publications. Papers published in ADVANCES IN CHEMISTRY SERIES are original contributions not published elsewhere in whole or major part and include reports of research as well as reviews since symposia may embrace both types of presentation.

CONTENTS

Preface ix

1. Elementary Reactions in Radical and Anionic Polymerizations 1
   G. V. Schulz

2. Monomer Constitution and Stereocontrol in Free Radical Polymerizations 21
   H-G. Elias, P. Goeldi, and B. L. Johnson

3. Design of Large Polymerization Reactors 37
   G. Beckmann

4. Behavior of Viscous Polymers during Solvent Stripping or Reaction in an Agitated Thin Film 51
   F. Widmer

5. Mechanochemical Polycondensations and Polycomplexations 68
   C. Simionescu and C. Vasiliu-Oprea

6. Branching and Crosslinking in Styrene–Butadiene Polymerizations 102
   G. M. Burnett and G. G. Cameron

7. Popcorn Polymers 110
   J. W. Breitenbach and H. Axmann

8. Polyether Modifiers for Polyvinyl Chloride and Chlorinated Polyvinyl Chloride 125
   P. Dreyfuss, M. P. Dreyfuss, and H. A. Tucker

9. Developments in Vinyl Chloride Graft Copolymers 135
   F. Wollrab, J. Dumoulin, F. Declerck, P. Georlette, and M. Obsomer

10. The Copolymerization of Tetrachloroethylene and Ethylene 156
    H. Hopff and N. Balint

11. Radiation-Induced Chlorination of Polyisobutene 161
    C. Schneider and P. Lopour

12. Some Aspects of Vinyl Ester Emulsion Polymerization 170
    M. Litt, V. T. Stannett, and E. Vanzo

13. One-Step Synthesis of Cured Polyester 176
    R. D. Deanin and V. G. Dossi

14. Improved Process for Polycondensation of High-Molecular Weight Poly(ethylene terephthalate) in the Presence of Acid Derivatives . 183
    T. Shima, T. Urasaki, and I. Oka

15. Hexacyanometalate Salt Complexes as Catalysts for Epoxide Polymerizations 208
    R. J. Herold and R. A. Livigni

16. Preparation and Properties of Poly(arylene oxide) Copolymers .. 230
    G. D. Cooper, J. G. Bennett, Jr., and A. Factor

17. Use of Gel Permeation Chromatography to Study the Synthesis of Bisphenol-A Carbonate Oligomers 258
    A. B. Robertson, J. A. Cook, and J. T. Gregory

18. Kinetics and Mechanism of Urethane Formation in DMF. The Reaction of 4,4'-Diphenylmethane Diisocyanate and Alcohols Catalyzed by Dibutyltin Dilaurate 274
    G. Borkent and J. J. Van Aartsen

Index 281

PREFACE

At current projections the world's population will double in the next 30 years. World production of synthetic polymers, however, is doubling every five years according to an annual growth rate of 15.6%. During the 1960's, U.S. polymer production showed an annual gain of 14.5% while the rest of the chemical industry expanded at an annual rate of only 9%. This made synthetic polymers the fastest growing area in chemical industry. During the same period the growth rate of world polymer production exceeded the domestic rate and amounted to an average of 16.6% per year.

The growth rate of polymer production is determined by two factors: demand and technology. Demand depends on economy. In 1970 domestic production stayed at almost the 1969 level, as Figure 1 illustrates. In 1971 it grew 8%. These values represented a significant

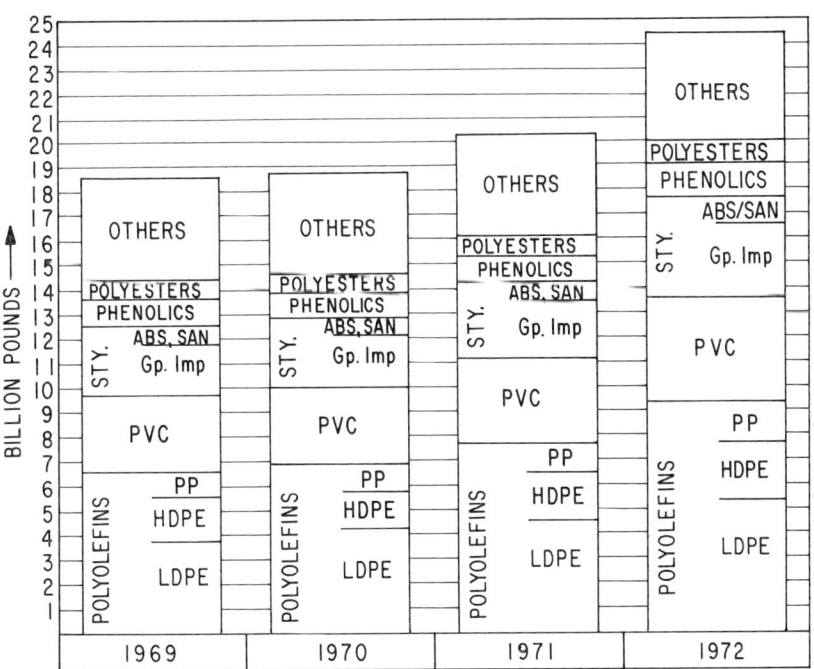

Figure 1. *U.S. production of synthetic thermoplastics and thermosets*

decrease, and the question arose as to whether the polymer industry had reached the plateau of saturation and maturity. The 1972 growth rate of more than 20% provided the answer: the slowdown was caused only by the general recession, and the demand for polymeric materials still exists and is rising.

Polyethylene, polypropylene, poly(vinyl chloride), styrene resins (including ABS), phenolics, and polyesters are the major commercial polymers and together represent over 80% of the total domestic production, also shown in Figure 1). The sharply increased demand in 1972 required manufacturing plants to operate at almost full capacities. Shortages may occur temporarily until new facilities are installed.

In 1950 the United States was the leading polymer manufacturer and contributed to 69% of the world's production. The economic boom in Western Europe and Japan of the 1950's and 1960's raised their productivity and reduced our world participation to 41% in 1960 and 27% today. The seven largest polymer producing countries are United States, Japan, West Germany, Italy, U.S.S.R., Great Britain, and France, as shown in Figure 2.

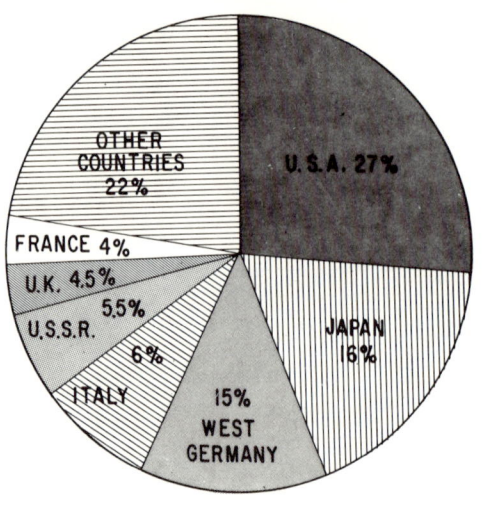

TOTAL 90 BILLION POUNDS

Figure 2. Participation of different countries in world production during 1972

Technology represents the second growth factor and depends on research. Poly(vinyl chloride) is the oldest of the synthetic commodity thermoplastics and has been used plasticized and rigid for 35 years. Its manufacturing technology by suspension and emulsion polymerization is known worldwide. The technology of high pressure polyethylene was developed later. Low pressure processes with Ziegler-Natta or

Phillips catalysts for linear polyethylene and polypropylene and the graft copolymerizations for impact polystyrene and ABS are even younger and have not yet spread into the less industrialized countries of world. The production of polyolefins, poly(vinyl chloride), and styrene resins on a worldwide basis as well as of all synthetic polymers is shown in Figure 3. A comparison of the U.S. production in Figure 1 and in Figure 3 demonstrates the effect of age and dissemination of technology. It shows that relatively more poly(vinyl chloride) but less polyolefins and styrene resins are produced worldwide than in this country.

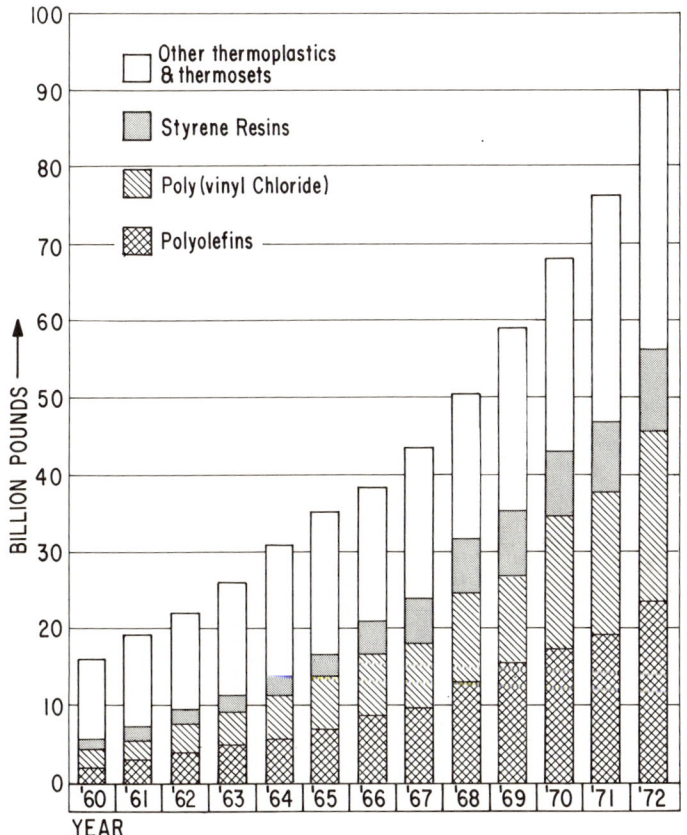

Figure 3. *World production of plastics*

With the growth of the polymer industry, competition has become fiercer and selling prices lower. In view of the competitive situation, priorities of industrial research have changed. It has become imperative to produce the same commodity polymers with better properties at reduced cost by improved processes and in larger units. To stay ahead, it is also necessary to understand the mechanism of polymerization more fully, to modify the existing commercial products, and to develop new ones.

This volume contains 18 papers on the kinetics and technology of addition and condensation polymerization processes. These papers were presented at the sixth symposium on this subject held by the Division of Industrial and Engineering Chemistry and the Division of Polymer Chemistry during the A.C.S. Meeting, Boston April 9-14, 1972. They are concerned with known commercial products. New polymers and novel polymerization reactions presented at the same symposium are collected in the companion volume, ADVANCES IN CHEMISTRY SERIES No. 129.

Polymerization Kinetics

To broaden our overall knowledge of process kinetics the first chapter of this volume deals with elementary reactions in radical and anionic polymerization; it was written by G. V. Schulz, the first recipient of the H. Staudinger Award. It is followed by a discussion on monomer constitution and stereocontrol in radical polymerization by H. G. Elias *et al.*

Equipment and Process Design

For commercial suspension and solution polymerization current batch reactors are generally 2000 to 10,000 gallons in size. G. Beckmann describes the design of larger, more economical polymerization kettles, 53,000 gallons in size, as installed at C. W. Huels, and the problems of heat removal, agitation, cleaning, and safety. The behavior of viscous melts during solvent or residual monomer stripping and the design of wiped-film devolatilizers is covered by F. Widmer. These devolatilizers can be used in mass or solution addition and condensation polymerization processes. Mechanochemical complexing is a novel method for condensation reactions developed by C. Simionescu and C. Vasilu-Oprea.

Addition Polymerization Products

To the synthetic thermoplastics and thermosets of Figure 3, the synthetic rubbers and elastomers may be added. In 1972 they amounted to more than 15 billion pounds worldwide. SBR has been the workhorse of the rubber industry since World War II. Being used in tires, it amounts to over half of all rubber production. G. M. Burnett and G. G. Cameron have investigated the way its crosslinking during polymerization affects its properties and processability. For years it has been the intention to raise the heat distortion temperature of polystyrene. One method is to copolymerize styrene with a small quantity of divinylbenzene. J. W. Breitenbach and H. Axman have studied the formation of popcorn polymers from styrene with less than 1% divinyl benzene.

Rigid impact poly(vinyl chloride) can be made either by polyblending or by grafting. P. Dreyfuss, M. P. Dreyfuss, and H. A. Tucker, discuss in their chapter polyblends of poly(vinyl chloride) with polyethers. F. Wollrab, F. Declerck, J. Dumoulin, M. Obsomer, and P. Georlette review grafting of vinyl chloride upon polyethylene, ethylene/propylene rubber, and chlorohydrin rubber.

H. Hopff and N. Balint developed a copolymerization process for tetrachloroethylene with ethylene. Radiation-induced chlorination of polyisobutylene is the subject of the chapter of C. Schneider and P. Lopour. M. Litt, V. T. Stannett, and E. Vanzo show that the polymerization of vinyl caproate follows the kinetics of styrene.

Condensation Polymerization Products

Unsaturated polyesters, primarily copolyesters of propylene glycol with maleic and phthalic acid, dissolved in styrene monomer, are used in the glass-fiber reinforced polyester market. They are cured commercially in a series of involved and expensive steps. R. D. Deanin and V. G. Dossi propose a simplified one-step curing method. Polyethylene terephthalate is manufatcured by a condensation reaction in which ethylene glycol is liberated. Diffusion and removal of the glycol from the viscous melt determine and decelerate the reaction rate. T. Shima, T. Urasaki, and I. Oka discovered that addition of a small amount of certain acid derivatives accelerates the condensation polymerization significantly, yielding a high molecular weight product of low free-acid content.

Propylene oxide is one of the raw materials used to manufacture rubbery and crystalline polyepoxides. R. J. Herold and R. A. Livigni describe propylene oxide polymerization with hexacyanometalate salt complexes as catalyst. Polyphenylene oxide is made by copper catalyzed oxidative coupling of 2,6-dimethylphenol. G. D. Cooper, J. G. Bennett, and A. Factor discuss the preparation of copolymers of PPO by oxidative coupling of dimethylphenol with methylphenylphenol and with diphenylphenol.

At our 1958 symposium, polycarbonate was introduced. Since then this polymer has reached a domestic production of 50 million pounds per year. J. T. Gregory, J. A. Cook, and A. B. Robertson report on the synthesis of polycarbonate oligomers useful in making random and block copolycarbonates.

Polyurethanes are manufactured by the reaction of diisocyanates with diols, diamines, or other organic compounds containing two or more active hydrogens. The reaction rate between a diisocyanate and alcohols catalyzed with dibutyltin dilaurate yielding urethanes was studied by G. Borkent and J. J. Van Aartsen.

The chapters in this volume will broaden our insight into polymerization kinetics and provide information technologically important on commercial processes and products. New polymers and novel polymerization reactions are covered in the 18 chapters of the companion volume.

NORBERT A. J. PLATZER

Longmeadow, Mass.
December 1972

Elementary Reactions in Radical and Anionic Polymerizations

G. V. SCHULZ

Institut für Physikalische Chemie der Universität Mainz, West Germany

> *Studies of the influence of solvents on polymerization processes by stationary and nonstationary methods give the following results. In radical polymerizations, solvent viscosity mainly influences diffusion-controlled elementary reactions. The effect on efficiency demands new considerations of the cage effect. The influence on chain termination provides interesting information on segment diffusion, mutual penetration of polymer coils, and internal viscosity. In anionic polymerizations, the polarity of solvents strongly influences equilibria between three forms of the active end groups—the two kinds of ion pairs and the free carbanion. The equilibria between these forms and the corresponding enthalpy and entropy differences were determined by kinetic and conductance measurements. The influence of solvents on the rate constants of the monomer addition to each form of the active chain ends is relatively small. Analysis of molecular weight distributions gives, in addition the activation enthalpies and entropies of the transitions between these forms.*

Solvents have different effects on polymerization processes. In radical polymerizations, their viscosity influences the diffusion-controlled bimolecular reactions of two radicals, such as the recombination of the initiator radicals (efficiency) or the deactivation of the radical chain ends (termination reaction). These phenomena are treated in the first section. In anionic polymerization processes, the different polarities of the solvents cause a more or less strong solvation of the counter ion. Depending on this effect, the carbanion exists in three different forms with very different propagation constants. These effects are treated in the second section. The final section shows that the kinetics of the transitions between these three forms of active chain ends can be measured by analyzing molecular-weight distributions on anionically prepared polystyrenes as a function of solvent and temperature.

Influence of Solvents on Radical Polymerization

The fundamental processes of initiation, propagation, and termination of a polymer chain can be formulated this way (transfer reactions are not considered):

Initiation
$$R_2 \xrightarrow{k_d} R \cdot$$
$$\left. \begin{array}{l} 2R \cdot \to R'_2 \\ R \cdot + M \to RM \cdot \end{array} \right\} \text{efficiency } f$$

Propagation $\quad RM_n \cdot + M \xrightarrow{k_p} RM \cdot_{n+1}$

Termination $\quad RM_n \cdot + RM_m \cdot \xrightarrow{k_t} \text{dead polymer}$

The least known of these reactions is chain initiation together with the efficiency of the primary radicals. This group consists of at least three elementary reactions. The so-called cage effect certainly plays an important role. On the other hand, the cage effect in its classical form cannot explain all phenomena sufficiently.

The propagation step is a special case of the reaction of a radical with an unsaturated compound. The process can be influenced by EDA-complexes of the radical with the solvent and the monomer, as Henrici-Olivé (1) have shown. These effects are not very large.

In contrast to the propagation step, the termination process strongly depends on solvent properties. As Dainton and North (2, 3) have shown, viscosity plays the most important role. The values of k_t (10^6 to 10^8 1/mole s) are additional proof for the termination-step as a diffusion-controlled reaction.

The data on k_p and k_t as reported in the literature differ considerably. Therefore, we conducted new studies on methyl methacrylate (MMA), benzyl methacrylate (BMA), and styrene (St) as monomers. The constants were obtained by applying the method of intermittent illumination (rotating sector) combined with stationary state methods. The viscosity of the solvents varied between 0.5 and 100 cP. No mixed solvents composed of low- and high-molecular components were used but pure solvents only, the molecules of which did not deviate very much from a spherical form (methyl formate, diethyl phthalate, diethyl malonate, dimethyl glycol phthalate, etc.).

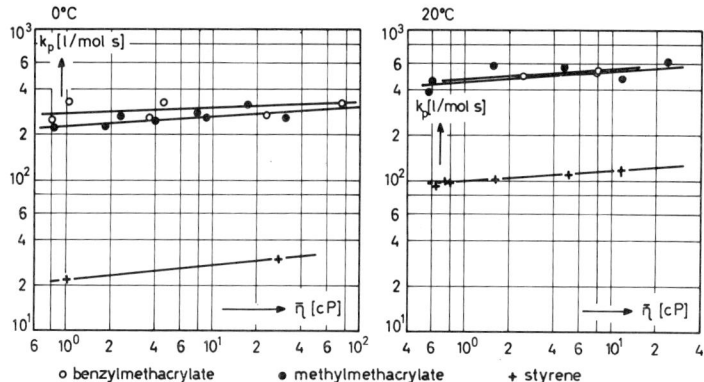

Figure 1. Propagation rate constants of three monomers as a function of solvent viscosity (including the monomer)

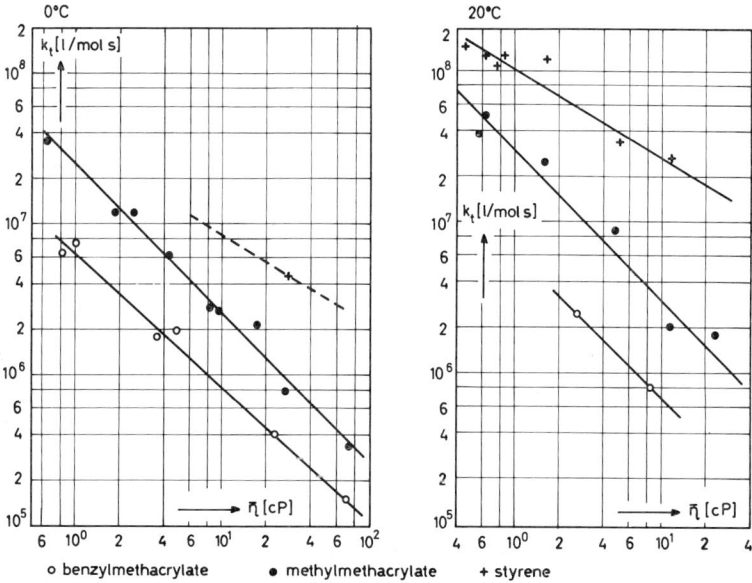

Figure 2. Termination rate constants of three monomers as a function of solvent viscosity

The result is shown in Figures 1 and 2 (4-7). In all cases, the propagation constant shows a small increase with increasing viscosity; the increase hardly exceeds the experimental error. We have no definite explanation for this small effect, which has been observed by other authors, too (8).

Figure 2 shows the strong dependence of the termination rate constant on the viscosity η of the solvent. For methyl methacrylate, k_t is exactly inversely proportional to η; the two other monomers depend less on η. In any case, this dependence leads to the conclusion that the termination is a diffusion-controlled reaction.

Diffusion of the macroradicals controls can be assumed to be the termination reaction. However, that is not the case; the termination rate constant is absolutely independent of the degree of polymerization, as shown in Table I. Therefore, the assumption must be that the diffusion of the segment at the end of the radical chain controls the termination process (as long as the Trommsdorff effect is not rate-determining).

Table I. Dependence of the Termination Rate Constant on Degree of Polymerization.

$DP \times 10^{-3}$	$k_t \times 10^{-7}$	$DP \times 10^{-3}$	$k_t \times 10^{-7}$
Methyl methacrylate in Diethyl phthalate at 0°C (4)			
3.24	1.03	17.2	0.98
5.50	1.00	23.3	1.01
8.75	0.98	30.0	1.21
11.6	0.98	33.3	1.03
15.0	1.00	62.8	0.98
Styrene in Bis(ethylhexyl phthalate) at 0°C (6)			
0.89	0.46	1.85	0.48
1.39	0.49	3.45	0.42

This assumption has been made by several authors (2, 3, 8, 10, 11); but the theoretical models proposed do not sufficiently account for the complete independence of the termination rate constant from the degree of polymerization. Therefore, we have developed a model of the termination process; the model seems to agree better with the experimental results (12). We assume that two encountering radical polymer chains partly permeate each other. So an overlapping volume builds up with a certain lifetime. During this overlapping time, the radical end groups can diffuse into that volume and terminate each other by combination or disproportionation (Figure 3).

Use of Smoluchowski's model gives these results:

(1) The diffusion of polymer chain radicals has no influence on the termination rate. This is because the frequency of encounter decreases—by decreasing diffusion constant—to the same extent as the lifetime of the permeation state increases.

(2) The probability of a termination step during an encounter depends on the lifetime and degree of overlapping; on the diffusion constant D_s of the segment of the radical chain end; and on a steric factor α, of the order of 10^{-2}. One obtains:

$$k_t = \text{const } D_s r_{12} \alpha \tag{1}$$

(const = 1.51×10^{22}, r_{12} = distance of encounter)

What is the size of the overlapping volume? The complete independence of the constant in Equation 1 from the degree of polymerization shows that the overlapping volume always consists of the same portion of the volume of the polymer coil (12). This can be easily understood by assuming that two polymer coils are able to migrate nearly unhindered through each other. Then the mean depth of permeation and, therefore, the time of overlapping is determined only by the statistics of the free Brownian motion. Equation 1 is based on this assumption.

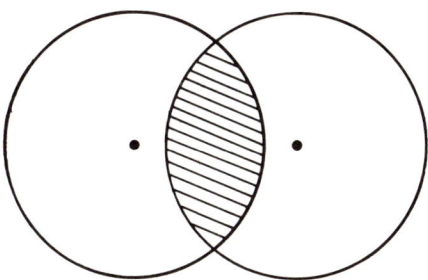

Figure 3. Overlapping volume of two radical polymer chains represented by equivalent spheres

The calculation proceeds as follows (13). First, we determine the frequency of encounter of two polymer chains following Smoluchowski's treatment. After two chains have come into contact with each other, we allow them to move freely in all directions so we obtain a mean volume and a mean lifetime of the overlapping. This mean volume is given by about 1/10 of the diameter of the polymer coil. This treatment is illustrated in Figure 4.

The overlapping occurs within a distance defined by 2ξ (depending on the angle γ). Taking the mean of ξ over all angles γ gives a mean overlapping volume.

The segmental diffusion constant is given by Einstein's formula (Equation 2):

$$D_s = \frac{kT}{6\pi\eta r_s} \tag{2}$$

Stokes' radius of a polymer segment = r_s. Applying our experimental data to Equation 2, we obtained $r_s \approx 10$ A. This value seems to be quite reasonable. On the other hand, we can set D_s in relation to experimentally determined diffusion constants of a certain polymer. This enables us to attribute a molecular weight of the segment to D_s. For MMA, we obtain $M_{s,\text{MMA}} \approx 2000$.

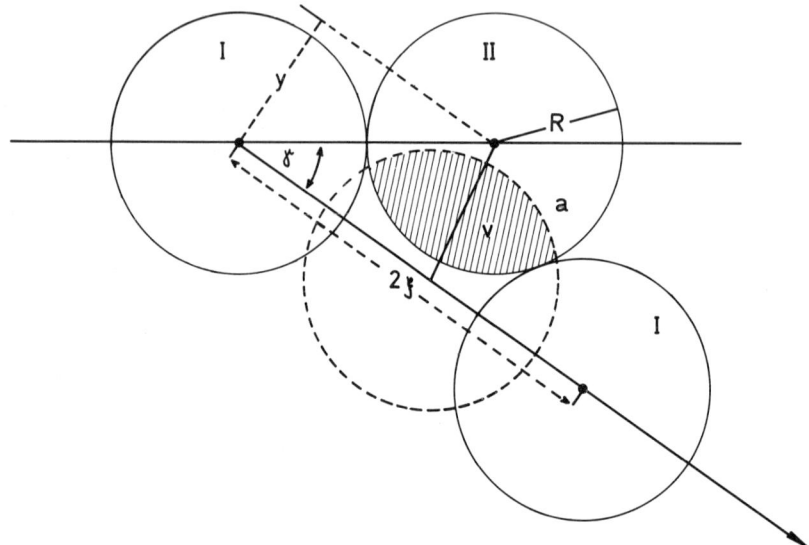

Figure 4. Model of encounter of two polymer coils represented by equivalent spheres; arrow indicates direction of diffusion of coil I relative to that of coil II after both coils have met.

Equation 2 can be checked by drawing a log-log-plot of the experimental data, obtained at different temperatures in various solvents, as shown in Figure 5. In the case of a pure segmental diffusion process, we find a straight line of slope 1. Figure 5 shows that this holds for MMA over a wide range of temperatures and viscosities.

This scheme does not hold for benzyl methacrylate and even less for styrene; k_t of these monomers is no longer inversely proportional to η. Maybe the theory of the so-called inner viscosity can help (as developed by Kuhn and Kuhn around 1950). According to that theory, the movement of a segment in a solvent of viscosity η_s is overlaid by an inner viscosity η_i, so that

$$\frac{1}{k_t} = C(\eta_i) + \text{const} \frac{\eta_s \times r_s}{r_{12} \alpha k T} \qquad (3)$$

holds instead of Equation 1; η_i is a constant depending on temperature and is characterized by the potential of the rotational movement of the segment. It is the object of our present work to find a suitable form of Equation 3 that corresponds to our data.

The above conclusions are valid only as long as the frequency of the encounter of the macromolecules is greater than the frequency of the termination reaction. That is the case at low degrees of conversion or in solution polymerization. With increasing conversion, the diffusion of the polymer radicals is increasingly hindered; then the frequency of encounter decreases and the lifetime of the overlapping increases accordingly. Finally, each encounter produces a termination reaction, and from that moment, the diffusion control passes from the segment to the polymer radical. This is the onset of the Trommsdorff-effect (gel effect).

Such investigations of elementary reactions lead to fundamental questions on the inner mobility of polymer molecules, their mutual penetration, and on general diffusion problems not yet solved.

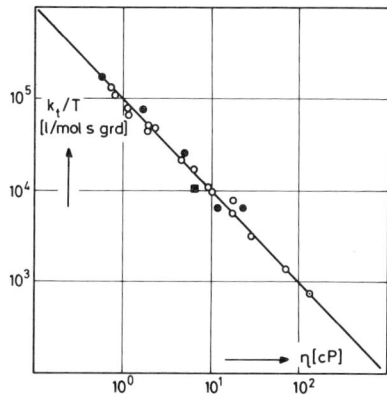

Figure 5. Segmental diffusion constant depending on temperature and solvent viscosity; data are calculated according to Equation 1 and the graph according to Equation 2

In the study of efficiency, another fundamental problem in radical kinetics comes up. The attempt is to use the so-called cage effect to explain the phenomenon that not all radicals produced start a reaction chain. One portion of the radicals generated in pairs in the solvent cage recombine there, while the other portion, which passes through the cage, starts a reaction chain. The higher the viscosity of the solvent, the more restricted is the movement out of the cage and, therefore, the lower the efficiency. This is true in a qualitative sense, as Figure 6 shows, but the quantitative results cannot be understood in terms of the theories so far developed. The strong influence of the type of monomer is especially strong evidence against the cage effect in its classical form. A portion of the radicals that escaped the cage will be necessarily deactivated by a twin termination reaction (4, 6, 7).

Influence of Solvents in Anionic Polymerization

Anionic polymerization differs from radical polymerization in that no chain termination of the propagating polymers with each other occurs ("living polymers"). Furthermore, the rate constant of the propagation k_p is not so high that this process is controlled by diffusion.

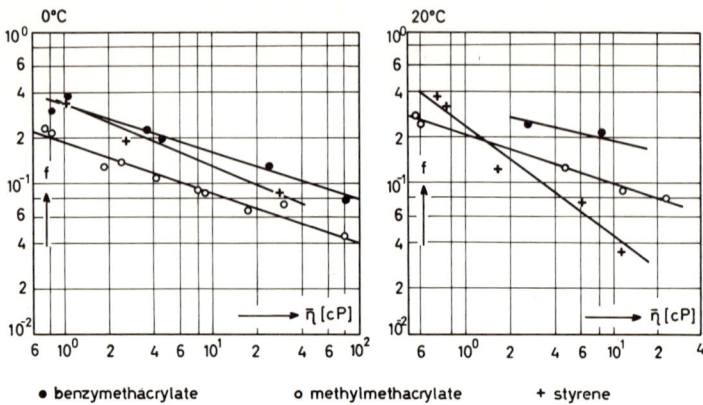

Figure 6. Efficiency as a function of solvent viscosity for three monomers (initiator is 1,1'-azo-biscyclohexanenitrile)

Nevertheless, the solvent shows a significant influence, mainly because of two reasons:

(a) The solvent enables the living polymer end to exist in different species that react with different propagation constants and whose concentrations are determined by a chemical equilibrium.

(b) The solvent may cause side reactions of the reactive carbanionic groups, resulting in an isomerization and deactivation of the "living end" before and during the polymerization process.

These effects of the side reactions are discussed in more detail elsewhere (15, 16).

The question of greater interest is: How can the solvent influence the state and the reactivity of the carbanionic end group?

Under good experimental conditions, initiation and chain termination within 1 msec can be achieved, the latter one by adding a proton donator. It is possible by developing the flow technique (17, 18) and by improving the normal reaction vessel technique (batch reaction) to measure the polymerization rate as a pure propagation rate. In the experimental results given here, the polymerization exactly follows the first order as to the monomer concentration, giving, for each experiment, the propagation rate constant according to:

$$\bar{k}_p = \frac{1}{c^* t} \ln \frac{[M]_o}{[M]} \qquad (4)$$

(t = reaction time, $[M]_o$ = initial monomer concentration, $[M]$ = monomer concentration after termination, c^* = concentration of the living ends).

Determination of the "living" end concentration c^* can be carried out by measuring the conversion and the degree of polymerization (17) or by spectroscopy (19-21).

The constant k_p strongly depends on the solvent and on the concentration c^* of the living ends. This concentration dependence is easy to understand, because the dissociation of weak electrolytes depends on the concentration according to Ostwald's law of dilution. When the ion pairs have a lower propagation rate constant than the free anions, $\bar{k}_p$ has two terms $\bar{k}_{(\pm)}$ and $k_{(-)}$. The first problem now is to determine the dissociation constant K_{Diss} and the two propagation rate constants. This is achieved by varying c^* and the gegen ion (e.g., Na^+), respectively. Applying the mass law, we obtain the following two equations (22-24)

$$\bar{k}_p = k_{(\pm)} + k_{(-)}(K_{Diss}/c^* f_\pm^2)^{1/2} \tag{5a}$$

and

$$\bar{k}_p = k_{(\pm)} + k_{(-)}(K_{Diss}/Na^+ f_\pm^2). \tag{5b}$$

The concentration of Na^+ may be changed by adding $Na\ BPh_4$ (Ph = phenyl). The plot according to equations 5a and 5b gives $k_{(\pm)}$ as intercept and $k_{(-)}(K_{Diss})^{1/2}$ and $k_{(-)}K_{Diss}$ respectively, for the slope. Figure 7 shows such a graph for polystyrylsodium in THF as an example (25).

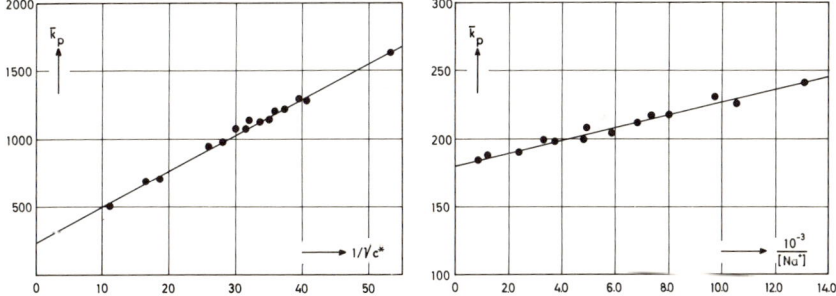

Figure 7. Propagation rate constant k_p of polystyrylsodium in THF plotted according to Equations 5a and 5b; T = 27°C

The value of the dissociation constant is confirmed by conductivity measurements (26).

Kinetic measurements in various solvents show that addition of the monomer to the free anion depends only on the temperature and not (or practically not) on the solvent; see Figure 8. The Arrhenius plot gives a straight line with the following (average) values of the parameters:

$$\log k_{(-)} = 8.0 - 3.9/(4.57T) \tag{6}$$

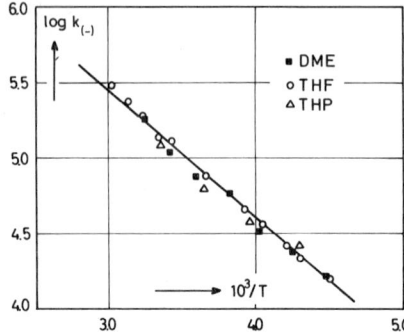

Figure 8. Arrhenius plot of the rate constants for the monomer addition to the free anion of polystyrylsodium in DME, THF, and THP (19, 25, 27)

The gegen ion has no influence on $k_{(-)}$ (28, 29).

The rate constant $\bar{k}_{(\pm)}$ for the monomer addition to the ion pair can be relatively easily determined in different ways: by extrapolation of Equations 5a and 5b (Figure 5a, b), or by kinetic measurements of the polymerization where the dissociation of the ion pairs is completely suppressed by the addition of a large enough excess of Na^+ ions. If the so-measured constants $\bar{k}_{(\pm)}$ are plotted according to Arrhenius' equation, the pattern shown in Figure 9 is obtained for five solvents of different dielectric constants.

This behavior can be understood by the assumption that two different types of ion pairs exist in a thermodynamic equilibrium and add the monomer with different rate constants. The less reactive contact ion pair (tight-ion pair) and the more reactive solvent separated ion pair (loose-ion pair), which is more stable at lower temperatures (*31*). The curved lines show the transition from one ion pair species to the other. Thus, the polymerization mechanism can be described by this scheme:

$$P_x^- Na^+ + mS \underset{K_{cs}}{\rightleftarrows} P_x^-(Sm\ Na^+) \underset{K^*_{Diss}}{\rightleftarrows} P_x^- + (Sm\ Na^+)$$

$$+M \downarrow k^{(\pm)}_c \qquad +M \downarrow k_{(\pm)s} \qquad +M \quad k_{(-)}$$

$$P_{x+1}^- Na^+ + mS \underset{K_{cs}}{\rightleftarrows} P_{x+1}^-(Sm\ Na^+) \underset{K^*_{Diss}}{\rightleftarrows} P_{x+1}^- + (Sm\ Na^+)$$

where P_x^- is a "living" polymer with DP = x; S = solvent, and M = monomer.

When we suppress the dissociation of the "living" ends by an excess of a dissociable sodium salt, the free anion cannot participate in the propagation, and the process is governed by the two species of ion pairs (two-way mechanism). In this case, this equation is ob-

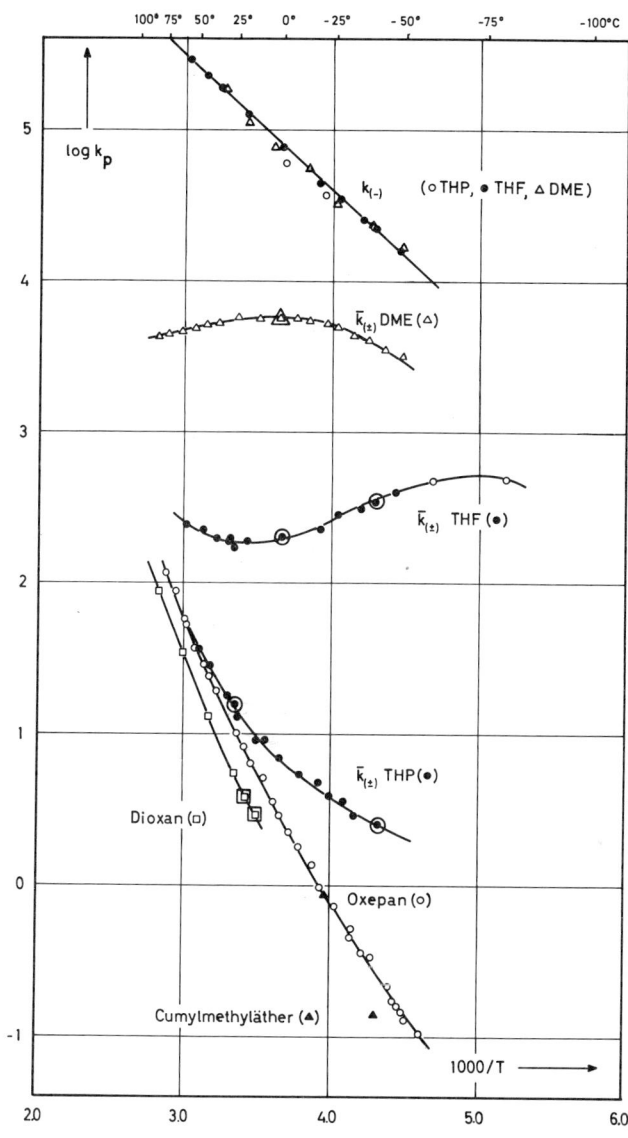

Figure 9. Arrhenius plot of rate constants for monomer addition to the ion pairs in five solvents of different polarity: DME (27), THF (25), THP (20), oxepane (30), and dioxane (21)

tained (the experimentally determined rate constant $\bar{k}_{(\pm)}$ of the ion pairs has to be split into two terms for the contribution of each ion pair species to the propagation):

$$\bar{k}_{(\pm)} = \frac{k_{(\pm)c} + k_{(\pm)s} K_{cs}}{1 + K_{cs}} \qquad (7)$$

The temperature dependence of $\bar{k}_{(\pm)}$ is given by two Arrhenius equations and one van't Hoff relation:

$$\log k_{(\pm)c} = A_c - E_c/4.57T \qquad (8)$$

$$\log k_{(\pm)s} = A_s - E_s/4.57T \qquad (9)$$

$$\log K_{cs} = \Delta S_{cs}/4.57 - \Delta H_{cs}/4.57T \qquad (10)$$

Considering Equations 7-10, the following expression is obtained for the experimentally determined rate constant of the ion pairs:

$$\log \bar{k}_{(\pm)} = \log[\exp(A_c - E_c/4.57T) + \exp(A_s + \Delta S_{cs}/4.57 - (E_s + \Delta H_{cs})/4.57T] - \log[1 + \exp(\Delta S_{cs}/4.57 - \Delta H_{cs}/4.57T)] \qquad (11)$$

When it is possible to measure $\bar{k}_{(\pm)}$ over a large enough temperature range, all parameters of Equations 8-10 can be calculated using the maximum, the minimum, and the turning point of the curve $\log k_{(\pm)} = f(1/T)$. When the temperature range is smaller, the "best" values for

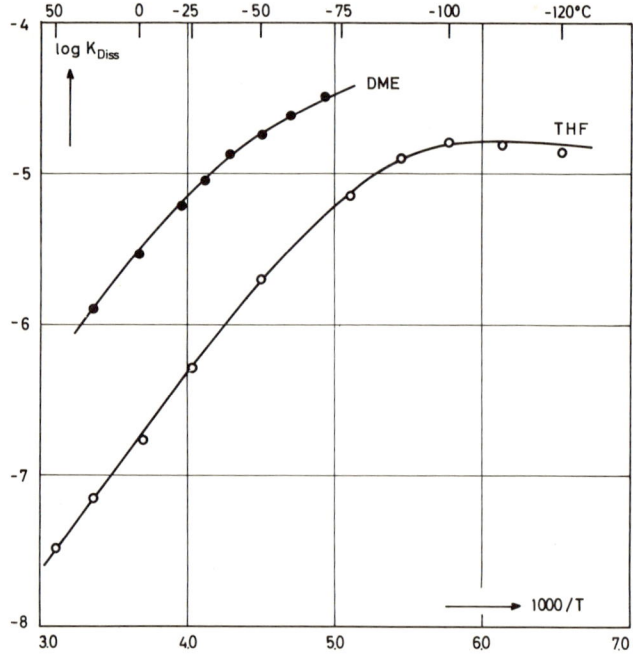

Figure 10. Dissociation constants of polystyrylsodium in dimethoxyethane and in tetrahydrofuran as functions of temperature (26)

the enthalpy and for the entropy can be approximated by using the least squares method.

A different and independent approach to determine the parameters of K_{cs} is made by conductivity measurements. The dissociation constant of carbanionic compounds displays no simple temperature dependence according to the van't Hoff equation as Worsfold and Bywater (*32*) and Szwarc (*33*) have already shown. When we assume two types of ion pairs, the experimentally measured dissociation constant is given by

$$K_{Diss} = \frac{K^*_{Diss} K_{cs}}{(1 + K_{cs})} \qquad (12)$$

K_{Diss} is controlled by two equilibrium-constants with two values for the enthalpy and the entropy, respectively:

$$\log K_{cs} = \Delta S_{cs}/4.57 - \Delta H_{cs}/4.57T \qquad (13)$$

$$\log K^*_{Diss} = \Delta S^*_{Diss}/4.57 - \Delta H^*_{Diss}/4.57T \qquad (14)$$

Therefore, the function $\log K_{Diss} = f(1/T)$ is a curve. The four parameters of Equations 13 and 14 can be calculated from this curvature (*34*). Figure 10 shows experimental points of two series measured at our institute (*26*).

The thermodynamic parameters of the equilibrium between the two ion pair species and those for the dissociation of the solvent separated ion pair, obtained from kinetic and conductivity measurements, are listed in Table II. The agreement of the values determined by the two independent methods is quite satisfactory.

Furthermore, the Arrhenius parameters can be calculated for the monomer addition to the different ion pairs by evaluating the experi-

Table II. **Enthalpy and Entropy of Transitions from Contact Ion Pair to Solvent-Separated Ion Pair, and of Dissociation of Solvent-Separated Ion Pair (K^*_{Diss}) of Polystyrylsodium**

Solvent	ΔH_{cs} kcal/mol		ΔS_{cs} eu		ΔH^*_{Diss} kcal/mol		ΔS^*_{Diss} eu	
	a[a]	b	a	b	a	b	a	b
Dimethoxiethane (*27*)	−5.5	−5.3	−22.5	−22	—	−1,2	—	−26
Tetrahydrofuran (*25*)	−6.5	−6.1	−34	−32	±0	+0.1	−26	−21
Tetrahydropyran (*19*)	−3.0	—	−28	—	±0.3	—	−27	—

[a] From kinetic measurements
[b] From conductivity measurements

mental results shown in Figure 9. Table III shows that the solvent has only a small influence on these parameters.

Table III. Arrhenius Parameters of Monomer Addition to Contact Ion Pair and to Solvent-Separated Ion Pair (log k = A − E/4.57 T)

Solvent	A_c	E_c kcal/mol	A_s	E_s kcal/mol
DME (27)	7.8	9.2	7.8	4.2
THF (25)	7.8	8.6	8.3	4.7
THP (19)	8.1	9.7	8.0	4.5
Oxepane (30)	8.2	9.8	—	—
Dioxane (21)	8.4	10.5	—	—

These results show that the big differences in the polymerization rate in the various solvents are caused mainly by the position of the equilibria, and only to a small extent by the direct interaction of the solvent. Figure 11 shows the equilibria as functions of temperature, calculated with the parameters of Table II and the corresponding rate constants for monomer addition.

Rate Constants of the Transitions Between the Three Forms of the Carbon-Sodium-Bond

To complete the three-way mechanism, the equilibrium constants are replaced by the corresponding rate constants:

$$K_{cs} = k_{cs}/k_{sc} \quad \text{and} \quad K^*_{\text{Diss}} = k_d/k_a \qquad (15a, b)$$

This gives:

$$P_x^-\text{Na}^+ + mS \underset{k_{sc}}{\overset{k_{cs}}{\rightleftharpoons}} P_x^-(mS\,\text{Na})^+ \underset{k_a}{\overset{k_d}{\rightleftharpoons}} P_x^- + (Sm\,\text{Na})^+$$

$$\downarrow k_{(\pm)c} \qquad\qquad \downarrow k_{(\pm)s} \qquad\qquad \downarrow k_{(-)}$$

The rate constants of these transitions can be estimated by determining the molecular-weight distribution of anionic polymer samples prepared under well-defined kinetic conditions. When all polymer molecules are formed under equal conditions during an equal time of propagation, a very narrow molecular weight distribution, a so-called Poisson distribution, is produced (Flory).

A broadening of the distribution occurs when the polymerization proceeds *via* different ways, and when the rate constants of the transitions are comparable to or smaller than the propagation rates.

To treat these effects quantitatively, it is useful to introduce a new parameter, the ununiformity (36):

$$U = (\overline{P}_w/\overline{P}_n) - 1 \tag{16}$$

where $\overline{P}_w$ and $\overline{P}_n$ are the normal average values of the degree of polymerization.

A one-way-mechanism produces the Poisson distribution with

$$U_{Pois} = 1/\overline{P}_n \qquad (\overline{P}_n \gg 1) \tag{17}$$

With a two- or three-way mechanism, the experimentally determined ununiformity is larger: $U_{exp} \geq 1\overline{P}_n$. The individual contributions to the value of U, which are caused by several simultaneous processes, can be summed to a first approximation.

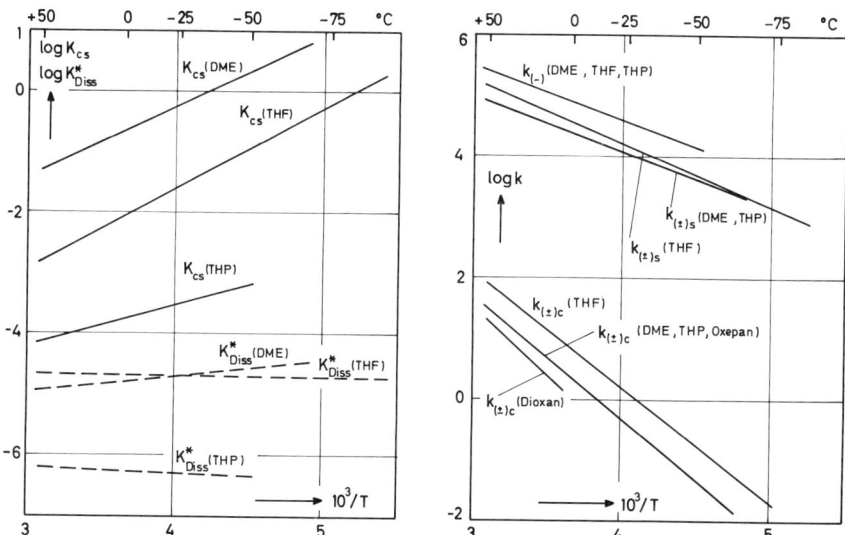

Figure 11. van't Hoff plots of the equilibrium constants K_{cs} and K_{Diss} of polystyrylsodium in three solvents and Arrhenius lines of the propagation constants of the three species

When the polymerization proceeds via two different ion pairs, the experimentally measured value of U has two (additive) terms:

$$U_{exp} = U_{Pois} + U_{sc} \tag{18}$$

In the case where the free anion also contributes to the propagation, a further broadening of the molecular-weight distribution results, and U_{exp} has to split into three terms:

$$U_{exp} = U_{Pois} + U'_{sc} + U_{Diss} \tag{19}$$
$$(U_{sc} \neq U'_{sc}).$$

These effects are shown in Figure 12. At higher temperatures in the presence of excess Na⁺, only the contact ion pair exists. The polym-

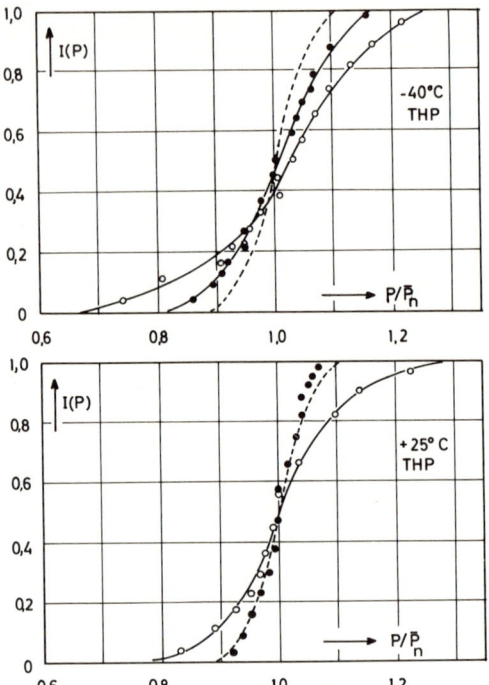

Figure 12. Molecular weight distribution of polystyrene prepared anionically in THP

- - - - - - - - Poisson distribution (one-way)

———●——— Addition of 4×10^{-3}M NaBPh$_4$ (two-way)

———○——— Without additional Na⁺ (three-way)

erization then is a one-way mechanism, and, consequently a Poisson distribution is produced in a good approximation. When no sodium salt is added, the "living" ends dissociate to a very small extent. Nevertheless, the result is a significant broadening of the distribution, because the propagation rate constant of the free anion is about 10^4 times higher than that of the contact ion pair. At lower temperatures, two species of ion pairs exist; their equilibrium cannot be influenced by adding a sodium salt. Therefore, in any case, an additional broadening is obtained. When no salt is added, a further broadening occurs.

The values of U_{sc} and U_{Diss}, which can be determined from the distribution curves, are approximately related to the rate constants for the transition, according to Figini (37) and Böhm (38), by the following equations:

$$k_{sc} = \frac{k_{(\pm)s}^2 K_{cs} c^*}{U_{sc} \overline{k}_{(\pm)}} \frac{2 - x_p}{2x_p} \qquad (20)$$

and

$$k_a = \frac{c^{*1/2}}{U_{\text{Diss}} k_p (K_{\text{Diss}} * K_{cs})^{1/2}} \left[k_{(\pm)}^2 K_{cs} + k_{(-)}^2 \left(\frac{K_{\text{Diss}} * K_{cs}}{c^*} \right)^{1/2} \right] : \frac{2 - x_p}{2x_p} \qquad (21)$$

where x_p = degree of conversion; the equation is valid for bifunctional polymer chains.

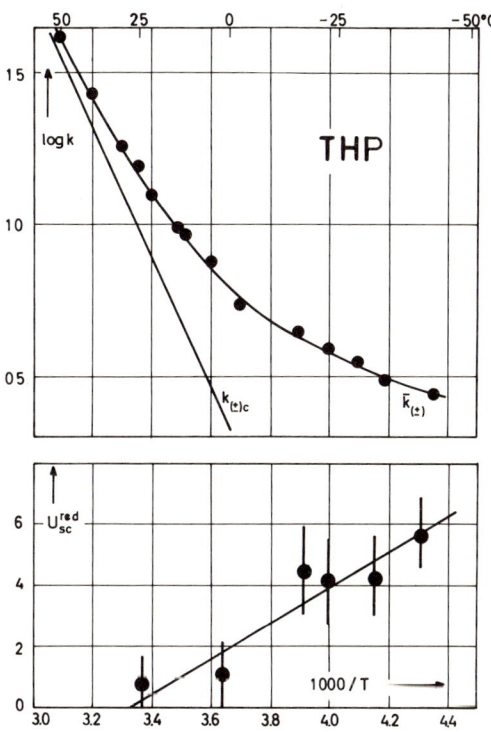

Figure 13. A: —•— $\overline{k}_{(\pm)}$ as a function of temperature (measured points); ——— Arrhenius line of the contact ion pair. B: experimentally determined U_{sc} (normalized)

Figure 13 shows the increase of U with decreasing temperature, because the equilibrium is shifted in the direction of the solvent-separated ion pair; no free anion is present. The rate constants k_{sc} and k_{cs}, respectively, can be calculated from these effects using Equations 20 and 15a. Figure 14 shows an Arrhenius plot of these constants.

Similar experiments were carried out to determine the values of the rate constants k_a and k_d according to Equations 15b and 21. These data are less reliable, but the obtained values are not unreasonable.

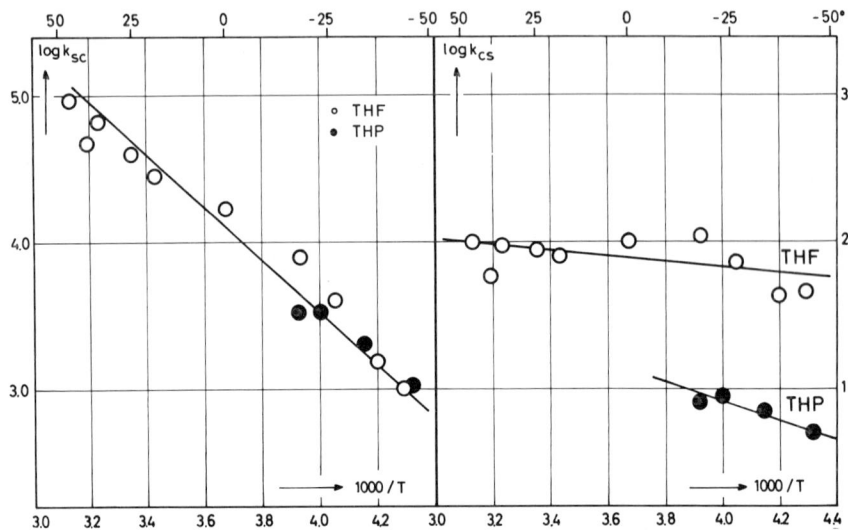

Figure 14. Arrhenius plots of the transition rate constants between the two forms of ion pairs for polystyrylsodium in THF and THP

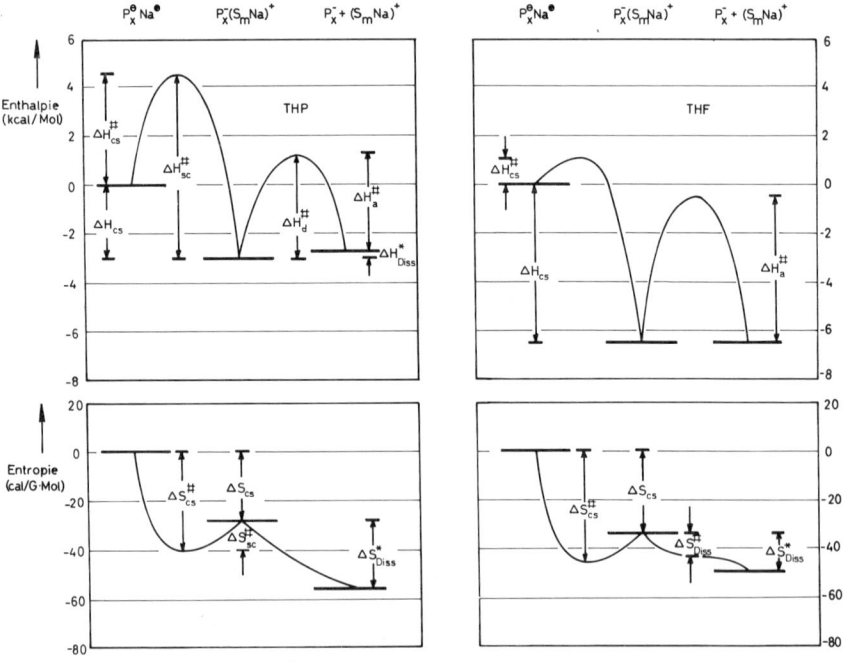

Figure 15. Scheme of enthalpies for the transition between the three forms of "living" chain ends for polystyrylsodium in THP and THF

All these results permit us to suggest a complete scheme for the enthalpy and entropy of the transition between the three states of the carbanionic end group in ethereal solvents (Figure 15).

Literature Cited

1. Henrici-Olivé, G., Olivé, S., *Makromol. Chem.* (1963) **68**, 219; *Z. Phys. Chem.* (Frankfurt am Main) (1965) **47**, 286.
2. Benson, S. W., North, A. M., *J. Amer. Chem. Soc.* (1959) **81**, 1339; (1962) **84**, 935.
3. North, A. M., Reed, G. A., *Tr. Faraday Soc.* (1961) **57**, 859.
4. Fischer, J. P., Mücke, G., Schulz, G. V., Ber. Bunsenges. (1969) **73**, 154.
5. Fischer, J. P., Schulz G. V., *Ber Bunsenges.* (1970) **74**, 1077.
6. Adam, H., unpublished results.
7. Mayer, G., unpublished results.
8. Burnett, G. M., "Kinetics and Mechanism of Polyreactions," *IUPAC Intern. Symp. Macromol. Chem.*, Budapest (1969).
9. Allen, P. E. M., Patric, C. R., *Makromol. Chem.* (1964) **72**, 106.
10. Burkhart, R. D., *J. Polym. Sci.* (1965) **A3**, 883.
11. North, A. M., *Makromol. Chem.* (1965) **83**, 15.
12. Schulz, G. V., Fischer, J. P., *Makromol. Chem.* (1967) **107**, 253.
13. Moroni, A. F., Schulz, G. V., *Makromol. Chem.* (1968) **118**, 313.
14. Schulz, G. V., *Z Phys. Chem.* (Frankfurt am Main) (1956) **8**, 290.
15. Schulz, G. V., Böhm, L. L., Chmelir, M., Löhr, G. Schmitt, B. J., "States and Reactions of the Carbanion in the Anionic Polymerization of Styrene," *IUPAC, International Symposium on Macromol. Chem.*, Budapest (1969).
16. Schmitt, B. J., Schulz, G. V., *Makromol. Chem.* (1969) **121**, 184.
17. Figini, R. V., Hostalka, H., Hurm, K., Löhr, G., Schulz, G. V., *Z. Phys. Chem.* Frankfurt am Main (1965) **45**, 269.
18. Löhr G., Schmitt, B. J., Schulz, G. V., *Z. Phys. Chem.* (Frankfurt am Main) (1972) **78**, 177.
19. Bhattacharyya, D. N., Lee, C. L., Smid, J., Szwarc, M., *J. Phys. Chem.* (1965) **69**, 612.
20. Böhm, L. L., Schulz, G. V., *Makromol. Chem.* (1972) **153**, 5.
21. Komiyama, J., Böhm, L. L., Schulz, G. V., *Makromol. Chem.* (1971) **148**, 297.
22. Hostalka H., Figini, R. V., Schulz, G. V., *Makromol. Chem.* (1964) **71**, 198; *Z. Phys. Chem.* (Frankfurt am Main) (1965) **45**, 286.
23. Bhattacharyya, D. N., Lee, C. L., Smid, J., Szwarc, M., *Polymer* (1964) **5**, 54.
24. Shimomura, T., Tölle, K. J., Smid, J., Szwarc, M., *J. Amer. Chem. Soc.* (1967) **89**, 796.
25. Schmitt, B. J., Schulz, G. V., *Makromol. Chem.* (1971) **142**, 325.
26. Chmelir, M., Schulz, G. V., *Ber. Bunsenges.* (1971) **75**, 830.
27. Alvariño, J., Chmelir, M., Schmitt, B. J., Schulz, G. V., in press.
28. Shimomura, T., Tölle, K. J., Smid, J., Szwarc, M., *J. Amer. Chem. Soc.* (1967) **89**, 5743.
29. Parry, A., Roovers, J. E. L., Bywater, S., *Macromolecules* (1970) **3**, 355.
30. Löhr, G., Bywater, S., *Can. J. Chem.* (1970) **48**, 2031.
31. Barnikol, W. K. R., Schulz, G. V., *Z. Phys. Chem.* (Frankfurt am Main) (1965) **47**, 89.
32. Worsfold, D. J. Bywater, S. J., *Chem. Soc.* (London) (1960) 5234.

33. Shimomura, T., Tölle, K. J., Smid, J., Szwarc, M., *J. Amer. Chem. Soc.* (1967) **89**, 5544.
34. Böhm, L. L., Schulz, G. V., *Ber. Bunsenges.* (1969) **73**, 260.
35. Schulz, G. V., *Z. Phys. Chem.* (1940) **B 47**, 155.
36. Böhm, L. L., Chmelir, M., Löhr, G., Schmitt, B. J., Schulz, G. V., *Fortschr. Hochpolymeren Forsch.* (1972) **9**, 1.
37. Figini, R. V., *Makromol. Chem.* (1967) **107**, 170.
38. Böhm, L. L., *Z. Phys. Chem.* (Frankfurt am Main) (1970) **72**, 199.

RECEIVED April 1, 1972.

2

Monomer Constitution and Stereocontrol in Free Radical Polymerizations

HANS-GEORG ELIAS, PAUL GOELDI, and BRIAN L. JOHNSON

Midland Macromolecular Institute, 1910 West St. Andrews Dr., Midland, Mich. 48640, and Department of Industrial and Engineering Chemistry, Swiss Federal Institute of Technology at Zurich, Universitatstrasse 6, CH-8006 Zurich, Switzerland

Stereocontrol of free radical polymerization is influenced by monomer constitution, solvent, and temperature. Most polymerizations seem to follow at least a Markov first-order one-way mechanism. Ratios of the four possible rate constants $k_{i/i}$, $k_{i/s}$, $k_{s/i}$, and $k_{s/s}$ can be calculated from the experimentally accessible concentrations of configurational triads and diads. With increasing temperature, more heterotactic triads are formed at a syndiotactic radical whereas the monomer addition at an isotactic radical favors isotactic and not heterotactic triads. Compensation effects exist for the differences of activation enthalpies and activation entropies for each of the six possible combinations of modes of addition. The compensation temperature is independent of the mode of addition whereas the compensation enthalpies are not.

The free radical polymerization of vinyl and acryl monomers normally does not lead to a true atactic polymer (1). A true atactic polymer is defined in this context as a polymer consisting of 50% isotactic and syndiotactic diads each, 25% iso- and syndiotactic triads each, and 50% heteroactic triads, etc. Furthermore all diads, triads, tetrads, etc., must be distributed at random.

Most free radical polymerized polymers exhibit a preponderance of syndiotactic diads. The syndiotacticity normally increases with decreasing temperature. This phenomenon has lead some authors to conclude that the stereocontrol is governed by the bulkiness of the substituent. Implicit in this argument are the assumptions that only repulsive forces

Table I. Markov First Order Process Probabilities for Diad Formation at Existing Diads in the Methacrylate Series CH=C(CH$_3$)COOR at −78°C in Toluene[a]

Substituent, R	Probabilities				Data from:
	$p_{s/s}$	$p_{s/i}$	$p_{i/i}$	$p_{i/s}$	
Methyl [b]	0.929	0.765	0.235	0.071	(2)
sec-Butyl	0.874	0.811	0.189	0.126	(3)
Glycidyl	0.973	1.000	0.000	0.027	(4)
tert-Butyl	0.834	0.812	0.188	0.166	(3)
α-Methylbenzyl	0.880	0.844	0.156	0.120	(3)
L-Menthyl	0.652	0.584	0.416	0.348	(5)

[a] Probabilities calculated from data of different authors. The ratio $p_{s/s}$ is the probability of formation of a syndiotactic diad at a radical attached to a syndiotactic diad; $p_{i/s}$ is the formation probability of a syndiotactic diad at a radical attached to an isotactic diad, etc. Two of the four probabilities are not independent because $p_{i/i} + p_{i/s} \equiv 1$ and $p_{s/s} + p_{s/i} \equiv 1$.
[b] At −35°C

are working between the substituents and that these repulsive forces are the same (or nearly so) in the transition state and in the dead polymer. Furthermore, the polymerization process is considered as taking part in a "quasi-vacuum"—i.e., a solvent influence on stereocontrol in free radical polymerization is denied.

Intuitively, the idea of an influence of the size of the substituent on the stereocontrol in free radical polymerization is very appealing. However, it is not supported by quantitative data. Table I shows the probabilities p of the formation of isotactic and syndiotactic diads at existing iso- and syndiotactic diads for the free radical polymerization of the methacryl type monomers CH$_2$=CH(CH$_3$)COOR in toluene at −78°C. These probabilities have been calculated from published mole fractions of iso- and syndiotactic diads (X_i, X_s) and iso-, syndio-, and heterotactic triads (X_{ii}, X_{ss} and X_{is}) via

$$p_{i/i} = X_{ii}/X_i \; ; \qquad p_{i/s} = 1 - p_{i/i}$$
$$p_{s/s} = X_{ss}/X_s \; ; \qquad p_{s/i} = 1 - p_{s/s} \qquad (1)$$

They correspond to a first-order Markov process for the stereocontrol—i.e., a penultimate effect of the last diad on stereocontrol.

Table I shows some peculiarities. In contrast to what has been claimed in literature, the probabilities of forming an isotactic diad at an existing isotactic diad ($p_{i/i}$) do not always equal the probabilities of forming an isotactic diad at an existing syndiotactic diad ($p_{s/i}$). The free radical polymerization of the six methacrylate monomers mentioned is thus not always Bernoullian, at least not at the temperature of −78°C and in toluene as a solvent.

An influence of the size of the substituent on the probabilities $p_{s/i}$ and $p_{s/s}$ is hardly detectable (with the possible exception of the L-menthyl ester). From the present data, it cannot be said with certainty that the probability $p_{i/i}$ runs through a minimum (and consequently $p_{s/i}$ through a maximum) with increasing bulkiness of the substituent. However, it is clear that there is no monotonic increase or decrease of each of the four probabilities with increasing size of the substituent.

These findings imply that the use of probabilities for i-ad formation at a given temperature in a given solvent is insufficient to describe the monomer constitution's influence on the stereocontrol in free radical polymerizations. The lack of correlation is either the result of the combined action of more than one parameter (size of substituent, resonance stabilization and/or structure of propagating radicals, etc.) or the result of noncomparable experimental conditions.

The search for stereocontrol parameters which depend on monomer constitution only has recently lead to the discovery of a compensation effect between the differences of the activation enthalpies and entropies, respectively (6). The calculated compensation temperature T_o and the compensation enthalpy can be used to evaluate the influence of the monomer constitution on the stereocontrol (6, 7, 8). The foundations and merits of this approach are discussed below.

Rate Constants of Diad Formation

With the experimental techniques available at present, rate constants of diad formation cannot be determined directly. There is however a way to calculate the rate constants from the experimentally determined triad, diad, etc. fractions if the rate constant of propagation and the statistical model are known (e.g., a one-way mechanism, a two-way mechanism, enantiomorphic site model) (9, 10). Very few rate constants of propagation are available, however.

More convenient and entirely sufficient for the present purpose is the calculation of *ratio* of rate constants. The calculation will be reviewed for a one-way first-order Markov process. A one-way mechanism is chosen because it is intuitively the most appropriate model for a free radical mechanism. Furthermore it has some experimental support. The assumption of a first-order Markov process does not rule out higher Markov processes. The differentiation between a first-order Markov process and higher order Markov processes is however possible experimentally in very rare cases because it involves the determination of tetrad, pentad, etc. fractions (11, 12, 13, 14). A Bernoullian process is ruled out by the analysis of the data of Table I.

The calculation of ratios of rate constants of diad formation starts with two assumptions. The first assumption is a steady state condition for the mole concentration of both isotactic and syndiotactic radicals

$$d[P_i^{\cdot}]/dt = k_{s/i}[P_s^{\cdot}][M] - k_{i/s}[P_i^{\cdot}][M] = 0$$
$$d[P_s^{\cdot}]/dt = k_{i/s}[P_i^{\cdot}][M] - k_{s/i}[P_s^{\cdot}][M] = 0 \qquad (2)$$

and consequently (10):

$$[P_s^{\cdot}] / [P_i^{\cdot}] = k_{i/s}/k_{s/i}$$

or for the instantaneous fraction of for example, isotactic diads:

$$[P_i^{\cdot}] / ([P_i^{\cdot}] + [P_s^{\cdot}]) = k_{s/i}/(k_{s/i} + k_{i/s}) = [P_i^{\cdot}] / [P^{\cdot}] \qquad (4)$$

The final fraction of isotactic diads is given by

$$X_i = \frac{k_{s/i}(k_{i/i} + k_{i/s})}{k_{s/i}(k_{i/i} + k_{i/s}) + k_{i/s}(k_{s/s} + k_{s/i})} \qquad (5)$$

Instantaneous and final mole fractions of isotactic (or syndioactic) diads are thus identical only if:

$$k_{i/i} + k_{i/s} = k_{s/s} + k_{s/i} \qquad (6)$$

If one assumes steady state conditions for each type of radical, then of course the instantaneous individual radical concentration cannot change with time and must thus equal the final diad concentration.

From the relationships derived (10), the six possible ratios of rate constants for a Markov first-order polymerization can be calculated from the corresponding diad and triad fractions via:

$$\frac{k_{i/i}}{k_{i/s}} = \frac{2X_{ii}}{X_{is}} \qquad (7)$$

$$\frac{k_{i/i}}{k_{s/i}} = \frac{2X_s X_{ii}}{X_i X_{is}} = \frac{1 + (2 X_{ss}/X_{is})}{1 + (0.5 X_{is}/X_{ii})} \qquad (8)$$

$$\frac{k_{i/i}}{k_{s/s}} = \frac{X_s X_{ii}}{X_i X_{ss}} = \frac{1 + (0.5 X_{is}/X_{ss})}{1 + (0.5 X_{is}/X_{ii})} \qquad (9)$$

$$\frac{k_{i/s}}{k_{s/i}} = \frac{X_s}{X_i} = \frac{X_{ss} - 0.5 X_{is}}{X_{ii} - 0.5 X_{is}} \qquad (10)$$

$$\frac{k_{i/s}}{k_{s/s}} = \frac{X_s X_{is}}{2 X_i X_{ss}} = \frac{1 + (0.5 X_{is}/X_{ss})}{1 + (2 X_{ii}/X_{is})} \qquad (11)$$

$$\frac{k_{s/i}}{k_{s/s}} = \frac{X_{is}}{2 X_{ss}} \qquad (12)$$

From the ratio of iso- and syndiotactic diad fractions, one thus does not get the ratio of the sum of rate constants leading to iso- and syndiotactic diads (Equation 11). Instead one gets only the ratio of rate constants of processes leading to the formation of heterotactic triads. In a Bernoulli process, the ratio X_i/X_s is however directly equal to the ratio k_i/k_s of the rate constants for the formation of iso- and syndiotactic diads (1, 16).

Temperature Control

A temperature influence on the stereocontrol of polymerization reactions was postulated by Huggins (17) as early as 1944 and backed by the experimental results of Fordham (16) and Bovey (18). A quantitative description of the temperature dependence was tried by Fordham (16) on the basis of the transition state theory. For a Bernoulli process, he obtained:

$$\frac{k_i}{k_s} = \frac{X_i}{X_s} = \exp\left(\frac{\Delta S_i^{\neq} - \Delta S_s^{\neq}}{R}\right) \exp\left(\frac{-(\Delta H_i^{\neq} - \Delta H_s^{\neq})}{RT}\right) \quad (13)$$

In analogy, we can write for any ratio of two rate constants:

$$\frac{k_a}{k_b} = \exp\left(\frac{\Delta S_a^{\neq} - \Delta S_b^{\neq}}{R}\right) \exp\left(\frac{-(\Delta H_a^{\neq} - \Delta H_b^{\neq})}{RT}\right) \quad (14)$$

where a, b are i/i, i/s, s/i, or s/s in the case of a Markov first-order trial.

The discussions in the literature concentrate on the problem of whether Equations 13 and 14 really reflect activation processes (16, 19) or conformational effects (20, 21) or both (8). It is agreed however that the linearity between $\ln (k_a/k_b)$ and $(1/T)$ as demanded by Equations 13 and 14 will be fulfilled for the normally accessible temperature range. Equations of the type of 14 may thus be used at least to interpolate data. The quantities $(\Delta S_a^{\neq} - \Delta S_b^{\neq})$ and $(\Delta H_a^{\neq} - \Delta H_b^{\neq})$ have at least some diagnostic value whatever their true meaning is. For convenience, we treat them here as differences of activation entropies and activation enthalpies, respectively.

In a Markov first-order process, the six possible ratios of rate constants can be calculated according to Equations 7–12 if the three triad fractions X_{ii}, X_{is}, and X_{ss} are known. We thus get six differences of activation enthalpies and activation entropies. Only two of them, however, are really independent because by definition:

$$X_{ii} + X_{is} + X_{ss} \equiv 1 \quad (15)$$

and

$$X_i = X_{ii} + 0.5\, X_{is} \quad (16)$$
$$X_s = X_{ss} + 0.5\, X_{is}$$

Table II. Activation Enthalpy Differences

$\Delta H_a^{\pm} - \Delta H_b^{\pm}$ (cal/mole) for

Monomer	Solvent	$a = i/i$ $b = i/s$	s/i i/i
Methyl methacrylate/ZnCl$_2$			
9.34/0.369 mole/mole	bulk	663±28	−(30±26)
1.10/9.6 mole/mole	70% ZnCl$_2$ in water	161±57	518±43
3.19/3.19 mole/mole	ethyl acetate	541±56	440±44
1.00/1.00 mole/mole	bulk	−(5±39)	−(5±26)
2.00/1.00 mole/mole	bulk	−(3±27)	−(24±18)
Methacrylic acid	methanol	5890±330	8980±590
(20 vol proc.)	1-propanol	3440±350	4890±420
	2-propanol	6410±440	7380±610
Glycidyl methacrylate	toluene	—	—
	DMF	—	—
	acetone	7860±550	—
	bulk	—	—
Vinyl formate	bulk	608±20	−(393±24)
	acetone	159±6	−(252±21)
	CHCl$_3$	396±43	−(756±46)
	DMF	0±36	−(59±45)
Vinyl chloride	water	101	288
	water/methanol	859±15	−(293±13)
Vinyl chloride β,β-d_2	bulk	760±11	129±11

Solvent Influence

Early studies of the free radical polymerization of methyl methacrylate did not show a solvent influence (*18, 22, 23, 24*) and consequently no solvent dependent influence of the conversion on the tacticity (*23*). A solvent dependence on stereocontrol in methyl methacrylate polymerization was however found by Watanabe and Sono (*25*) as early as 1962. Apparently, their paper has been overlooked. A literature search and a recalculation of most of the published data showed solvent influences on stereocontrol to be the rule and not the exception (*6*). Later experimental data on methyl methacrylate in about 50 solvents (*7*) and in 14 solvents (*8*) confirmed the earlier findings of Watanabe and Sono (*25*).

If a solvent effect is present, one would expect an influence of the initial solvent/monomer ratio although weak. This was indeed found by Goeldi and Elias (*8*). The effect of monomer conversion is apparently too small to be detected by the present accuracy of tacticity determinations (*8, 23*).

Calculated from Data of Various Authors

$\Delta H_a^{\pm} - \Delta H_b^{\pm}$ (cal/mole) for

i/i s/s	s/i i/s	i/s s/s	s/i s/s	Reference
580±14	633±5	−(81±14)	552±13	(26)
331±36	680±17	168±22	848±9	(26)
630±41	981±17	89±15	1071±13	(26)
−(7±30)	−(10±31)	−(2±14)	−(11±32)	(27)
−(10±21)	−(27±19)	−(7±7)	−(34±18)	(27)
1550±180	750±12	100±19	783±26	(28)
74±37	921±18	−(297±18)	957±23	(28)
1327±2	1854±41	36±7	1900±52	(28)
—	2069±33	—	2410±240	(4)
—	1867±87	—	1381±8	(4)
—	1691±24	—	1702±27	(4)
—	2358±26	—	2215±34	(4)
155±6	214±4	−(453±25)	−(217±30)	(29)
−(100±1)	−(93±4)	−(254±22)	−(514±22)	(29)
−(125±11)	−(84±10)	−(810±46)	−(876±59)	(29)
−(42±11)	−(63±10)	−(44±47)	−(104±56)	(29)
410	353	235	598	(30)
1056±7	1151±13	198±20	1348±33	(30)
682±5	−(633±1)	−(78±7)	554±6	(31)

The diad fractions have been determined as a function of polymerization temperature for many vinyl and acryl polymers. Many values of ($\Delta H_{s/i}^{\neq} - \Delta H_{i/s}^{\neq}$) and ($\Delta S_{s/i}^{\neq} - \Delta S_{i/s}^{\neq}$) can thus be calculated (see the compilation of Elias and Goeldi for 85 monomer/solvent pairs). It is interesting that many negative differences ($\Delta S_{s/i}^{\neq} - \Delta S_{i/s}^{\neq}$) can be found for a given system monomer/solvent (Table III), but only a few negative values of ($\Delta H_{s/i}^{\neq} - \Delta H_{i/s}^{\neq}$). Most negative is the ($\Delta H_{s/i}^{\neq} - \Delta H_{i/s}^{\neq}$) for the polymerization of vinyl chloride in bulk (see Table II). All other values of ($\Delta H_{s/i}^{\neq} - \Delta H_{i/s}^{\neq}$) are higher than zero or at least nearly zero.

A positive value of ($\Delta H_{s/i}^{\neq} - \Delta H_{i/s}^{\neq}$) means that the formation of a heterotactic triad is easier with increasing temperature if isotactic diads are formed at syndiotactic ones and not vice-versa. More interesting are the differences of activation enthalpies and activation entropies for the other five ratios of rate constants. Because very few data for the temperature dependency of triad fractions have been reported in literature, the calculation of the corresponding activation enthalpies and en-

Table III. Activation Entropy Differences

$\Delta S_a^{\pm} - \Delta S_b^{\pm}$ (cal mole^{-1} deg^{-1})

Monomer	Solvent	$a = i/i$ $b = i/s$	s/i i/i
Methyl methacrylate/ZnCl$_2$			
9.34/0.369 mole/mole	bulk	0.95±0.08	−(0.69±0.07)
1.10/9.6 mole/mole	70% ZnCl$_2$ in water	−(0.73±0.17)	1.25±0.13
3.19/3.19 mole/mole	ethyl acetate	0.33±0.21	0.92±0.17
1.00/1.00 mole/mole	bulk	−(1.24±0.15)	−(0.41±0.09)
2.00/1.00 mole/mole	bulk	−(1.68±0.11)	−(0.66±0.07)
Methacrylic acid	methanol	12.9±1.0	29.9±2.0
(20 vol. proc.)	1-propanol	7.6±1.7	17.2±1.9
	2-propanol	17.8±1.8	23.8±2.6
Glycidyl methacrylate	toluene	—	—
	DMF	—	—
	acetone	20.4±2.1	—
	bulk	—	—
Vinyl formate	bulk	2.66±0.06	−(1.82±0.06)
	acetone	1.22±0.06	−(1.33±0.07)
	CHCl$_3$	1.83±0.13	−(3.10±0.15)
	DMF	1.09±0.11	−(0.98±0.14)
Vinyl chloride	water	−0.31	1.10
	water/methanol	2.80±0.06	−(0.97±0.05)
Vinyl chloride β,β-d_2	bulk	1.99±0.04	0.50±0.04

tropies is possible for relatively few monomer/solvent pairs (see Tables II and III).

The result is interesting in many respects. First of all, it shows that a certain order of ratios of rate constants found for a given monomer in a certain solvent may be changed if another solvent is used. It is not surprising therefore that no correlation between the differences of activation enthalpies ($\Delta H_a^{\neq} - \Delta H_b^{\neq}$) and macroscopic properties of the solvents have been found so far. Secondly, an increase of heterotacticity with increasing temperature is not a general rule for all addition processes, contrary to what has been said in the literature. The change of rate constants with temperature is governed by the sign of ($\Delta H_a^{\neq} - \Delta H_b^{\neq}$). A positive difference ($\Delta H_a^{\neq} - \Delta H_b^{\neq}$) leads to an increase of k_a/k_b with temperature, independent of the sign of ($\Delta S_a^{\neq} - \Delta S_b^{\neq}$).

Table II shows that all differences ($\Delta H_{i/i}^{\neq} - \Delta H_{i/s}^{\neq}$) are positive. In all of these systems, there is a tendency to form more isotactic triads than heterotactic triads (coming from an s to an i addition) with increasing temperature. With the exception of vinyl formate, all differences ($\Delta H_{s/i}^{\neq} - \Delta H_{s/s}^{\neq}$) are positive however. A formation of heterotactic triads (from an i to an s addition) is thus favored over the formation of syndiotactic triads with increasing temperature. Because the syndio-

Calculated from Data of Various Authors

$\Delta S_a^{\pm} - \Delta S_b^{\pm}$ (cal mole^{-1} deg^{-1})

i/i s/s	s/i i/s	i/s s/s	s/i s/s	Reference
0.42±0.04	0.27±0.01	−(0.53±0.04)	−(0.27±0.03)	(26)
−(0.36±0.10)	0.53±0.05	0.37±0.06	0.90±0.03	(26)
0.43±0.15	1.26±0.05	0.09±0.06	1.35±0.03	(26)
−(1.43±0.11)	−(1.65±0.11)	−(0.20±0.05)	−(1.84±0.10)	(27)
−(1.91±0.09)	−(2.34±0.07)	−(0.36±0.03)	−(2.57±0.07)	(27)
−(0.31±0.54)	−(1.01±0.04)	0.55±0.06	−(0.67±0.08)	(28)
−(3.77±0.13)	−(0.43±0.07)	−(0.81±0.06)	−(0.21±0.09)	(28)
−(0.44±0.01)	2.43±0.14	0.17±0.03	2.66±0.18	(28)
—	3.43±0.11	—	4.06±0.93	(4)
—	3.37±0.28	—	1.39±0.03	(4)
—	2.30±0.09	—	2.47±0.10	(4)
—	4.20±0.09	—	3.84±0.13	(4)
0.61±0.02	0.84±0.01	−(2.05±0.08)	−(1.20±0.09)	(29)
−(0.18±0.01)	−(0.11±0.01)	−(1.38±0.07)	−(2.21±0.07)	(29)
−(0.25±0.03)	−(0.08±0.03)	−(3.32±0.14)	−(3.33±0.18)	(29)
−(0.07±0.03)	0.11±0.03	−(1.03±0.14)	−(0.92±0.17)	(29)
1.07	0.68	0.90	1.62	(30)
3.66±0.13	4.20±0.05	0.87±0.08	4.91±0.13	(30)
1.68±0.02	−(1.50±0.01)	−(0.31±0.02)	1.19±0.02	(31)

tacticity is higher in all systems monomer/solvent, this leads to the result that heterotacticity is normally increasing with temperature.

A similar result has been recently found for the free radical polymerization of methyl methacrylate in 14 solvents (32). All differences ($\Delta H^{\neq}_{s/i} - \Delta H^{\neq}_{s/s}$) were found to be positive, but only three of the 14 differences ($\Delta H^{\neq}_{i/s} - \Delta H^{\neq}_{i/i}$). Again, isotactic triad formation is favored over heterotactic triad formation and heterotactic triad formation over syndiotactic with increasing temperature as long as the individual modes of addition are considered and not the net result. Except for methacrylic acid in alcohols (cf. Lando et al. (28)) no model is known which shows why a certain solvent acts differently from another one with respect to stereocontrol in free radical polymerization.

Compensation Effects

Compensation effects are well known in physical organic chemistry for both equilibrium and rate processes. A compensation effect has also been found between the activation enthalpies and activation entropies of stereocontrol for a given mode of addition for a monomer in different solvents (6)

$$(\Delta H_a^{\neq} - \Delta H_b^{\neq}) = \Delta \Delta H_o^{\neq} - T_o(\Delta S_a^{\neq} - \Delta S_b^{\neq}) \tag{17}$$

a, b may be i/i, i/s, s/i, or s/s. T_o will be called the compensation temperature (sometimes called isokinetic temperature in low molecular weight chemistry). $\Delta \Delta H_o^{\neq}$ is the compensation enthalpy.

The results can be summarized as follows:

(1) A linear relationship between $(\Delta H_a^{\neq} - \Delta H_b^{\neq})$ and $(\Delta S_a^{\neq} - \Delta S_b^{\neq})$ is obeyed for all but one system checked so far. The compensation effect holds for different solvents and different initial monomer/solvent compositions, as Figure 1 shows for the s/i and i/s additions of the free radical polymerization of methyl methacrylate.

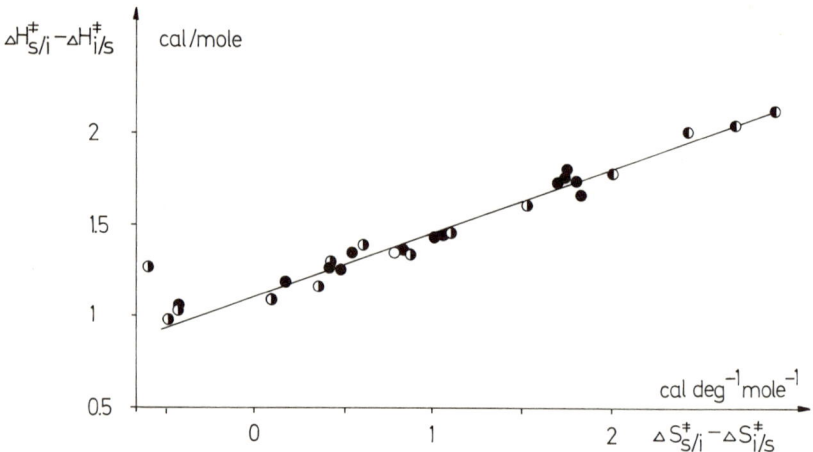

Figure 1. Compensation plot for the formation of diads leading to heterotactic triads. Free radical polymerization of methyl methacrylate in bulk (o), in 9 mole % solution in different solvents (●), and in acetone (o) and dimethyl sulfoxide (o) in different concentrations. Data from Goeldi and Elias (8).

(2) The exception to the rule is the system methyl methacrylate/$ZnCl_2$ in various solvents (Figure 2). Except for higher values of $(\Delta S_a^{\neq} - \Delta S_b^{\neq})$, the values of $(\Delta H_a^{\neq} - \Delta H_b^{\neq})$ seem to level off at about zero.

(3) Within the limits of error, the slope of the function $(\Delta H_a^{\neq} - \Delta H_b^{\neq}) = f(\Delta S_a^{\neq} - \Delta S_b^{\neq})$ is constant for a given monomer (Figure 3). The compensation temperature T_o is thus independent of the mode of addition. The compensation enthalpy $\Delta \Delta H_o^{\neq}$ differs however (Table IV).

Much has been said for and aginst the existence of compensation effects [for a recent literature review *see* Lumry and Rajender (33)]. In low molecular weight chemistry, activation enthalpies and activation entropies are calculated occasionally from rate constants at two different temperatures only. It has been claimed that many reported compensation effects are spurious because the compensation temperature equals

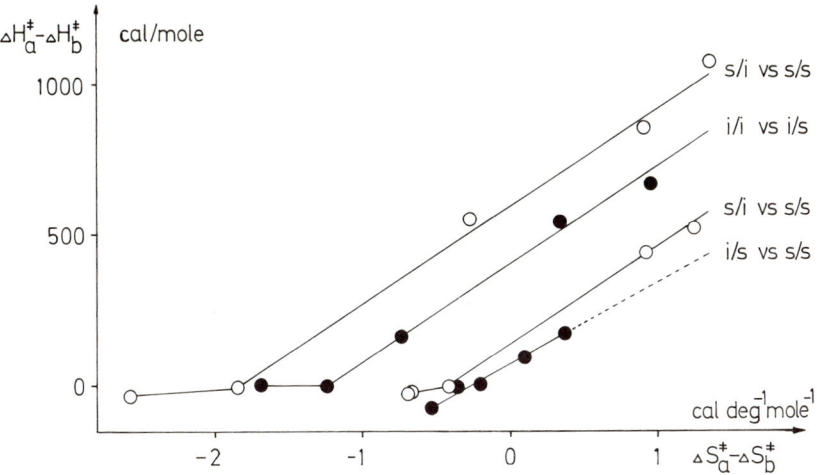

Figure 2. Compensation plot for different modes of addition in the free radical polymerization of methyl methacrylate/$ZnCl_2$ systems. Data from Otsu et al. (26) and Okuzawa et al. (27).

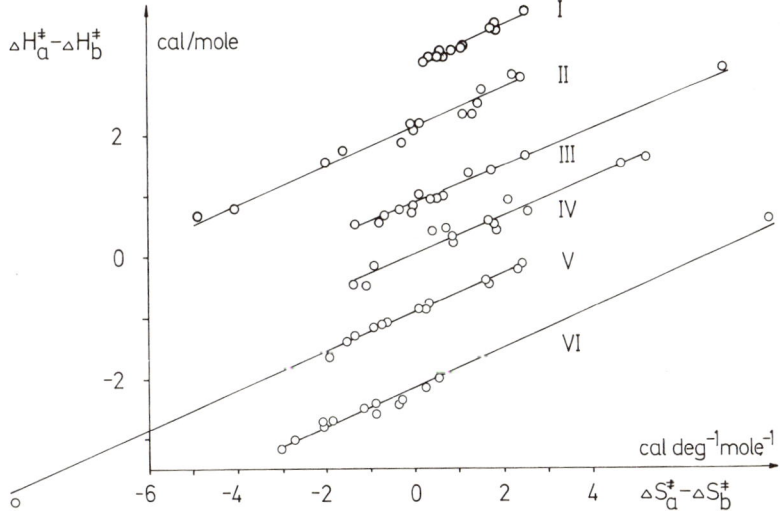

Figure 3. Compensation plot for different modes of addition in the free radical polymerization of methyl methacrylate. For a clearer picture, the data of $(\Delta H^{\neq}_a - \Delta H^{\neq}_b)$ have been shifted vertically by -2500 $(\Delta H^{\neq}_{i/s} - \Delta H^{\neq}_{s/s})$, $+1000$ $(\Delta H^{\neq}_{i/s} - \Delta H^{\neq}_{s/s})$, $+1500$ $(\Delta H^{\neq}_{s/i} - \Delta H^{\neq}_{i/i})$, and $+2500$ cal/mole $(\Delta H^{\neq}_{s/i} - \Delta H^{\neq}_{i/s})$.

more or less the error slope (34). The error slope is defined as the geometric mean of the most far-apart experimental temperatures. Because most free radical polymerizations are carried out in the temperature

Table IV. Compensation Temperatures (T_o) and Compensation Enthalpies ($\Delta\Delta H_o^{\neq}$) of Various Monomers[a]

Monomer	Rate Constants k_a	k_b	T_o, K	$\Delta\Delta H_o^{\neq}$, cal/mole
Methyl methacrylate	i/i	i/s	315±49	29±37
	s/i	i/i	310±20	614±30
	i/i	s/s	323±124	344±20
	s/i	i/s	362±15	590±12
	i/s	s/s	338±51	334±22
	s/i	s/s	334±54	957±18
Methyl methacrylate/ZnCl$_2$	i/i	i/s	314±39	394±16
	s/i	i/i	321±9	129±8
	i/i	s/s	331±24	461±10
	s/i	i/s	336±21	537±12
	i/s	s/s	304±10	58±1
	s/i	s/s	328±42	604±19
Methacrylic acid/alcohols	i/i	i/s	294±26	1500±300
	s/i	i/i	323±5	−(546±122)
	i/i	s/s	403±137	1591±49
	s/i	i/s	323±10	1069±5
	i/s	s/s	302±74	−(44±15)
	s/i	s/s	333±23	1016±6
Glycidyl methacrylate	s/i	i/s	345±15	889±48
	s/i	s/s	373±11	829±29
Vinyl formate	i/i	i/s	362±23	−(325±30)
	s/i	i/i	315±16	204±19
	i/i	s/s	323±27	−(48±6)
	s/i	i/s	330±95	−(69±10)
	i/s	s/s	320±18	232±23
	s/i	s/s	316±5	177±5
Vinyl chloride	i/i	i/s	283±25	196±31
	s/i	i/i	282±6	−(18±4)
	i/i	s/s	235±28	215±38
	s/i	i/s	305±124	−(53±100)
	i/s	s/s	248±9	−(3±8)
	s/i	s/s	219±10	270±14

[a] Data recalculated from papers of various authors

range −20° to 120°C, an error slope of approximately 300 K can be expected. As Table V shows, many cases exist where the compensation temperature is either considerably higher or significantly lower than the error slope. By analogy, the compensation temperatures around 300 K should have a physical significance also. The compensation temperature is the temperature at which a solvent does not influence the stereocontrol of the polymerization mechanism.

Other criteria developed to prove the existence of compensation effects (35) also favor a physical significance of compensation. Compensation has been said to be spurious if the ratio of the two extreme experimental temperatures is 0.9 and higher. Goeldi and Elias (8) polymerized methyl methacrylate in the temperature range $-5°$ to $120°C$, so that $T_{min}/T_{max} = 268/393 = 0.68$. The same reasoning applies to the polymerization data of other workers.

Table V. Compensation Temperature (T_o) and Compensation Enthalpy ($\Delta\Delta H_o^{\neq}$) for the Various Monomers

Monomer	T_o, K	$(\Delta\Delta H_o^{\neq})_{s/i-i/s}$ cal/mole	Data Calculated from:
Acrylonitrile	1406	0	(36, 37)
Methacrylic acid	455	664	(7)
Methacrylic acid/alcohols	318	1069	(28)
Methacrylic acid isobutyl ester	294	789	(7)
Methacrylic acid n-butyl ester	236	854	(7)
Methacrylic acid p-carboxyphenol ester	481	790	(38)
Methacrylic acid glycidyl ester	325	970	(4)
Methacrylic acid hexyl ester	303	778	(7)
Methacrylic acid methyl ester	347	778	(7)
	344	930	(25)
	362	590	(8)
Methacrylic acid methyl ester/$ZnCl_2$	342	530	(26, 27)
Methacrylic acid isopropyl ester	230	817	(7)
Methacrylic acid n-propyl ester	248	779	(7)
Vinyl acetate	1100	-490	(39)
Vinyl chloride	237	187	(30, 40, 41)
Vinyl formate	344	-74	(29)
Vinyl trichloroacetate	300	100	(39)
Vinyl trifluoroacetate	340	40	(39)

[a] T_o is averaged for the various modes of addition. $\Delta\Delta H_o^{\neq}$ refers to the ratios of the rate constants $k_{s/i}/k_{i/s}$.

Influence of Monomer Constitution

Based on a compilation of few data, Yamada and Yanagita (7) concluded that $\Delta\Delta H_o^{\neq}$ (for the s/i vs. i/s case) was independent of the monomer constitution and that T_o varied unsystematically with monomer constitution and of the mode of addition.

Because $\Delta\Delta H_o^{\neq}$ is independent of temperature and solvent effects, it can serve for a quantitative evaluation of the effects of mode of addition and monomer constitution. Even more convenient however may be the ratio k_a/k_b of rate constants at the compensation temperature T_o. That ratio can be calculated from

$$\frac{k_a}{k_b} = \exp\left(\frac{-\Delta\Delta H_o^{\neq}}{RT_o}\right) \exp\left(\left(\frac{\Delta H_a^{\neq} - \Delta H_b^{\neq}}{R}\right)\left(\frac{1}{T_o} - \frac{1}{T}\right)\right) \qquad (18)$$

which reduces for $T = T_o$ to:

$$\frac{k_a}{k_b} = \exp\left(\frac{-\Delta\Delta H_o^{\neq}}{RT_o}\right) \qquad (19)$$

From the ratio $(k_a/k_b)_{T_o}$ in Table VI, one can conclude:

(1) Within the limits of error, the ratio $k_{s/s}/k_{i/i}$ is always greater than 1—i.e., the formation of syndiotactic triads is preferred over the formation of isotactic triads (column I). The system methacrylic acid/alcohol with $(k_{s/s}/k_{i/i}) = 7.3$ is outstanding in this respect.

(2) At existing syndiotactic diads, synoditactic diads are always formed more easily than isotactic diads (column II). At existing isotactic diads, syndiotactic diad formation is however preferred to isotactic diad formation (column III) if one considers the limits of error.

(3) With the exception of methacrylic acid in various alcohols, a homotriad (iso- or syndiotactic triad) formation is always preferred to the formation of heterotactic triads (columns IV and V).

(4) Heterotactic triads are formed more easily by a formation of isotactic diads at existing syndiotactic diads in the case of methacrylic monomers than vice-versa (column VI). The probability for the formation of heterotactic triads is however independent of the mode of addition for vinyl polymers.

Table VI. Ratios of Rate Constants of Various Monomers at Their Compensation Temperatures[a]

	Ratios of Rate Constants at $T = T_o$					
	I $k_{s/s}/k_{i/i}$	II $k_{s/s}/k_{s/i}$	III $k_{i/i}/k_{i/s}$	IV $k_{i/i}/k_{s/i}$	V $k_{s/s}/k_{i/s}$	VI $k_{s/i}/k_{i/s}$
Methacrylic acid/alcohols	7.30	4.64	0.08	0.43	0.93	0.19
Methacrylic acid methyl ester	1.71	4.23	0.95	2.71	1.64	0.44
Methacrylic acid methyl ester/ZnCl$_2$	2.02	2.53	0.53	1.22	1.10	0.45
Methacrylic acid glycidyl ester	—	3.06	—	—	0.83 [b]	0.27
Vinyl chloride	1.59	1.86	0.71	1.00	0.99	1.09
Vinyl formate	0.93	1.33	1.57	1.39	1.44	1.11

[a] Data from Table IV. Because different values of T_o have been used for the different modes of addition, the values of the various columns cannot be used for rechecking.
[b] Calculated from other k_a/k_b data.

These findings suggest a strong influence of conformational effects on the stereocontrol in free radical polymerizations. Conformation analyses are lacking however for the monomers investigated.

Conformational effects may also influence T_o values. The T_o values of vinyl esters CH=CH–OCO–R fall with increasing van der Waals radii of substituents

R =	CH$_3$	CF$_3$	CCl$_3$
$T_o =$	1100°	340°	300 K

The T_o = 344 K of vinyl formate (R = H) is interesting. Hydrogen bonds are likely to participate in the transition state so that the substituent OCOH works *via* attraction forces and not repulsion forces. This may influence the conformation in the transition state. Intramolecular hydrogen bonds are known to occur in solid poly(vinyl formate) according to infrared measurements (*42*). It is also known that intramolecular hydrogen bonds exist in poly(vinyl alcohol) in an all-trans-conformation (*43*). If similar intramolecular hydrogen bonds exist in poly(vinyl formate), an all-trans-conformation might be preferred. It may well be that this is most easily achieved by an addition step leading to an isotactic diad. A $k_{i/i}/k_{i/s}$ = 1.57 for vinyl formate could be an indication for this.

Literature Cited

1. Coleman, B. D., *J. Polymer Sci.* (1958) **31**, 155.
2. Kato, Y., Nishioka, A., *Bull. Chem Soc. Japan* (1964) **37**, 1614.
3. Matsuzaki, K., Ishida, A., Tateno, N., Asakura, T., Hasegawa, A., Tameda, T., *Kogyo Kagaku Zasshi* (1965) **68**, 852.
4. Iwakura, Y., Toda, F., Ito, T. and Aoshima, K., *Makromol. Chem.* (1967) **104**, 26.
5. Sobue, H., Matsuzaki, K., Nakano, S., *J. Polymer Sci.* (1964) **A2**, 3339.
6. Elias, H.-G., Goeldi, P., Kamat, V. S., *Makromol. Chem.* (1968) **117**, 269.
7. Yamada, A., Yanagita, M., *IUPAC, Intern. Symp. Macromol. Chem., Budapest 1969*, Preprint 5/60.
8. Goeldi, P., Elias, H.-G., *Makromol. Chem.* (1972) **153**, 81.
9. Bovey, F. A., Tiers, G. V. D., *Fortschr. Hochpolym.-Forschg. Adv. Polymer Sci.* (1963) **3**, 139.
10. Elias, H.-G., *Makromol. Chem.* (1970) **137**, 277.
11. Coleman, B. D., Fox, T. G., *J. Polymer Sci.* (1963) **A1**, 3183.
12. Coleman, B. D., Fox, T. G., Reinmöller, J., *J. Polymer Sci.* (1964) **B4**, 1029.
13. Frisch, H. L., Mallows, C. L., Bovey, F. A., *J. Chem. Phys.* (1966) **45**, 1565.
14. Bovey, F. A., "Polymer Conformation and Configuration," Academic, New York, 1969.
15. Johnson, B. L., Elias, H.-G., *Makromol. Chem.* (1972) **155**, 121.
16. Fordham, J. W. L., *J. Polymer Sci.* (1959) **39**, 321.

17. Huggins, M. L., *J. Amer. Chem. Soc.* (1944) **66**, 1991.
18. Bovey, F. A., *J. Polymer Sci.* (1960) **46**, 59.
19. Bawn, C. E. H., Janes, W. H., North, A. M., *J. Polymer Sci* (1963) **C4**, 427.
20. Fischer, H., *Kolloid-Z. Z. Polymere* (1965) **206**, 131.
21. Luisi, P. L., Mazo, R. M., *J. Polymer Sci. Pt. A-2*, (1969) **7**, 775.
22. Bawn, C. E. H., Janes, W. H., North, A. M., *J. Polymer Sci* (1962) **58**, 335.
23. Fox, T. G., Schnecko, H. W., *Polymer (London)* (1962) **3**, 575.
24. Schroeder, G., *Makromol. Chem.* (1966) **97**, 232.
25. Watanabe, H., Sono, Y., *Kogyo Kagaku Zasshi* (1962) **65**, 273.
26. Otsu, T., Yamada, B., Imoto, M., *J. Macromol. Chem.* (1966) **1**, 61.
27. Okuzawa, S., Hirai, H., Makishima, S., *J. Polymer Sci., A-1* (1969) **7**, 1039.
28. Lando, J. B., Semen, J., Farmer, B., *Polymer Preprints* (1969) **10**, 586; *Macromolecules* (1970) **3**, 524.
29. Elias, H.-G., Riva, M., Goeldi, P., *Makromol. Chem.* (1971) **145**, 163.
30. Bargon, J., Hellwege, K.-H., Johnsen, U., *Makromol. Chem.* (1966) **95**, 187.
31. Cavalli, L., Borsini, G. C., Carraro, G., Confalonieri, G., *J. Polymer Sci., Pt. A-1* (1970) **8**, 801.
32. Elias, H.-G., Goeldi, P., *Makromol. Chem.* (1971) **144**, 85.
33. Lumry, R., Rajender, S., *Biopolymers* (1970) **9**, 1125.
34. Blackadder, D. A., Hinshelwood, C., *J. Chem. Soc. (London)* **1958**, 2720 and 2728.
35. Exner, O., *Nature* (1964) **201**, 488; *Coll. Czech. Chem. Commun.* (1964) **29**, 2.
36. Svegliado, G., Talamini, G., Vidotto, G., *J. Polymer Sci., Pt. A-1* (1967) **5**, 2875.
37. Murano, H., Yamadera, R., *J. Polymer Sci.* (1967) **B5**, 333.
38. Amerik, Y. B., Konstantinov, I. I., Krentsel, B. A., *IUPAC, Intern. Symp. Macromol. Chem., Tokyo 1966*, Preprint 1.1.09.
39. Uoi, M., Sumi, M., Nozakura, S., Murahashi, S., cited in Murahashi, S., *IUPAC, Macromol. Chem.* (1967) **3**, 435; (*Pure Appl. Chem.* (1967) **15**, (3-4)).
40. Bovey, F. A., Hood, F. P., Anderson, E. W., Kornegay, R. L., *J. Phys. Chem.* (1967) **71**, 312.
41. Talamini, G., Vidotto, G., *Makromol. Chem.* (1967) **100**, 48.
42. Fujii, K., Imoto, S., Mochizuki, T., Ukida, J., Matsumoto, M., *Kobunshi Kagaku* (1962) **19**, 587 (cited in Fujii, K., Michizuki, T., Imoto, S., Ukida, J., Matsumoto, M., *J. Polymer Sci.* (1964) **A2**, 2327).
43. Elias, H.-G., "Makromoleküle," p. 111, Hüthig and Wepf, Basel and Heidelberg, 1971.

RECEIVED February 2, 1972.

Design of Large Polymerization Reactors

GUENTHER BECKMANN

Chemische Werke Hüls A. G., 4370 Marl, West Germany

> *A new type of giant reactor has been developed at Chemische Werke Hüls and tested on a large scale. This reactor has some remarkable properties—a wide range of applicability for different polymerizations and capacities on the order of 20,000 to 50,000 tons per year. Polymerizations which have been successfully tested are suspension and emulsion PVC, polystyrene, polyolefin, SBR-latices, and solution rubber.*

Forecasts indicate that polymer consumption will increase rapidly during the decades ahead. The development of polymer plants with capacities far exceeding those in use today, therefore, demands attention. The heart of a polymer plant is the polymerization reactor section. These reactors are often agitated vessels in which the polymerization takes place in a continuous or batchwise operation. Monomers that are polymerized include vinyl chloride, styrene, olefins, butadiene, and the like. The most important features of a polymerization reactor are its agitation and, since the polymerization is an exothermic process, its cooling systems. Today the volume of such reactors is usually between 5 and 30 m^3, corresponding to annual unit capacities of 5,000 to 10,000 tons of plastic or rubber polymer. Recently, several companies have started to develop much larger polymerization reactors with volumes between 50 and 200 m^3 and capacities of 10,000 to 50,000 and possibly 100,000 tons per year (Figure 1). The design of such equipment entails some entirely new problems. The most important basic question is whether it is possible to stir and to mix sufficiently such amounts of viscous and rheological complex liquids at a reasonable cost—*i.e.*, using technical means which are not highly sophisticated.

Agitation

Polymerization in agitated reactors is carried out in solution, emulsion, or suspension. Water, organic solvent, or the liquid monomer is

Figure 1. A 200-m^3 PVC reactor installed at Chemische Werke Hüls with a capacity of 30,000 to 50,000 tons/year of PVC grade suspension or emulsion

used as diluent. Common to all these processes is the aim for a high polymer content in the liquid; it is often so high that the reactor content becomes viscous and close to the stirring limit. Reactors of conventional size and design. (Figure 2) can handle liquids with viscosities up to 20,000 cp. If one intends to design a larger reactor—reactor volume scaled from 10 m^3 to 100 m^3—it is advisable to know beforehand whether the large scale still permits sufficient mixing. In other words, under what conditions is it possible to obtain the same mixing in two geometrically similar reactors of different size (1). Here, "same mixing" means of similar geometric agitation in both reactors (Figure 3).

From the Navier-Stokes' equations it follows that three criteria of fluid motions (*i.e.*, small arrows in Figure 3) are to be kept constant to obtain geometrically similar agitation patterns for both reactor sizes.

The criteria are (1) the Froude number, (2) the Power number, and (3) the Reynolds number; these are shown below in their well-known form.

$$\frac{d_s n_s^2}{g_s} = \frac{d_b n_b^2}{g_b} = \text{const}_1 = \text{Fr}$$

$$\frac{N_s}{\rho_s n_s^3 d_s^5} = \frac{N_b}{\rho_b n_b^3 d_b^5} = \text{const}_2 = P$$

$$\frac{d_s^2 n_s \rho_s}{\eta_s} = \frac{d_b^2 n_b \rho_b}{\eta_b} = \text{const}_3 = \text{Re}$$

Subscripts s and b stand for small and big reactor, d is typical diameter, n is revolutions of agitator per unit time, N is power input of agitator, η is viscosity, ρ is density, and g is gravity constant. Using this set of equations, it is impossible to calculate the power input (N_b) and the

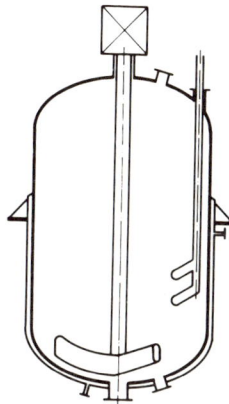

Figure 2. *Conventional polymerization reactor with top-driven agitator, baffle, and wall-cooling (volumes up to 40 m³)*

revolutions of the agitator (n_b) of the big reactor from the data of the small reactor and the liquid properties η and ρ, and the gravity g. The three equations above are overdeterminate because there are only two variables N_b and n_b. A solution is possible only if a third variable is created—i.e., if a constant is made a variable. The viscosity is the best parameter for this purpose. Then the solution to the three equations can easily be calculated as:

$$\eta_b = \eta_s \left(\frac{d_b}{d_s}\right)^{1.5}$$

[This calculation can be found in a paper by Rushton (2).] This result means that in the big reactor the viscosity must be higher than it is in the small one to achieve geometrically similar agitation patterns. From this, another conclusion may be drawn: if one does not vary the viscosity in two different reactor sizes, no geometric similarity of the fluid motion is possible. The viscous fluid in the big reactor then behaves differently—as if its viscosity were lower than it actually is.

A practical example will illustrate this point. Imagine a thimble and a bucket, both full of the same viscous honey. When both receptacles are simultaneously turned upside down, the honey will flow more rapidly from the bucket than from the thimble. One might think that the honey in the bucket is less viscous than that in the thimble, but it isn't. It is the size of the bucket which accounts for the difference.

A general rule may now be set up: the greater the volume of a vessel, the less viscously the same liquid behaves when being stirred in that vessel. Using a more complicated derivation it can also be shown

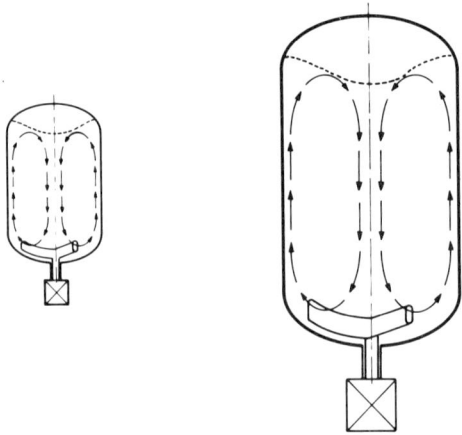

Figure 3. Geometrically similar flow patterns in two reactors of different size

that a departure of the viscosity behavior from the Newtonion viscosity behavior, as encountered with polymer solutions, for example, having structural viscosity, has a lesser effect on the large scale than it has on the small one. Thus, from the hydrodynamic point of view no fundamental difficulties are expected with mixing in large reactors.

In constructing agitators for large reactors, however, practical difficulties are inherent in the design. These agitators cannot be driven from above by a shaft entering the reactor hood. The agitator shaft needed for 100-m^3 or even 200-m^3 reactors would be too long, too thick, and too expensive since the mixing element must be close to the reactor bottom for effective mixing while the reactor contents are discharged.

A possible solution to this problem is provided by what may be called "bottom entering agitator." In this case the mixing element

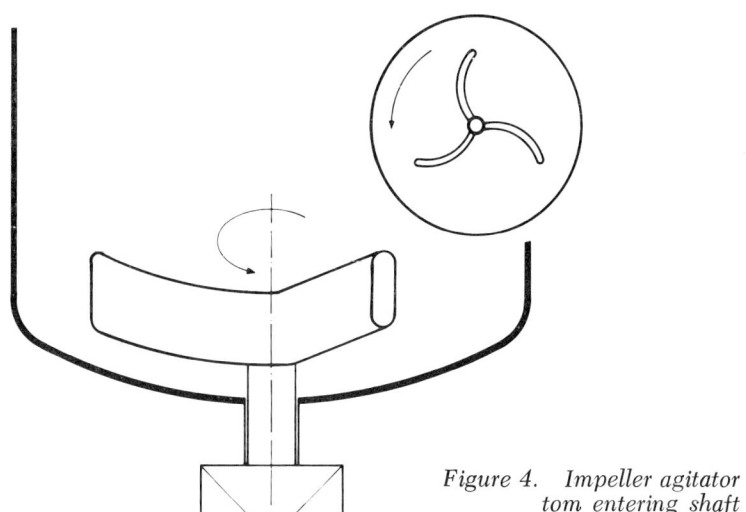

Figure 4. Impeller agitator with bottom entering shaft

arranged above the reactor bottom is driven through a short and relatively thin shaft from a drive unit located below the reactor bottom. Impeller-type or helical screw agitators are used as the mixing elements. Big, three-vane impeller agitators, as shown in Figure 4 are recommended for reactors requiring frequent cleaning (3). Helical screw-types (Figure 5) are suitable for mixing very viscous fluids. The mixing action of these agitator types is based on a typical circulation in the reactor—overturn—consisting of an upward stream at the periphery and a downward stream at the center as shown in Figure 6. Of course, a rotation flow is superposed on these streams, which results in the streamlines shown in Figure 7. For the three-vane impeller agitator the peripheral upward stream is generated by the difference of the centrifugal pressures in the lower and upper areas of the reactor. In the lower area the impeller causes high rotation whereas in the upper part the baffles brake the rotation. In a well mixed reactor, the time for one overturn top-bottom-top should be short and in accordance with the heat generated by the reaction. Recommended time intervals are 30 to 100 seconds. These times can be checked with a dye or ink which is introduced into the vortex of the liquid surface and which reappears after a prescribed time at the periphery of the surface.

A substantially new and critical feature in bottom-entering agitators is the sealing of the shaft; double mechanical seals are used here. On modern designs, they are provided with additional "standstill" seals for emergencies. The standstill seal allows the mechanical seal to be replaced while the reactor is full and under pressure. In a number of polymerizations, particularly in vinyl-chloride polymerization, the rotating elements must be protected by additional water flushing (4); other-

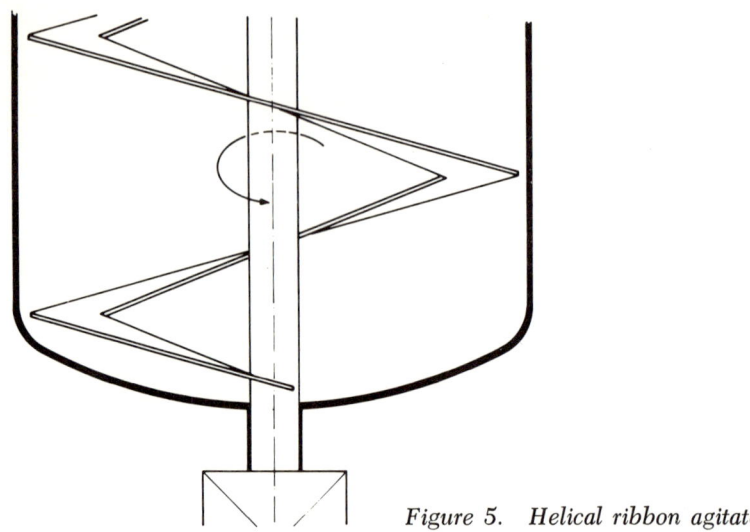

Figure 5. Helical ribbon agitator

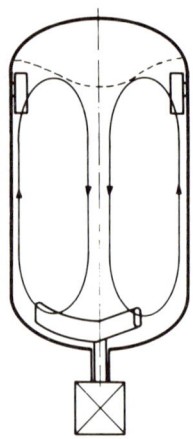

Figure 6. Overturn motion in reactor, consisting of a peripheral upward stream and a downward stream in the center part

wise the vinyl chloride will diffuse against the sealing oil stream between the rotating elements and polymerize there with the rotating ring lifting off as a consequence.

Cooling

Polymerizations are exothermic processes. The specific heats of polymerization of some important monomers are shown in Table I. Note that the heat of polymerization of vinyl chloride is about 400 whereas that of styrene is only 160 kcal/m^{2}hour °C. It is useful to visualize the great amount of heat generated in a large reactor (*e.g.*, 200 m³ volume) during polymerization. Table II shows two examples. Polymerization of styrene in a 200-m³ reactor batch results in a heat load of 1.6 million kcal per hour, corresponding to a reaction rate of 10 tons/hour of polymer during the reaction peak. Vinyl chloride poly-

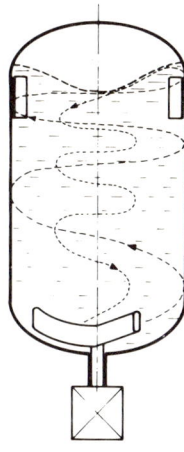

Figure 7. Stream lines in impeller agitated reactor which result from the superposition of rotational and overturn motion

merization in a 200-m³ batch gives even higher heat loads of 6 or 7 million kcal/hour since the reaction rate can be as high as 16 to 17 tons/hour at a maximum. Let us compare the heat evolved with the heat rates removable through the wall. The jacket cooling area of a completely jacketed 200-m³ reactor amounts to about 160 m². Table III shows the heat rates dissipated through the reactor wall at two possible Δt values and two heat transfer values, which are estimated for typical polymerizations. More wall cooling will suffice only for a comparatively slow reaction like that of styrene when the most favorable Δt and heat transfer values are given (*see* last line). In other cases, where conditions are not as favorable, additional cooling equipment is needed. The most important additional means of cooling are discussed below.

Cooling coils or tube bundles can be installed in the reactor (Figure 8). This additional equipment is effective only when the cooling areas are relatively large compared with the wall surface. Since the tubes interfere with mixing, they can be used only where relativly low viscous fluids are handled. Also, the tubes are an obstacle to reactor cleaning, and they are suitable only for reactions where no wall caking occurs. If these rules are disregarded, an inhomogeneous polymer containing "fish eyes" will result.

A similar method is the circulation of the reactor contents through externally arranged tube bundles with help of a circulation pump. It

Table I. Polymerization Heat of Monomers

Monomer	Specific Polymerization Heat, kcal/kg	State
Styrene	160–170	liquid
Vinyl chloride	390–440	liquid
Ethylene	930	gas
Propylene	590	gas
Butene	300	liquid
Butadiene	310	dissolved
Isoprene	230	dissolved

Table II. Polymerization Heats Generated in a 200-m³ Reactor [a]

	Specific Polymerization Heat, kcal/kg	Max. Rate in 200 m³ Reactor, tons/hr	Polymerization Heat, million kcal/hr
Styrene	160	10	1.6
Vinyl chloride	400	16	6.4

[a] Operating conditions are those for smaller reactors

Table III. Removable Heat through Cooling Jacket of 200-m³ Reactor [a]

Δt	$\dfrac{kcal}{m^2\ hr\ °C}$	Removable Heat, million kcal/hr
10	100	0.16
	200	0.32
50	100	0.8
	200	1.6

[a] Cooling area = 160 m². Data for typical heat transfer values and Δt values

can be very effective if the polymerization is free of wall build-up and if a suitable pump for the active polymerization liquid is available.

A second method involves the installation of cooling cylinders and wall scrapers (Figure 9). On large reactors this is complex and expensive. Great problems will ensue if the scraped-off material remains stuck to the scraper blades. Cleaning and dismanteling are extremely difficult.

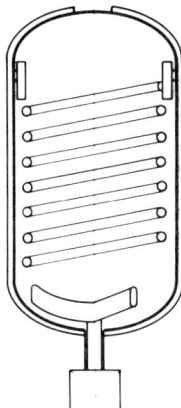

Figure 8. Reactor with cooling coil

A third method is reflux cooling by a condenser mounted on top (Figure 10). This type of cooling is very effective. On the 200-m³ reactor heat rates of up to 15 million kcal/hour can be removed. Boiling conditions must prevail in the reactor, and some restrictions must be observed:

(a) Inert gases (nitrogen) will interfere since they tend to form insulating gas cushions in the condenser. In this case either the gas must be bled from the top of the condenser or the vapors in the condenser must be kept in turbulent motion by a circulation fan.

(b) The catalyst used must be non-volatile; otherwise polymerization will result in the condenser.

(c) Reflux cooling can be used on foaming batches to only a limited extent.

(d) In a number of polymerizations remixing of the refluxed condensate involves problems.

Feed cooling is a fourth method of additional cooling. Cold liquid or ice—water ice or benzene ice or the like—is fed to the reactor during polymerization. Heat is consumed by melting and heating to reaction temperature. This type of cooling may prove favorable primarily in continuous polymerization in single reactors. The ice must be available as a pumpable mass. This technique is not optimal with regard to thermodynamics since refrigeration is required. A special case of this type of cooling is what might be called "racing polymerization." The reaction is allowed to "race," that is, to increase in temperature, until all reaction heat is consumed, with only part of the reaction heat being removed by cooling. This method can only be applied to special cases.

Temperature control is important in cooling. Basically, temperature control of large reactors operates the same way as in small reactors. However, if mixing in the reactor is poor—*i.e.*, if the overturn is too

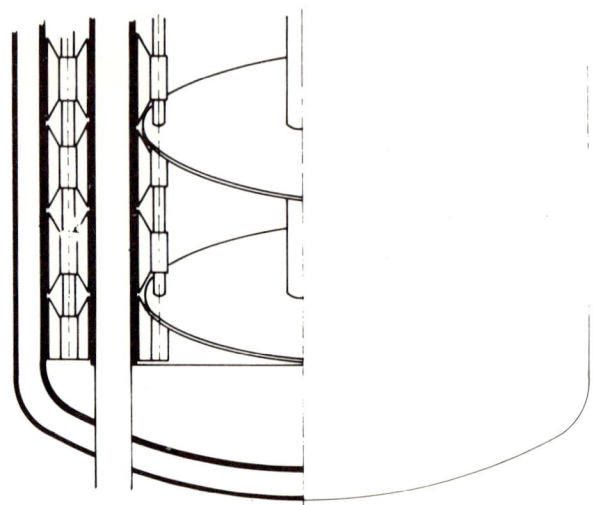

Figure 9. Reactor with cooling cylinder and scraper blades

slow—serious interference is caused by delays in boiling; the content bubbles up rhythmically. This is caused by the great hydrostatic head of the reactor content—8 to 9 m—and the difference in boiling point associated with it. A special case is represented by suspension PVC. Here a peculiar phenomenon has recently been observed with small reactors (5) which probably will also have to be considered in the case of big reactors. Towards the end of reaction the temperatures in different regions of the reactor differ by as much as 10 or 20°C. Not being able to look inside the reactor one might speculate that ineffective stirring is the reason, perhaps caused by the fibrous constitution of the suspended polymer or a gel effect. For large reactors with reflux condensers this phenomenon could be particularly adverse because during the critical phase towards the end of polymerization a pressure drop occurs as a result of vinyl chloride depletion and affects the condenser performance. In view of the complex control problems connected with this, it seems advisable for now to use digital control (computer control) since this permits automatic variation of control dynamics during the reaction.

Reactor Shell

Several problems are associated with the reactor shell. Few workshops have annealing furnaces suitable for large reactors. Thus, to avoid annealing after welding and to keep the weight for transportation within limits, we tried to use a wall thickness as small as possible. Using high strength, special steel, a transportation weight of less than 100 tons is achieved on 200-m³ reactors of 16 atm pressure rating and a height vs. diameter ratio of 1.8 : 1 as shown in Figure 11. The wall thickness is so small that the reactor does not require annealing after welding. However, special welding methods must be used. The rigidity of the reactor shell is inferior to that of small reactors. Therefore, the load-bearing capacity of the cooling jacket is used to advantage in supporting the shell.

Large reactors lend themselves well to outdoor installation with four, six, or eight supports resting on a foundation. The support attachment to the shell must be designed carefully to prevent the shear stresses

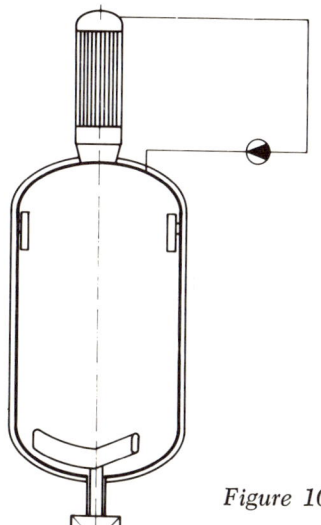

Figure 10. *Reactor with reflux cooling and vapor circulation*

from denting the shell. It is important to avoid shell vibrations, which, because of the thin wall, are liable to occur more readily here than in small reactors. Bracing of shell, proper arrangement of supports, and minimization of agitor unbalance and vibrations provide the proper solution of this problem.

If the inside surface of a large reactor is stainless steel, clad steel plate is the material of construction for the reactor shell. The extra cost for stainless steel cladding is relatively low. Highly polished surfaces in such reactors can best be obtained by electric polishing, which we effect by a special method. At present it is not possible to construct 100-m³ or 200-m³ reactors with enamelled inside walls. If sufficient protection could be provided against damage to enamel, which on large reactors entails high repair cost, this application would be desirable in certain polymerizations.

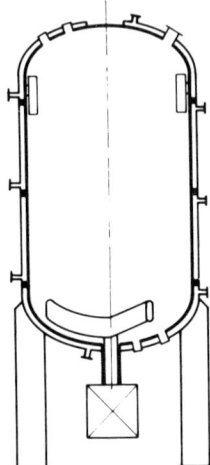

Figure 11. Hull of a large reactor with cooling jacket

A difficult problem is the transportation of the reactor shell from the workshop to site. Road transportation of the shell, about 6 m in diameter, is possible only if there are no obstacles such as bridges, tunnels, or bottlenecks. Transportation by rail is prohibited because clearance requirements are not satisfied. On the other hand, transportation by ship—also by river barge—can be done without difficulty.

Reactor Cleaning

With many polymerizations reactor cleaning is very important. On large reactors manual cleaning of the inside walls is less economical than mechanical or chemical cleaning. In manual wall cleaning with scraper tools a scaffold must be erected in the reactor, or a raft must be introduced to float on water, on which the workmen can stand and lower the level from which to work by draining water away.

Mechanical cleaning, which can also be automated, can be done by using high-pressure water pumps (50 to 300 atm gage) and rotating spray heads introduced into the reactor which partly move up and down (similar to familiar tank cleaning equipment). The pressurized water jet will loosen and break up even tenacious wall deposits. However, the reactor must be designed to prevent dead spaces that cannot be swept by the water jet. The pressurized water may also contain additives which, at the end of the cleaning cycle, condition the reactor wall for the next polymerization cycle. Polymerizations sensitive to traces of water require that the inside wall (of that the reactor) be thoroughly dried after cleaning.

Recently, methods for cleaning polymerization reactors contaminated with caked wall deposits have been developed. These methods also lend themselves to automation. Hot solvents which can dissolve or swell caked material are being used to squirt or steam out the reactor. However, they must not attack the sealing elements of the reactor. Squirting pressures can be lower than those in pressurized water cleaning. The solvent is used until it is highly viscous as a result of its solids content. Then it must be treated in separate equipment—*i.e.*, freed from solids. For this purpose distillation or precipitation and filtration may be applied, depending on thermal stability and solubility. The toxicity and flammability of some solvents present problems since it is desirable to avoid expensive inert gas purging. Traces of solvent must not interefere with or affect the subsequent polymerization.

Safety Features

Because of the huge amounts of monomer handled in a large reactor, special safety measures are required to prevent or to minimize the hazard of a reaction getting out of control. Obviously, a prerequisite for the safe operation of a large reactor is the liberal design of the cooling system and the agitator. Nonetheless, one must take into account the fact that under abnormal operating conditions either cooling or agitation or both may fail or prove inadequate. For example, the agitator and the cooling water pump can come to a standstill in the event of a power failure. Similarly, the agitator shaft may break, or a coolwater supply line may burst. In such cases the reaction must be controlled or rising pressure and temperature could cause the reactor to explode.

Two measures can be taken to ensure safety:

(1) Emergency Shutdown. In an emergency a chemical is added to the reactor to stop the reaction chemically. The addition has to be done so that the stopping agent mixes immediately with total contents of the reactor, even with the agitator at a standstill. For example, this can be achieved by using a gaseous stopping agent. It is introduced into the reactor from below and then mixed in by gas bubble stirring.

(2) Emergency Venting. Emergency venting is primarily used if the cooling system fails. In the event of an uncontrolled rise in temperature, a valve on the reactor hood will open and gas will be vented from the reactor. Thus, the heat of evaporation is being removed from the vent gas since the reactor contents are boiling. In most cases the vent gas (order of magnitude of several tons in 10 minutes), for environmental reasons, can not be vented to a flare but must be blown down

into a gas tank or a condenser of sufficient size. The use of gas balloons has been discussed. It is important that the venting operation be controlled to prevent solids from being entrained from the reactor. For this reason it is better to use a control valve rather than a spring-loaded valve.

The high production capacity of large reactors justifies the use of expensive additional equipment. Generally, the large reactors can be equipped with more instruments and accessories than smaller reactors. Therefore, a higher reliability in service is to be expected in return.

Literature Cited

1. Beckmann, *Chem.-Ing.-Tech.* (1964) **36**, 169–174.
2. Rushton, J. H., Chem. Engng. Progr. (1951) **47**, 485.
3. Deutsches Patent **OLS 2, 032,700**.
4. French Patent **7,112,877**.
5. Hedden, H., Ph.D. Dissertation, Universität Stuttgart, p. 22–24 (1969).

RECEIVED September 29, 1972.

4

Behavior of Viscous Polymers during Solvent Stripping or Reaction in an Agitated Thin Film

FRITZ WIDMER

Swiss Federal Institute of Technology, Zurich, Switzerland

> *A specially designed thin-film machine can be used to process very viscous, non-Newtonian materials. The apparatus can also be used to remove solvents from polymers and polycondensation processes having viscosities exceeding 10,000 poises. The Luwa thin-film machine has a small clearance between the heated wall and rotor blade. This clearance results in high shear gradients and considerably reduces apparent viscosity. The increased turbulence and improved surface renewal that ensue improve reaction velocities and aid the required forced product flow on the walls of the apparatus.*

Liquid films, whether in the form of natural falling films or in the more advanced form of mechanically agitated films in thin-film apparatus, offer convenient conditions for the solution of many difficult processing problems because of the films' favorable surface-to-volume ratio and good heat transfer. This is especially so with an agitated film between heating wall and film. The agitated-film principle has previously found its broadest application in thin-film evaporators.

These thin-film evaporators are equipped with rotating elements that create a thin, liquid film of high turbulence along the inner surface of the heated tube (*see* Figure 1). Consequently, favorable heat and mass-transfer conditions (1), (2) and short residence times result owing to the small holdup (3, 4).

The favorable working conditions of the thin-film evaporator enabled the equally successful application of the thin-film principle for fractionation (5), absorption, chemical reactions (6), and drying (7, 8). In these processes and applications, the thin-film apparatus is used mainly to treat heat sensitive and lower-viscosity products that flow on the influence of gravity alone. This particular type of apparatus in a special design can also be used to process highly viscous products to

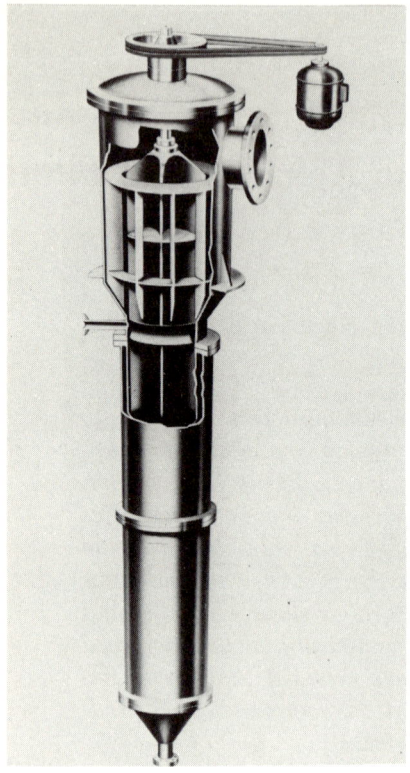

Figure 1. Thin-film evaporator equipped with a fixed blade rotor (Luwa)

remove solvents or monomer components from polymer solutions and polymers and for polycondensation processes up to 15,000 poises (P) and more.

Film Formation in Thin-Film Apparatus

The influence of viscosity on the formation and behavior of a mechanically agitated film is based on the theory of the natural falling film along a wall. The mean film thickness $\bar{\delta}$ of a falling film under laminar conditions is determined according to Nusselt (9):

$$\bar{\delta} = \left(\frac{3\nu^2}{g}\right)^{0.33} \mathrm{Re}^{0.33} \qquad (1)$$

Work done by Brauer (10) pointed out that the waves arising in a falling film at low viscosity are about 2.6 times higher than the mean film thickness defined by Equation 1. In Figure 2, the mean thickness of the film according to Equation 1 is expressed as a function of the peripheral load for different product viscosities (curve 1: water 1 CST, curve 2, 3: higher viscosities.

By comparison, the conditions in a thin-film apparatus equipped with a rigid blade rotor can now be examined. In this apparatus, the rotor blades move along the heating wall at distance s of about 1.5 mm. This gap—large, compared with the mean thickness of a low-viscosity liquid—is still small enough to enable the rotor blade to dip into the waves of the film, even for low liquid loadings (compare section A, Figure 2). The resulting measured mean thickness for the water film is shown by curve 1° in Figure 2. As soon as the gap is filled with liquid, the mean film thickness increases considerably as a result of the wave that forms in front of the blade (detail B, Figure 2).

Assuming the gap s between rotor blade and heating wall is not enlarged, the rotor of the thin-film apparatus for processing liquids of

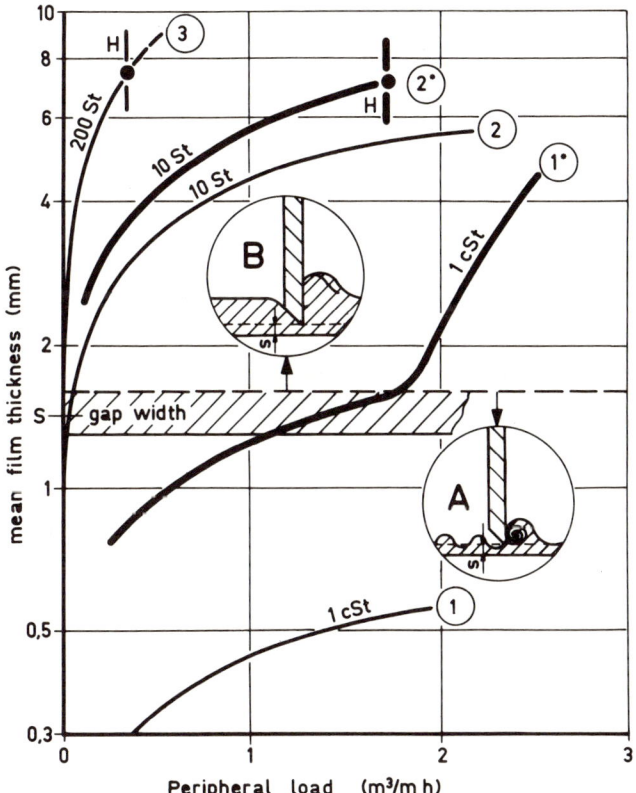

Figure 2. Comparison between mean film thickness of falling and agitated films in a thin-film apparatus, expressed as a function of peripheral load and viscosity. Curves 1, 2, and 3, falling film according to Equation 1; curve 1° and 2°, agitated film in the thin-film apparatus

higher viscosity will almost always operate within the filled gap zone, even at reduced loads (curve 2, 2*, and 3). For every level of product viscosity, it is possible to indicate a maximum load *H* for a given thin film apparatus. Above this limiting load, the influence of gravity alone is no longer enough to move the agitated film downward fast enough. This maximum load decreases rapidly with increasing viscosity and rotor speed, (*see* Figure 3). To prevent a possible overflow of viscous products at the desirable high peripheral speed, it is essential that the rotor be equipped with additional elements; these help to accelerate film flow axially.

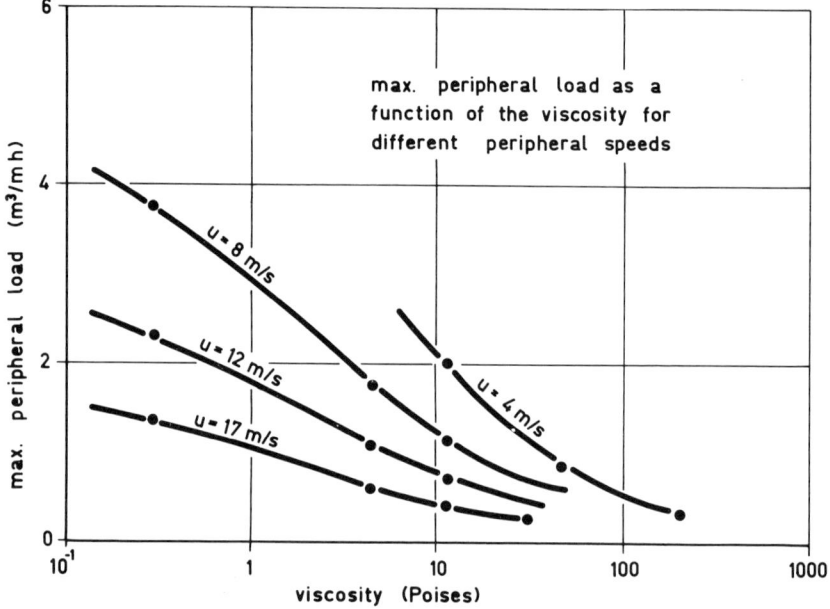

Figure 3. *Maximum peripheral load as a function of the viscosity for different peripheral speeds*

When the straight rotor blades are replaced by slightly pitched blades (design A in Figure 4), which accelerate the axial film flow, it is possible to reduce considerably the mean thickness of the film even with high viscosities. The mean thickness as measured in the apparatus is practically independent of the viscosity. However, in the case of design A, there is a risk that the film will not be renewed at a high enough rate and that the product will be turned downward as a spiral wave. The smaller film thickness resulting from higher speed confirm this assumption.

If we were to provide the rotor with additional distributing elements, as in design B, a more efficient film formation would occur and

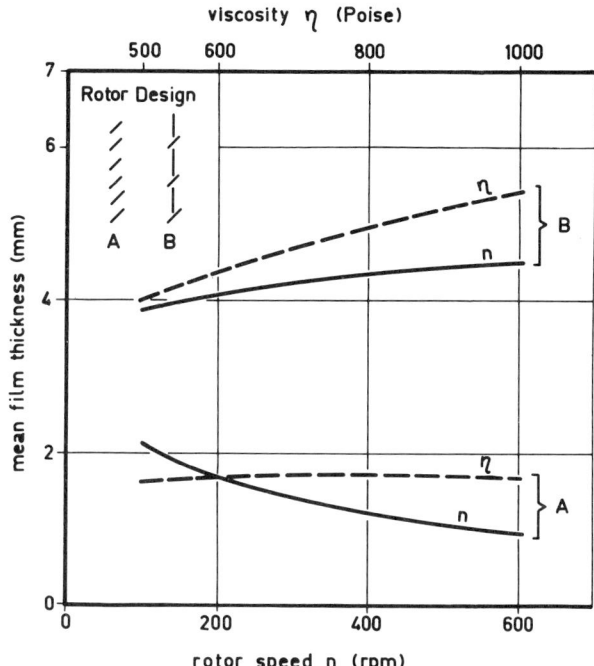

Figure 4. Influence of different rotor designs A and B in thin-film apparatus on the mean thickness, expressed as a function of rotor speed and viscosity

more favorable processing conditions would emerge despite the increase in the mean thickness of the agitated film. Figure 4 shows that the rotor speed has little effect on film thickness or residence time and, therefore, may be adjusted for optimal heat and mass transfer conditions.

Comparative measurements of residence time for rotor designs A and B showed relatively small differences in the width of the residence-time distribution characteristics. In Figure 5, the modified Pe number [according to Levenspiel (11)] is used to demonstrate the residence-time distribution. A lower Pe number corresponds to an increased degree of mixing and will be accompanied by a broader residence time distribution.

The difference in processing behavior between rotors provided with straight blades and those with pitched blades is shown in Figure 6 by an application referring to solvent recovery from a polymer solution of low initial viscosity. The evaporation capacity of a thin-film evaporator equipped with straight blades decreases considerably as soon as the concentrated polymer reaches a viscosity between 1000 and 2000 P. The greatly increased mean film thickness that characterizes this vis-

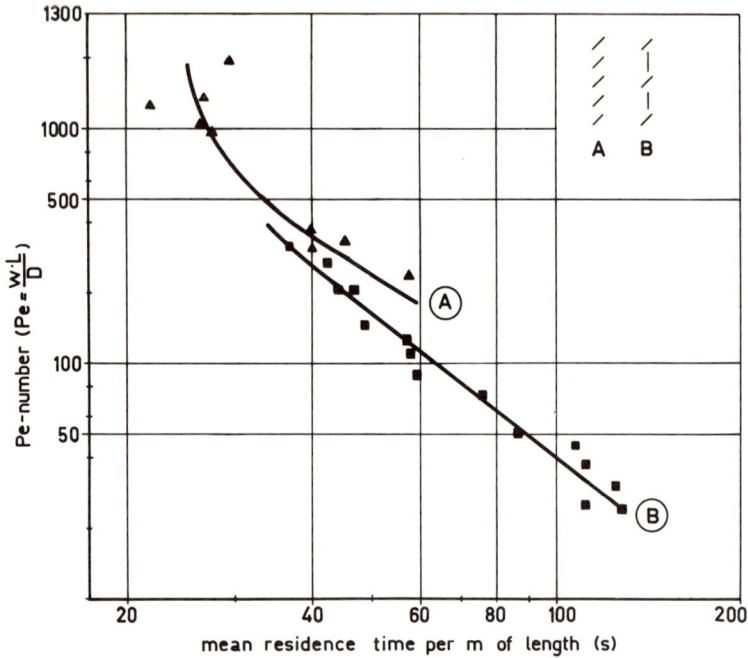

Figure 5. Residence time distribution, characterised by the Pe number, expressed as a function of the mean residence time for rotor design A and B (viscosity of 600—800 P).

cosity range prevents evaporation and, as a result of the limiting load, the throughput must be reduced. By contrast, in the Luwa high-viscosity machine in which the film is accelerated partially by the pitched-blade elements, only a very slight reduction in evaporation capacity can be measured up to viscosities of 20,000 P and more.

Design of The Luwa High-Viscosity Machine (Luwa-Filmtruder)

Figure 7 shows schematically the layout of the Luwa high-viscosity machine. The rotor designed to create the thin film is provided with a conventional bearing on the drive side and is supported at the opposite end by a guide bearing against the heating wall. The rotor itself is equipped with elements for distributing the film and for producing a forced thin-film flow along the wall. A short distance from the rotor is the discharging device, in this case a worm-type pump with a drive that is independent of the rotor drive.

The close proximity of processing and discharging zone ensures that the product to be processed in the thin-film machine will be subjected to a high shear rate from the moment of entry into the apparatus until final discharge. This high rate of shear creates favorable processing conditions for handling viscous non-Newtonian products, whose apparent viscosity decreases with increasing shear gradients.

The viscosity curve of a typical non-Newtonian product such as polystyrene is plotted in Figure 8 as a function of the shear gradient. Because of the small gap between rotor and heating wall, together with the relatively high rotor speed, the thin-film machine works within the limits of shear gradient of 1000 to 10,000 sec $^{-1}$.

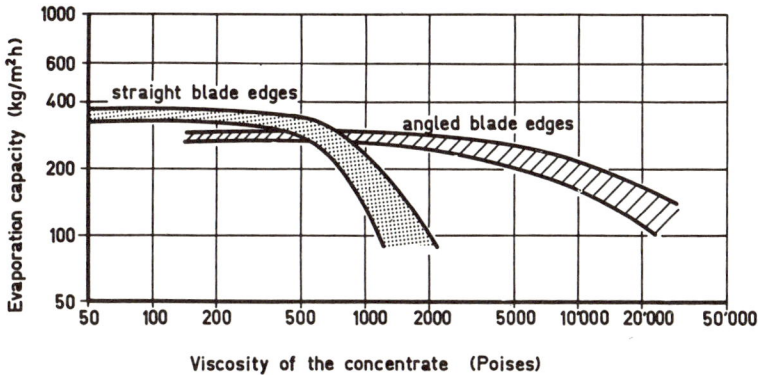

Figure 6. *Comparison of evaporation capacity between thin-film machines equipped with straight-blade rotors and the high viscosity machine (angled blades), expressed as a function of the concentrated polymer (polypropylene-heptane-solution)*

The reduction in apparent viscosity, particularly in the intake region of the rotor blade, results in a considerable improvement of the flow, and greatly accelerates heat- and mass-transfer processes compared with the same processing at low shear rates or in unstressed conditions. It should therefore be obvious that these machines, which produce an agitated film mechanically, are specially suitable for processing non-Newtonian viscous materials.

All measurements described here were carried out on a Luwa high-viscosity machine having a heating area of 0.5 square meter (Figure 9). A Luwa high-viscosity machine of industrial size (heating area of 12 square meters), used for solvent stripping from a viscous polymer, is shown in Figure 10.

Solvent Stripping of Polymer Solutions

To characterize heat transfer of solvent stripping processes in the Luwa high-viscosity machine, two different heat transfer coefficients have been defined:

$$\alpha = \frac{H_E^*}{F(T_w - T)} \qquad (2)$$

$$\alpha_{th} = \frac{H_E^* - H_R^*}{F(T_w - T)} \qquad (3)$$

Coefficient α is calculated from the total amount of energy H_E^* that is fed into the product by heat transfer through the heating wall and by internal friction between the rotor blades and the process fluid. (F indicates the surface area, and $T_w - T$ the temperature difference between heating wall and product).

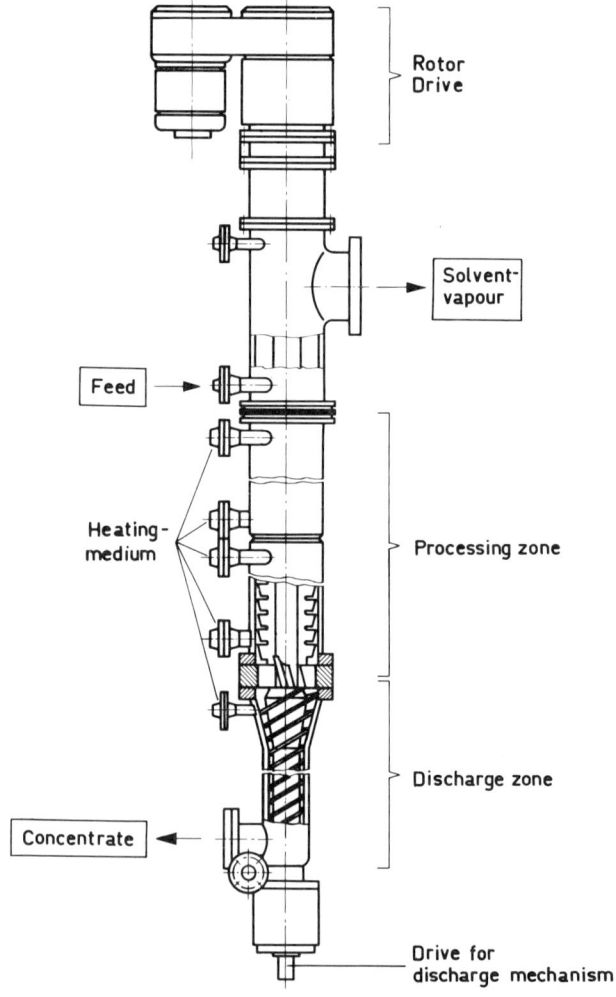

Figure 7. Schematic layout of high viscosity machine (Luwa Filmtruder) with discharge mechanism

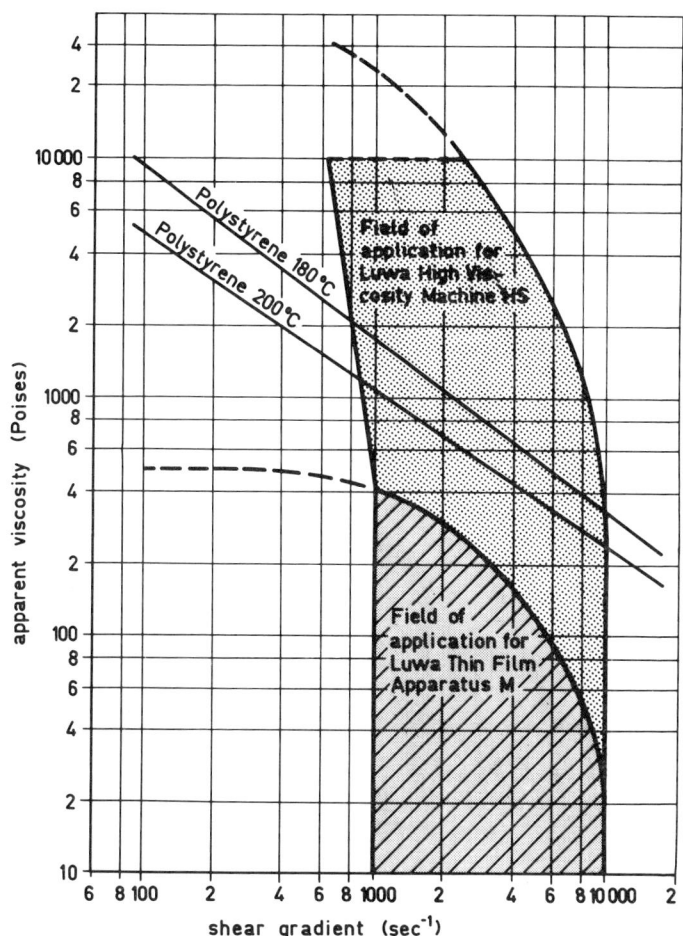

Figure 8. Viscosity curves for polystyrene expressed as a function of shear gradient. Fields of applications for Luwa thin-film machines type M and type HS (Luwa Filmtruder)

Assuming the dissipation (friction) energy is equal to the rotor drive energy H_R^* the coefficient α_{th} refers only to the amount of heat transferred directly from the heating wall into the film. For low-viscosity products, some correlations for this coefficient in agitated film equipment are known (14).

The heat-transfer coefficients are represented as a function of the peripheral speed of the rotor in Figure 11. In the evaluated tests, a polymer solution having an initial viscosity of approximately 300 P was concentrated down to a final product of about 3000 P. The higher peripheral speed produces only a small increase in the amount of heat

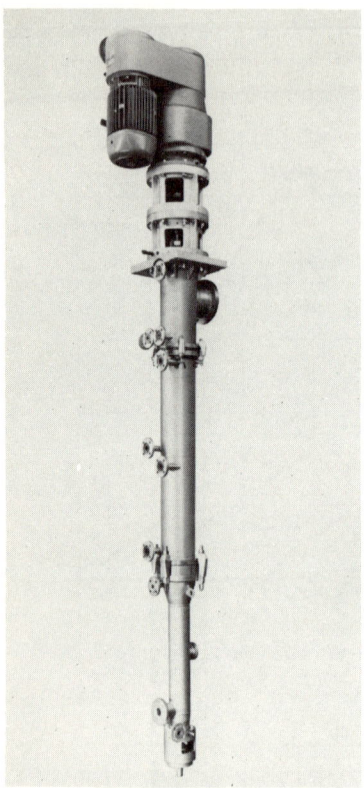

Figure 9. High viscosity machine (Luwa Filmtruder) with rotor drive and discharging device (heating area=0.5 square meter)

transferred between heating wall and film. However, a much higher increase in the total amount of energy fed into the product results because of the considerable rise in the energy dissipated by internal friction. The fact that the increase in α is proportional to the square root of the peripheral speed can be explained by the behavior of the agitated film corresponding to the penetration theory model (*12, 15, 16*). According to this theory, the heat-transfer coefficient α (as well as the mass-transfer coefficient) depends on the contact time Δt of every film element on the wall or on the film surface.

$$\alpha \sim (\Delta t)^{0.5} \qquad (4)$$

If we were to suppose that, in the agitated film, the surface of the film will be renewed after every passage of a blade, we would find:

$$\Delta t \sim \frac{1}{uz} \quad \text{and} \quad \alpha \sim (uz)^{0.5} \qquad (5)$$

(z = number of blades; u = peripheral speed)

The validity of the penetration theory points out that heat transfer in an agitated viscous thin film (even in the case of evaporation) and the mass transfer are mainly effected by forced convection and continuous surface renewal.

Figure 10. Industrial size high viscosity machine (Luwa Filmtruder) (heating area=12 square meters)

The residual solvent content of a highly viscous polymer can be reduced to less than 1%. (Figure 12). Because of the thermodynamic equilibrium conditions, it is necessary to increase considerably the product temperature to obtain low residual solvent contents. The higher product temperature can offer, at the same time, more favorable operating conditions, as shown in Figure 13. Tests that were carried out with a PVA-toluene solution enabled us to determine the course taken by the important processing variables, such as shear gradient S, film temperature T, viscosity η, and concentration X along the entire length of the processing zone.

Because of solvent removal, the concentration and apparent viscosity rise considerably with increasing length of the processing zone. Within the limits of 90-95% polymer, the influence exerted by the great increase

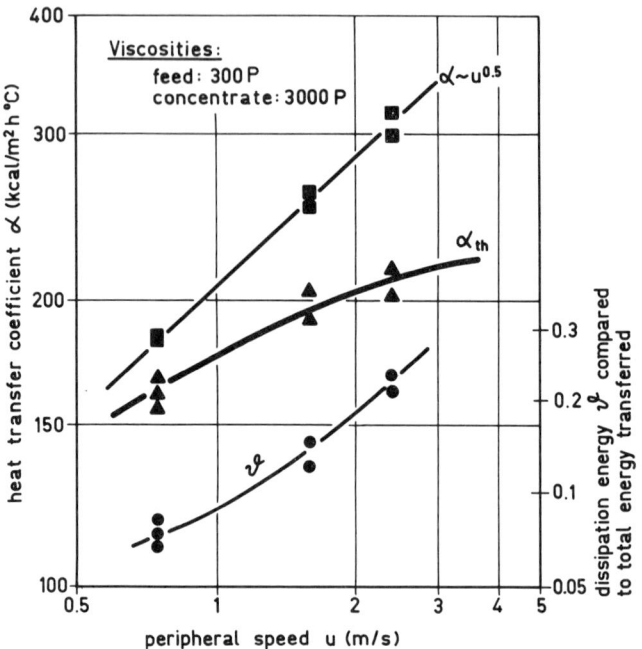

Figure 11. Heat transfer in a high viscosity machine expressed as a function of peripheral speed of the rotor (polypropylene-heptane-solution)

in the product temperature T (which has a much larger effect upon viscosity than does the small change in concentration) the apparent viscosity drops again toward the end of the concentration. In processes where polymer solutions are concentrated down to minimal solvent residual contents, it is evident that the maximum viscosity is reached somewhere in the processing zone. From the standpoint of equipment operation, the lower viscosity between rotor and discharge device aids the flow into the discharge pump.

Polycondensation of Polymers

For polycondensation reactions such as the final polycondensation of polyester, the reaction equilibrium and the reaction rate are mainly influenced by the removal of ethylene glycol (*13*):

$$\text{DGT} + [\text{BM}]_n \rightleftarrows [\text{BM}]_{n+1} + \text{G}$$
$$[\text{BM}]_n + [\text{BM}]_m \rightleftarrows [\text{BM}]_{m+n} + \text{G}$$

where DGT = diglycol therephthalate; BM = basic molecule; and G = ethylene glycol.

Polycondensation will be more complete as ethylene glycol is more efficiently removed from the reaction mixture. At higher viscosities, the mass transfer of the ethylene glycol is controlling the reaction. For this reason, a reactor must be designed to operate under very high vacuum to give a high specific surface area with respect to the product volume and to have very efficient agitation. Therefore, the agitated thin-film principle offers favorable operating conditions for this mass-transfer process.

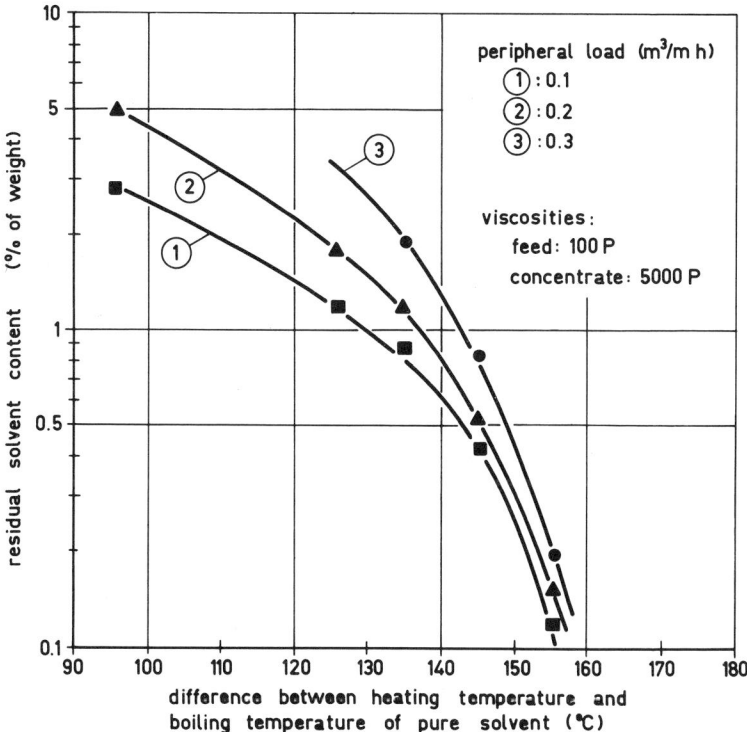

Figure 12. Residual solvent content expressed as a function of the temperature difference between heating medium and boiling temperature of the pure solvent (polypropylene-heptane-solution)

Figure 14 shows the measured increase of the intrinsic viscosity of polyester as a function of the residence time in an agitated thin-film reactor. This is compared with a reactor operating with free-falling films at the same operating pressure. Within this relatively low viscosity range, the higher rate of reaction in the agitated film is caused by the extremely high continuous surface renewal.

Figure 15 shows the same characteristics in an advanced polycondensation step of polyester. The differences in the reaction rates are

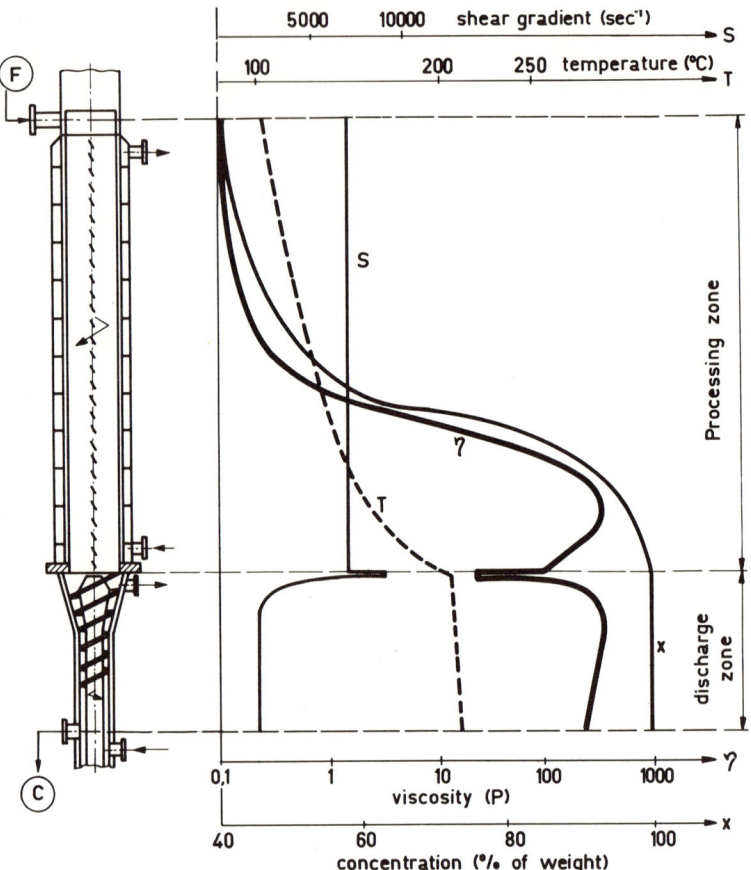

Figure 13 Course taken by shear gradient S, concentration X, viscosity η, and product temperature T along the length of the processing and discharge zone of a high viscosity machine (Luwa Filmtruder)

even more pronounced because of the increasing influence of the reduced apparent viscosity in the intake region of the rotor caused by the high shear rate. The importance of efficient agitation combined with rapid surface renewal is shown by the difference in reaction rates for a horizontal rigid film (with no surface renewal at all), in a free-falling film with slow surface renewal, and in the agitated film with high surface renewal and additional reduction of the apparent viscosity.

Summary

Mechanically agitated liquid films in thin-film machines are commonly used for continuous processing of heat sensitive products of low viscosity, which flow readily by gravity alone. In specially designed equipment for materials of higher viscosity, the agitated thin-film principle can be used to process very viscous non-Newtonian materials, solvent removal from polymers and polycondensation processes having viscosities exceeding 10,000 P under operating conditions.

In this thin-film machine, the small clearance between heated wall and rotor blade, together with the high peripheral blade velocity, results in high shear gradients, whereby the apparent viscosity in the film is considerably reduced. The resulting increased turbulence and better surface renewal improve heat transfer, increase reaction velocities, and aid the required forced product flow on the wall. On the basis of test

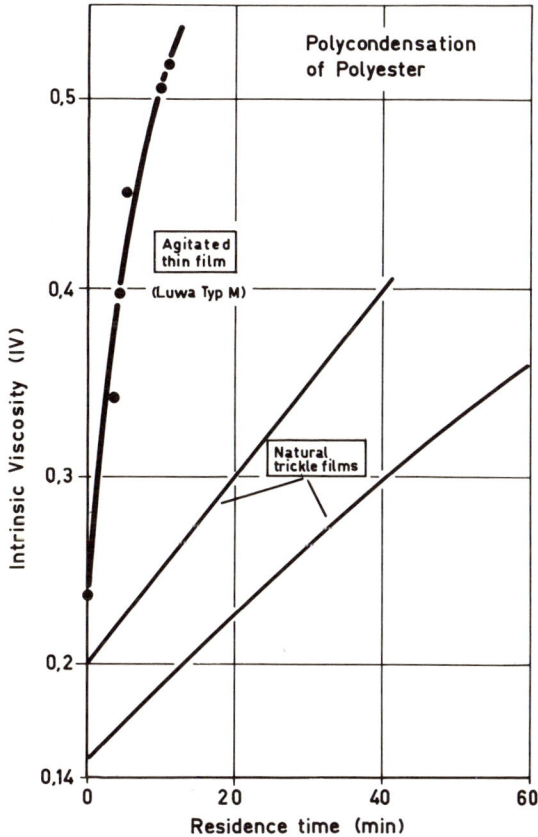

Figure 14. Increase of the intrinsic viscosity during the polycondensation reaction of polyester as a function of the residence time. Comparison between free-falling-film reactor (13) and agitated thin-film machine

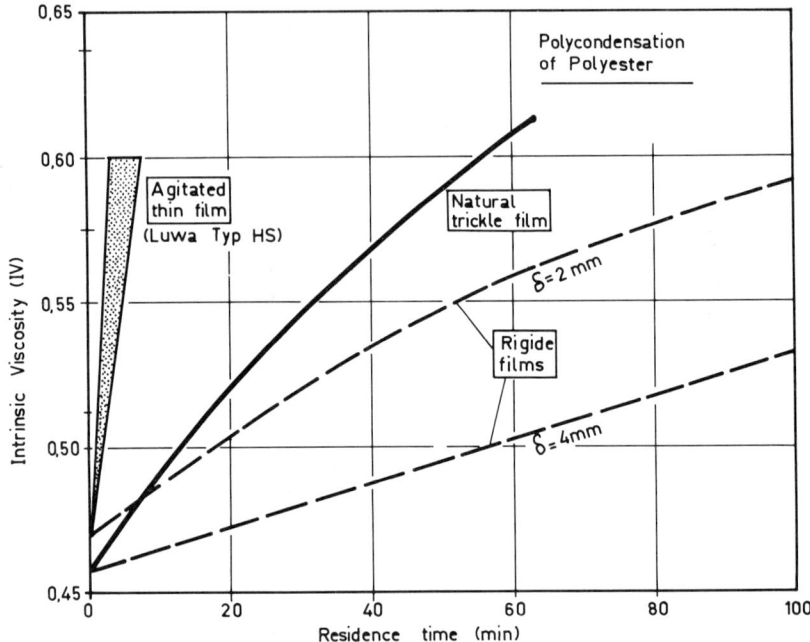

Figure 15. Final polycondensation of polyester. Comparison between horizontal rigid film, free-falling film, and agitated thin film in a high viscosity machine (Luwa Filmtruder)

results for various rotor configurations, the relation of heat transfer and residence time is discussed as a function of shear gradient and product viscosity. The effect of the agitated film on the reaction velocity of polycondensation reactions in comparison with the falling film is also covered.

Acknowledgment

The author thanks Luwa, Ltd., for permission to publish this paper, which is based on his work while he was R&D manager of Luwa's process equipment division.

Literature Cited

1. Dieter, K., *Chem.-Ing.-Tech.* (1960), **32**, 521.
2. Bott, T. R., Azoory, S., *Chem. Process Eng.* (1969), **50**, 185.
3. Mützenberg, A. B., *Chem. Eng.* (1965), **72**, 175.
4. Widmer, F., Giger, A., *Chem. Process Eng.* (1970), **51**, 11.
5. Stevanovic, V., thesis, ETH Zurich (1961).
6. Mützenberg, A. B., Giger, A., *Trans. Inst. Chem. Eng.* (1968), **46**, T 187.

7. Widmer, F., *Vakuum Technik* (1969) **18**, 147.
8. Hauschild, W., *Chem. Process Eng.* (1969) **50**, 83.
9. Nusselt, W., *Z.V.D.I.* (1923) **67**, 206.
10. Brauer, H., *VDI-Forschungsheft* (1956) 457.
11. Levenspiel, O., *Chem. Reaction Engineering* (1962).
12. Highbie, R., *Trans. Amer. Inst. Chem. Eng.* (1935) **31**, 365.
13. Dietze, M., *Chemiefasern* (1969) **3**, 194.
14. Penney, W. R., Bell, K. J., *Ind. Eng. Chem.* (1967) **59**, 4, 47.
15. Harriott, P., *Chem. Eng. Progr. Symp. Ser.* (1959) **29**, 137.
16. Latinen, G. A., *Chem. Eng. Sci.* (1959) **9**, 263.

RECEIVED March 29, 1972.

5

Mechanochemical Polycondensations and Polycomplexations

CRISTOFOR SIMIONESCU and CLEOPATRA VASILIU-OPREA

Polytechnic Institute of Jassy, Department of Natural and Synthetic Macromolecules, Jassy, Romania

> *Mechanochemical condensation of polyethylene terephthalate with aliphatic and aromatic diamines has been achieved by vibratory milling. The main factors influencing this process were the diamine's structure and the relation between reagents and mechanical parameters. The products of the polycondensation contain atoms (oxygen and nitrogen) with nonparticipating electrons, and were capable of being mixed with various metals. Crystalline metallic salts (ferric chloride, manganese chloride, and manganese acetate) added to the reaction gave polychelates. The products exhibit both thermostability and semiconductivity.*

Mechanical activation of compounds such as polymers consists primarily of changing valence angles and intermolecular distances, and in changing the potential energy of the materials. The changes give structural modifications at both inter- and intramolecular levels, resulting in the appearance of mechanically excited structures as the stress acts up to the moment of bond breaking. These structures are characterized by a greater reactivity than that possessed by the compound in its normal state. The potential energy accumulated in the fragments (radicals, ions, or mixed) of the deformed chain may be compared with the energy required to break the original chain.

The distortion of the valence angles and the increase in interatomic distances are phenomena produced simultaneously during mechanical stress. The one determines the other, and their effects are summed. The overall result is activation of the backbone and transition of the material to a mechanically excited (tensioned or stressed) state exhibiting a higher sensitivity to chemical agents. This result is mainly related to the structure of the polymer and the chemical in contact with the polymer during mechanical stress. Under stress (such as impact or high-frequency waves), active intermediate states appear. These are characterized by the weakening of numerous bonds (not only in the polymer's backbone, but also in the pendant groups); in other words, stress leads to increased reactivity. Subsequent transformations can therefore give active centers without chain scission (1).

The increase of interatomic distances by stretching the main chain and diminishing the energy of chemical bonds (that is, weakening them), is accompanied by redivision of the potential energy between the polymer backbone and that of its partner in the working medium. This results in the formation of mechanoexcited complexes (2). From there, the reaction mechanism is determined by the value of the activation energy (E_a) of the reaction. In mechanochemical destruction by a radical mechanism, E_a equals the energy of formation of the active particles.

Because this kind of reaction takes place at low temperatures, thermal oscillations do not essentially contribute to backbone stretching. In fact, E_a is zero in this case. When E_a of bond scission is less than that required to form the active particles, cracking exhibits the character of a "mechanically activated" chemical reaction—chain scission in active particles takes place. There is a cause-and-effect relationship between the strain processes (which are cumulative in the deformed fragment to activate the backbone) and the destruction processes.

The problem of "interconditioning" destruction—that is, interrupting the continuity of a stressed body—and of the strain representing the sum of all the regrouping phenomena taking place at atomic, molecular, and supermolecular levels is of particular interest. It allows us to explain the mechanical properties of polymers and to elucidate the causes of the initial process. Kinetic studies showing that the strength criterion is the longevity of the body under tension and that the deformation characteristic is the accumulation rate of the strains, point to the interdependence between strain and destruction (3-6). Both processes are of a "thermofluctuation" nature, being characterized by equal activation energies, as shown by studies of different materials—metals (3), polymers (4, 5), semiconducting crystals (6), and others.

A strong interconditioning between destruction and strain occurs at the atomic and molecular levels. (Information concerning transformations at this level is given by the activation energy of the process calculated from phenomenological studies.) We must explain the causes of this interdependence and ascribe a primary role to one of them. It is possible that this may not, for the time being, offer a general solution to the problem. The elementary acts that are the basis of these processes are too intricate and there is too great a variety of solid bodies and strain conditions. Nevertheless, some special cases have been already solved so far—for example, monoaxial stretching of the oriented polymers.

UV irradiation of a polymer tensioned by stretching has resulted in a considerable increase of the stationary flow rate (4). However, this kind of irradiation results in scission of the bonds of the macromolecular chains. For such solids, therefore, the strain (in stationary flowing) is determined by backbone breaking—that is, by the destruction process, although interconditioning the strain and the destruction processes appear at different levels of the supermolecular structure (6,7).

Although there are many studies dealing with the supermolecular structure of polymers and the influence of this structure on physicomechanical properties, there is a lack of information concerning transformations that take place at that level in materials under stress. This is especially true concerning transformations dealing with the mechanism of plastic deformation and the factors controlling the strength of the materials. Using monoaxial stretching of oriented polymers as an example once again, important information has come from studies of the structure of defect energy, defects and their crystal distributions, and the nature of surfaces resulting from detachment of different supermolecular structures.

Stressed Polyethylene. Studies involving stressed polyethylene under the polarizing microscope have shown that as the load gradually increases, plastic deformation begins on that spherulitic fragment for which axial orientation of the constituent ribbons coincides, at least roughly, with the axis of stretching. In Figure 1, the dark parts represent the domains of the plastic deformations, allowing detection of the zones in which microstructural transformations took place. When the

Figure 1. Initial deformation stage of spherulites
(7)

ribbon type of deformations are set normally *vs.* the spherulite radius affected by mechanical force, the deformation is called tangential (7).

Under these conditions, even high plastic deformation values do not affect the formation of ribbons which keep their individuality (Figure 2). At small elongation values in the spherulite center, microcracks appear; as the strain increases, these microcracks merge, becoming larger and forming a main crack that brings about destruction.

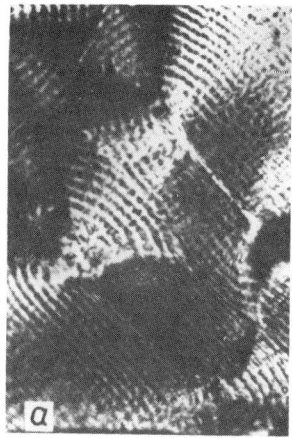

Figure 2. *Initial deformation stage of the crack at its center* (7)

Such information offers an opportunity to study details of the fibrillation mechanism. The fibers formed by stretching the spherulitic polymer representing nothing other than ribbon formations plastically deformed and oriented towards the mechanical stress that is released by comparatively weak mutual interconditions existing in an earlier formation (Figure 3). This behavior points to the existence of some weak surfaces in the crystalline polymers. Elements of the supermolecular structure detached by action of the external mechanical forces can slide on the weak surfaces. Evidence for the strain-destruction relationship must come from studies of the modification of the contact surfaces of two neighboring spherulites under mechanical stress.

Rupture occurs without important plastic deformation when the ribbons of some neighboring spherulitic formations show a parallel orientation in the contact zone. Conversely, considerable elongation occurs when the orientation is normal at the contact limit because of plastic deformation of the ribbons (Figures 4 through 6).

Figure 5 shows interspherulite destruction initiated by a small microcrack appearing just at the starting point of the deformation as a result of lateral and local ribbon separation. The subsequent evolution of these two processes—deformation and destruction—is determined by the increase in the number and size of the cracks. Crack formation results

Figure 3. Detachment of fibriles during deformation.

from destruction of the continuity of the structure of a polymer's solid bodies. Studies of the newly formed surfaces give information not only about the distribution of these rupture centers in the stressed mass, but also about the prehistory of the destruction process.

When the cracks first appear, the propagation rate is small, and the stressed state is preserved until the fracture occurs. Mechanical force orients both the structural units and the macro- and supermolecular formations into directions parallel to the direction of the force.

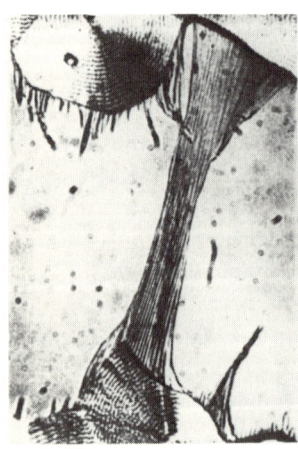

Figure 4. Character of interspherulite destruction, depending on orientation of ribbon formations at detaching limit (7)

The motion of the frontal part of the crack is thus determined by the boundary shift between the macromolecules containing broken bonds and their neighboring macromolecules, which are characterized by a reserve of energy and "ready" for their own bond scission. Hence, the destruction at this stage is preceded by orientation (8,9). The rate of crack development increases as the frontal part becomes more distant from the starting point.

Since there is a flexible zone before the breaking front (because of local tension fluctuations), even scission of chemical bonds is possible, as well as orientation processes and conformational changes. These scissions, in turn, accelerate the displacement of the breaking front during subsequent stages of crack development.

Figure 5. Interval destruction initiated by crack formed in spherulite center (7)

Displacement of the crack front—from the start of crack formation until final destruction—takes place at a variable rate. For the crack to overcome impediments (such as macromolecules, chain bundles, supermolecular structure formations, inclusions, and micropores), it needs varying time lengths, and the fissure perimeter takes on a sinuous form. The limit between different formations on the entire perimeter of the crack front is evidently determined by the equilibrium set up between elastic mechanical forces and bond forces; displacement of the crack front results from this equilibrium.

Let us assume a situation in which the elastic forces along the entire perimeter of the crack are in equilibrium, being at the end of a just-finished destruction act. To eliminate any possible impediment, a specified time is necessary to achieve a selective shift of the crack front. However, the crack front is hindered by neighboring difficulties that prevent the elastic forces from breaking the structural elements. Redistribution of tension on the perimeter of the crack then paves the way for sudden "annihilation" of all the impediments. Thus, elastic energy along the entire crack perimeter accumulates and, at the crack-polymer limit, an overloading occurs and becomes a force impulse. The field of tension and the impulse energy are, in a fragile or quasifragile way, able

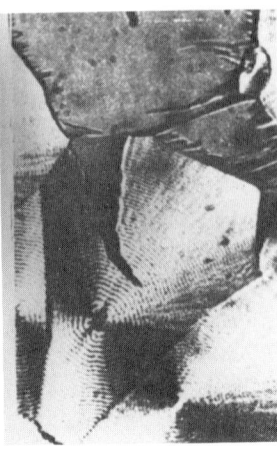

Figure 6. *Local tangential deformation of spherulite* (7)

to shift the crack perimeter with some value determined by the total energy reserve existing at the moment the crack "leap" is released. The process is successively repeated further on.

This point of view concerning the mechanism of mechanochemical destruction of macromolecular compounds is based on two factors. One is the gradual character of this process, as reported by us earlier (*10*). The other factor is the increase in potential energy and, consequently, the chemical reactivity of polymer systems that are placed in a stressed state mechanically.

Stages of Transformation. The gradual character apparently represents stages of the transformation initiated by mechanical energy at different structural levels of the polymer. These stages are:

- Polymer modification at the macroscopic level (either as a material subjected to mechanical processing or as a "running" object) consists of initiating the destructive phenomena at microdefects—that is, at submicroscopic cracks, statistically distributed on the surface or within the body of the stressed material. These cracks become centers where a detachment of intermolecular bonds occurs. This process might be called a "mechanical disaggregation," the opposite of "aggregation," a term that expresses (in this context) the assembly of various structural elements into polymers.

- Exertion of mechanical energy at the atomic/molecular level transforms the polymer into a precursor; at this stage, backbone breaking occurs as a change in interatomic distances and valence angles, or mechanical activation.

- Scission of chemical bonds results from the concentration of mechanical energy of the structural inhomogeneities; this represents mechanocracking or mechanochemical destruction.

At all levels of supermolecular structure, the transformations caused by mechanical stress stem from the anisotropy of the mechanical properties—by the existence of defects at the macroscopic level; by anisotropic properties of the crystal lattice at various levels of the supermolecular structure; and by chain asymmetry at the molecular level.

The transformations initiated by mechanical energy do not uniformly and necessarily pass through all three stages. The number and duration of the stages are determined by the nature and intensity of the mechanical stress.

When force is applied very intensely or very rapidly, the activation stage does not occur, and the polymer is affected at the atomic/molecular level by the direct breaking of chemical bonds. This bond breaking simultaneously causes, at the other structural levels, corresponding changes in the morphological formations according to the principle of crack emergence, development and propagation.

By contrast, when stress is applied slowly, transformations depend on how long it takes for deformations to accumulate. The transformations occur at the macroscopic level; the mechanical energy is first located on the body inhomogeneities and then on the structural ones, leading ultimately to bond breaking. At other times, when the stress is less than that needed for bond breaking, the polymer remains in an activated or mechanoexcited state, characterized by the peculiarities described elsewhere in this paper.

A frequent and well-known case involves stretching elastomers to their breaking points, thus increasing their chemical reactivity and making them more susceptible to the action of oxygen or ozone (11-13). In another case, some polymers with heteroatoms in their backbones (gelatin, for example) can be mechanically destroyed at low temperatures. Destruction is easier in the presence of some inactive saponification agents that otherwise act only at high temperatures. Actually, the transformation stages are inseparable and only the rates at which they occur differ, regardless of the level at which structure transformation begins—the level primarily "stressed" under the given conditions.

A body (polymeric material) inevitably involves some inhomogeneities at various structural levels. These represent concentration points of mechanical energy and, consequently, of destruction initiation. Bond breaking is unimaginable without corresponding transformations at other levels; conversely, it is also unimaginable that there can be a stress at macroscopic and supermolecular levels without scission, at least accidentally, of chemical bonds.

Another characteristic of stressed systems is that their energy levels rise in mechanoexcited states (that is, mechanoexcited complexes form), and intermediate active particles of all kinds (macroradicals, macroions, or macroion-radicals) appear at the last stage. That stage is mechanocracking, thus justifying the consideration of mechanochemistry as a general method for transforming macromolecular compounds.

Such a view leads to the idea that it is possible to extend mechanochemical synthesis, using practically all of the methods of macromolecular chemistry and mechanical energy as activation factors. With that end in mind, judicious elaboration of the reaction systems, the choice of an adequate kind of mechanical stress (which must agree with the physical state of the initial polymer), and the use of a macromolecular reaction partner are required.

The fundamental task, in our opinion, is to correlate the principles and methods of the proposed synthesis with those of mechanochemical synthesis. Thus, besides the destruction processes and mechanochemical synthesis discussed in the literature, other kinds of transformations sometimes occur as side reactions, or even as major processes. These include chemical fixation of small molecules (methyl chloride or butyl alcohol) on mechanically activated macromolecular backbones; grafting of inorganic surfaces (quartz, metals, metallic oxides, inorganic salts, etc.) dispersed by vibratory milling on polymerized fragments synthesized from monomers present in the reaction medium, and activated by centers on the inorganic surface (14); and the possibility of some reactions (such as nitration), achieved so far on macromolecular supports and only as side reactions.

Examples of the last possibility include: the use of vibratory milling and nitrogen oxides as a radical acceptors to nitrate the aromatic ring of polystyrene (15); celluluose xanthogenation (16); and dehydrochlorination and dehydrocyanation of some polymers.

The variety of reaction possibilities implies a new science—the science of mechanochemistry. Starting with just such results, we have achieved some new syntheses in the domain of macromolecular compounds—namely mechanochemical polycondensation and complexation.

Mechanochemical Polycondensation

The idea of this synthesis was suggested by a phenomenon observed during the mechanochemical destruction of polyamides. In addition to the main reaction of homolytic fragmentation of macromolecular backbones, a polycondensation occurred, involving the interaction of the amino and carboxylic end groups. These were condensed until they disappeared from the reaction medium:

$$NH_2 \ldots\ldots CH_2 \overset{\downarrow}{-} CH_2—CO—NH—CH_2 \ldots\ldots COOH$$

$$\downarrow \text{Mechanochemical destruction} \qquad (1)$$

$$NH_2 \ldots\ldots CH_2\cdot + \cdot CH_2—CO—NH—CH_2 \ldots\ldots COOH$$

$$NH_2 \ldots CH_2-CH_2-CO-NH-CH_2 \ldots COOH + NH_2 \ldots CH_2\cdot$$

$$\downarrow \qquad (2)$$

$$NH_2 \ldots CH_2-CH_2-CO-NH-CH_2 \ldots CO-NH- \ldots CH_2\cdot + H_2O$$

This observation led to the idea that both the choice of some suitable reaction partners and judicious selection of working condition could transform such side processes into main reactions, thus opening a new way to modify macromolecular compounds and a new prospect for mechanochemistry.

The mechanochemical polycondensation reaction has been studied using heterochain polymer systems—polyethylene terephthalate poly-(ε-caprolactam), cellulose, etc.—characterized by end groups that can be activated to increase their own number by mechanochemical destruction of corresponding polymers. The mechanochemical destruction was done in the presence of some suitable condensing agents, such as aliphatic and aromatic diamines and fatty acid dichlorides.

Mechanical processing was carried out by vibratory milling in the absence of air, the condensing agent having been purified earlier by suitable methods. The reaction product was treated with solvents specific for the condensing agents to remove unreacted excess. The product was then dried in vacuum and analyzed with a view toward establishing the existence of chemical bonds between the support polymer and the reactant used.

Influence of Physical and Chemical Factors on the Mechanochemical Polycondensation

Chemical Nature of the Polymer. The mechanochemical polycondensation can take place two ways: by condensing polymer end groups (the preexisting end groups and those obtained by mechanochemical scission of the chains at their heteroatoms) with those of the partner, or by interacting active centers formed by mechanocracking with the partner.

An attempt to verify the two hypotheses and to obtain information regarding the nature of the active centers formed by mechanochemical destruction of the support polymer consisted of the selection of three kinds of polymers for study:

(1) Those containing end groups whose number can be increased by mechanochemical destruction—polyethylene terephthalate, poly(ε-caprolactam), and cellulose.

(2) Those containing pendant functional groups and only carbon atoms in the main chain (polyvinyl alcohol).

(3) Those without functional groups (polyethylene).

These polymers were chosen to obtain some information about the nature of the activation centers, which eventually appear during the mechanochemical polycondensation. Amines, used chiefly as condensing agents, can act as radical acceptors. If the proposed reaction takes place *via* that kind of active particle, the polyethylene destruction fragments obtained by a homolytic chain scission would fix at the ends of the diamine introduced into the working medium and thus verify the hypothesis. This last aspect was also studied by introducing, besides condensation agents, some more-active radical acceptors such as oxygen, β-naphthol, and hydroquinone.

The conditions used for a radical reaction could require a modification of the mechanochemical polycondensation. Such a polycondensation of poly(ε-caprolactam) was carried out using sebacic acid dichloride as a condensing agent. For the mechanochemical polycondensation of all the other polymers, aliphatic and aromatic diamines were used. In all cases, the reaction was followed by determining chemically linked nitrogen, which decreased in the first case (owing to the fixation of acid dichloride) and increased in the remaining ones because of the reactions with diamines.

The experiments were carried out in an inert medium with liquid condensation agents at $18° \pm 1°C$; results are presented in Table I. Among the polymers shown in that table, only polyethylene terephthalate was systematically studied. Some informative results, yet to be completed, were collected with the other polymers.

Although limited, the results are still quite convincing, and enable us to characterize mechanochemical polycondensation. Heterochain polymers, which give an increasing number of functional end groups by scission of heteroatom links in the mechanochemical cracking process, bind large quantities of condensing agents—polyethylene terephthalate, cellulose, and poly(ε-caprolactam).

A comparison between two hydroxy polymers—cellulose and polyvinyl alcohol—whose chains have many pendant functional groups, shows an essentially different behavior. Cellulose, a polymer that is mechanocracked at the glucosidic groups and gives aldehyde end groups, binds large diamine quantities (about 10% over 48 hours). During the same time, polyvinyl alcohol binds incomparably smaller diamine quantities (1.5%), indicating the secondary importance of pendant hydroxyl groups. Finally, polyethylene—a polymer free of functional groups—does not bind ethylenediamine under the conditions used, but binds *m*-phenylenediamine (about 1.9%) in the same time period.

Table I. Behavior of Some Polymers Produced by Mechanochemical Polycondensation

Polymer (10 grams)	Condensing Agent	Milling Duration (hours)	Chemically Linked Nitrogen in Polymer (%)
Polyethylene terephthalate	ethylenediamine	6	4.8
		9	10.4
		48	17.0
		96	18.43
Poly(ε-caprolactam)	—	—	12.36
Poly(ε-caprolactam)	Sebacic acid dichloride	48	9.74
Cellulose	ethylenediamine	15	5.4
Polyvinyl alcohol	ethylenediamine	48	1.5
Polyethylene	ethylenediamine	48	—
	m-phenylenediamine	48	1.9

The results so far suggest that to achieve this reaction, it is important to use polymers containing functional groups susceptible to mechanochemical destruction along the main chain. It remains to be proved by additional experiments whether the active centers obtained react directly with the condensing agents, or whether there first occurs a stabilization of the end groups, followed by condensation with the partner.

In the mechanochemical destruction, the fixation of small quantities of m-phenylenediamine at the polyethylene destruction fragments and the results shown in Figure 7 support the idea that the active centers contribute to the successful completion of the process. The figure shows a clear-cut difference in chemical binding of condensing agents as a function of the working medium. Thus, in an inert medium (purified nitrogen, curve 2) and after a mechanical working time of six hours, the nitrogen content of the sample was 4.8%. In the presence of pure oxygen for the same working time, the oxygen content was 8.4%. As described elsewhere (18), the number of intermediate active destruction particles in the presence of oxygen is greater than in an inert medium. Oxygen fixation at the active destruction fragments occurs after peroxidation. Macroradicals obtained by peroxidation were more active in the subsequent reaction than those that appeared in an inert medium. However, only a detailed study using a larger number of radical acceptors and having different activities (reaction accelerators or inhibitors) can

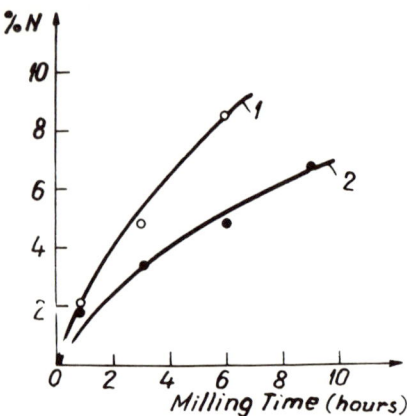

Figure 7. Mechanochemical polycondensation of polyethylene terephthalate with ethlylenediamine; 1, oxygen atmosphere; 2, nitrogen atmosphere

contribute to the elucidation of the mechanism of mechanochemical polycondensation.

Chemical Nature of the Condensing Agent. This problem was studied using systems formed from polyethylene terephthalate and different aliphatic and aromatic diamines. The selection of diamines was made to obtain information concerning the influence of their aliphatic and aromatic characters, the length of their hydrocarbon chains (for aliphatic amines) or the number of rings (for aromatic amines), and of the number and ring positions of amine groups. Since the majority of diamines answering the description in terms of structures are solids at room temperature, they were used either as solutions in tetrachlorethane, or as melts. The resulting data are in Table II.

As for the influence of the diamine structure, Table II shows that aromatic derivatives are quite reactive in polycondensation reactions when the ring is small and the number of amino groups large. In the case of aliphatic amines, the reaction ability is limited by an increase in the length of the hydrocarbon chain separating the two active groups.

The influence of the number of amino groups was pursued in the aromatic series using aniline and m- and p-phenylenediamine. Diamines are more active than monoamines, especially when the two amino groups are in the para position. The behavior of triamine is similar to that of diamine, perhaps because of steric hindrances.

Temperature. To determine the influence of temperature, milling of polyethylene terephthalate with ethylenediamine was done at $-50°$,

18°, and 75°C. All other conditions were held constant.

Two fractions—one soluble and the other insoluble—were obtained. The percentage of insoluble fraction decreased to a constant value after milling for 18 hours, with the largest percentage obtained at the lowest temperature (Figure 8, curve 1). The maximum percentage of soluble fraction corresponded to minimum values of insoluble fraction, with the largest amounts at the highest temperature; *see* Figure 9, curve 3.

The influence of temperature on nitrogen linking of insoluble fractions was also examined. The temperature was raised from $-5°$ to $120°C$ while maintaining the filling ratio (0.8%) and milling duration (three hours) constant. When data are plotted as in Figure 10, a direct correlation was obtained between temperature and nitrogen linking.

At higher temperatures, between $-5°$ and $40°C$, the reaction is essentially mechanochemical. At higher temperatures, participation of the thermal factor as a reaction activator is not ruled out.

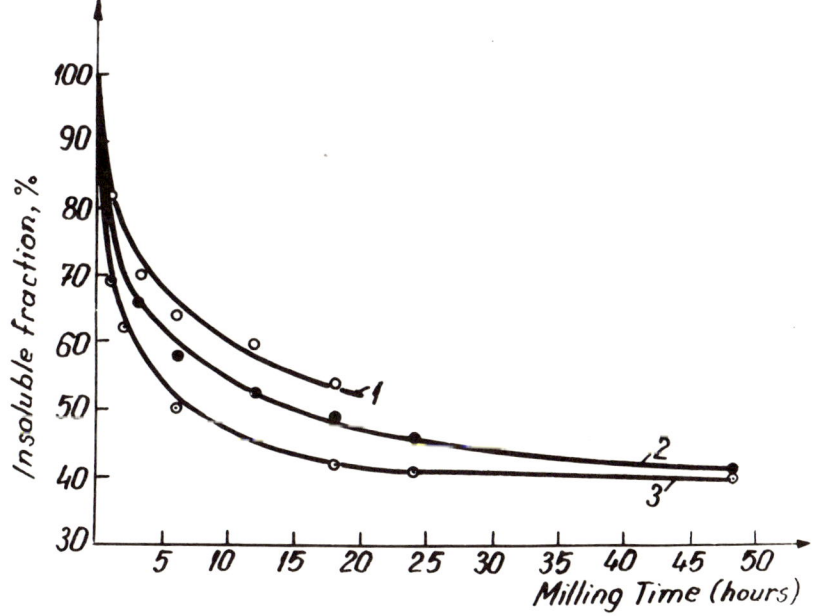

Figure 8. Variation of amount of insoluble fraction with milling time and temperature; 1, $-5°C$; 2, $18°C$; 3, $75°$

Mechanical Parameters

Filling Ratio and Milling Duration. The filling ratio, η, the ratio of the quantity of materials to be processed (polyethylene terephthalate and diamine) to the milling bodies (spherical particles 9 mm in

Table II. Influence of the Diamine Structure ($\eta = 0.8\%$;

Diamine	Structure of Diamine
Ethylenediamine p-Phenylenediamine	aliphatic-aromatic character
Ethylenediamine Hexamethylenediamine	length of hydrocarbon chain for aliphatic diamine
p-Phenylenediamine	aromatic ring
Benzidine	aromatic ring
Aniline m-Phenylenediamine p-Phenylenediamine	ring substitution

^a 75°C

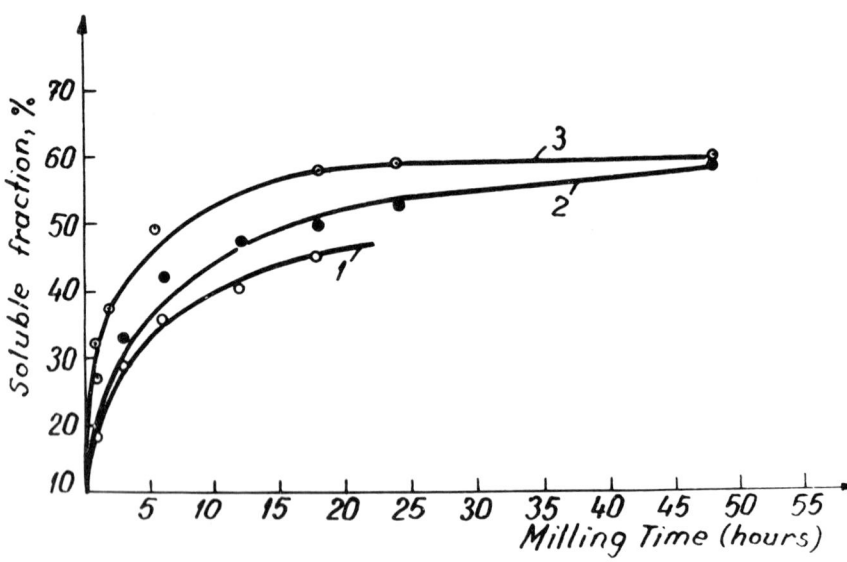

Figure 9. Variation of amount of soluble fraction with milling time and temperature; 1, –5°C; 2, 18°C; 3, 75°C

Milling Duration = 24 hours; Temperature = 18° ± 1°C)

Chemical Formula	Aggregation State	Chemically Linked Nitrogen (%)
$NH_2CH_2CH_2NH_2$	solution	2.00
NH_2—⟨C$_6$H$_4$⟩—NH_2	solution	3.75
$NH_2(CH_2)_2NH_2$	liquid	16.00
$NH_2(CH_2)_6NH_2$	solid	2.20
	melt	13.70[a]
NH_2—⟨C$_6$H$_4$⟩—NH_2	solution	3.75
NH_2—⟨C$_6$H$_4$⟩—⟨C$_6$H$_4$⟩—NH_2	solution	2.37
NH_2—⟨C$_6$H$_5$⟩	liquid	0.59
NH_2—⟨C$_6$H$_4$⟩—NH_2 (ortho)	solution	1.67
NH_2—⟨C$_6$H$_4$⟩—NH_2	solution	3.75

diameter) was also studied. Three successive millings were done at varying durations—1 to 24 hours for $\eta = 0.2\%$, 0.8%, and 3.2%. Results are given in Figure 11.

The polycondensation reaction was followed by determining the percentage of chemically linked nitrogen on support polymer by the Kjeldahl method. The correlation between the nitrogen percentage and the filling ratio leads to the conclusion that low values of the latter give the highest nitrogen percentage. The same plot shows the influence of the second parameter, milling duration. Nitrogen percentage increases with milling time and, consequently, the reaction develops along with the accumulation of mechanical energy in the system.

Influence of Building Materials. Influence of the building materials was determined by two successive millings, one using stainless-steel vessels and balls and the other using porcelain materials. Working temperature was about 28°C, and the filling ratio was 0.8%. The results are shown in Figure 12, which shows conclusive differences. Nitrogen percentages obtained with metallic equipment are plotted as an ascending line in that figure (curve 1) while porcelain equipment gives a descending line for similar milling times.

Starting with the idea that there are large differences in hardness between the building materials used in the two types of milling, we tried to obtain a correction factor *via* ash determinations. This was possible with metallic equipment although the results were only slightly affected by applying the factor. The ash percentage was small and nearly constant with respect to time (Figure 12, curve 2). When the porcelain mill was used, however, the polycondensation products gave, on combustion, increasing amounts of residue; the increase was proportional to milling duration, up to about 90% for a mechanical working time of 18 hours (Figure 12, curve 3).

The increase in the percentage of chemically linked nitrogen points to the existence of a reaction between the reactants and inorganic powder—that is, finely ground porcelain—inevitably present as a result of the balls' contact with the walls of the equipment. As a matter of fact, the possibility of a mechanochemical activation of inorganic materials by vibratory milling has been reported earlier (19-21).

Characterization the Polycondensation Products

To obtain some indication of the structure of the thermostable residue (undecomposed at 900°C) obtained in these reactions, its IR absorption was determined. Absorption maxima appear at 810 cm^{-1} and 1190 cm^{-1}, corresponding to $\equiv$C—, NH—, and $\equiv$C—N=bonds. The IR spectrum of the control sample lacks these bands, testifying to the existence of a reaction's having occurred either between the final polycondensation products and the active centers of the inorganic mass, or between the two compounds (polymer and diamine) and the active centers of the inorganic mass. The first of these possibilities is the more probable one.

Our assumption is that the lower values obtained for nitrogen linked by this fixing method perhaps cannot be determined by the Kjeldahl method. Also, the products obtained mechanochemically and those of the combustion residue may have different structures.

The IR spectrum recorded for an insoluble fraction obtained *via* mechanochemical polycondensation shows numerous new absorption bands, compared with the spectrum of the starting polymers. These bands are specific for groups formed by reaction with diamines. Thus, at 1470, 1640, 2990, and 3100 cm^{-1}, strong absorption maxima appear because of peptide groups, —CO—NH—. The absorption bands at 3300 and 920 cm^{-1} show amino and primary amino groups, respectively.

To establish the identity of the —CO—NH— links, a comparative recording was made between 2000 and 4000 cm^{-1}, a range that is specific for those groups. Samples of pure poly(ε-caprolactam) and a copolymer synthesized mechanochemically were used. The reaction time exerted a decisive influence on the polymer structure. The products obtained after short milling times (less than 20 hours) gave bands characteristic of the support polymer and a succession of new bands corresponding to groups formed by reaction with diamines. The products obtained after longer reaction times (96 hours, for example) did not give the most significant band of the support polymer—that is, the one appearing at 1720 cm^{-1} and corresponding to ester carbonyl in the starting polymer and in the IR spectrum of polycondensation products obtained after short reaction times.

We think that this change in the IR spectrum may be explained by a decarboxylation that takes place at long reaction times, followed by tautomerism of peptide groups:

$$-\underset{\underset{O}{\parallel}}{C}-\underset{\underset{H}{|}}{N}- \rightleftarrows -C=N- \atop |\atop OH$$

The insertion of peptide groups in the backbone of polycondensing products along the aromatic rings of the polyester chain provides the conjugation of ring electrons with those participating in the binding of oxygen and nitrogen. This idea is supported by properties such as electrical conductivity ($\sigma = 0.22 \times 10^{-12}$ to 0.55×10^{-8} ohm^{-1}cm^{-1} at 84° to 184°C), activation energy of the conductivity ($E_a = 0.98$ ev), and paramagnetism. The latter was observed both in the soluble and insoluble fractions by ESR spectra (Figure 13).

Mechanochemical Complexation

The observations recorded during mechanochemical polycondensation indicates that the fragments synthesized can react with the building materials of the milling vessel, thus suggesting a new kind of synthesis. Starting with the idea that the polycondensation products obtained mechanochemically contain, along with their macromolecular backbones, electron pairs that do not participate in binding, a complexing reaction was run using either trivalent iron (Fe^{3+} from crystalline ferric chloride) or manganese ion (Mn^{2+} from manganese chloride or acetate) as complexing centers. Thermostable semiconducting and chemically resistant polychelate compounds were obtained from these reactions. Thus, it is the first time polychelates have been synthsized by mchanochemical procedures. The polycondensing products of polyethylene terephthalate and ethylenediamine played the part of macromolecular ligands.

This synthesis was carried out by reaction of polyethylene terephthalate and ethylenediamine in the presence of the metallic salts. Mechanical activation was supplied by vibratory milling in a nitrogen atmosphere. Granular polyethylene terephthalate (supplied by U.F.S.-Jassy) was subjected to mechanical processing in powdered form. It was purified by dissolving in a 40/60 phenol/chloroform mixture and reprecipitating with methanol. After filtration, the polymer was extracted

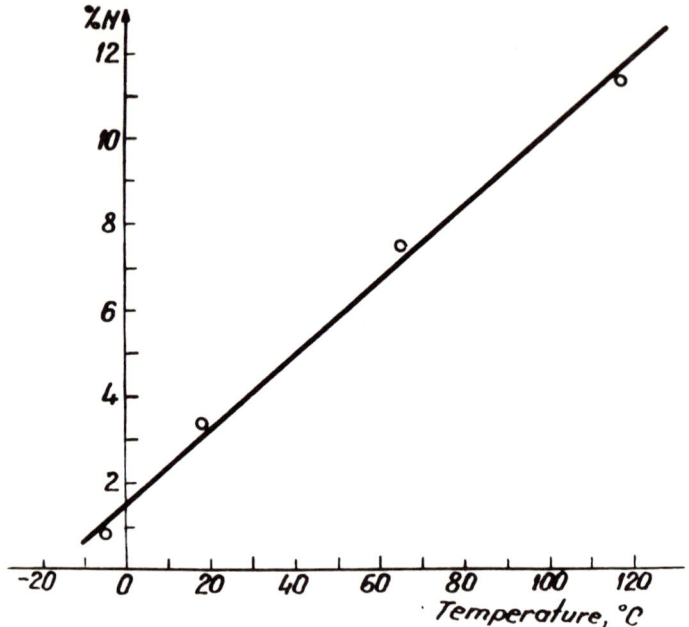

Figure 10. Influence of temperature on percentage of chemically linked nitrogen from insoluble fraction

with methanol until constant weight and dried at 105°C. Ethylenediamine was purified by distillation. The crystalline metallic salts, already of high purity, were used without further purification.

The reactants were introduced simultaneously into milling vessels. After the vessels had been hermetically sealed, air was removed with a vacuum pump, and residual air was removed by several flushings with purified nitrogen; milling was carried out in a nitrogen atmosphere. Milling was at a constant amplitude (4 mm) and frequency (1475 rpm, recorded by the rotation number of the electric motor).

Depending on the amount of amine used and on the milling time, the reaction mass either had a pastelike consistency or that of a fluid dispersion. The experiments were intended to establish some parameters (duration of mechanical processing, amount of diamine and complexing agent, etc.) and correlate them to characterize the polymers obtained, and to determine certain chemical and physical properties of the polymer. In all cases, the samples were purified by extraction in a Soxhlet apparatus with water or alcohol to remove unreacted ethylenediamine and metallic salts. The extractions were carried out until constant weight was obtained. Total removal of chloride was determined by silver nitrate. Purified samples were then washed with methanol, dried, and analyzed.

Some Determining Factors in Mechanochemical Complexation

Duration. The influence of time, which is actually the influence of the quantity of mechanical energy supplied to the system, was followed using a reaction mixture composed of support polymer (10 grams), ethylenediamine (30 grams), and crystalline metallic salt (6 grams). Vibratory milling was done over a broad time range, beginning with 15 minutes and ending with 96 hours.

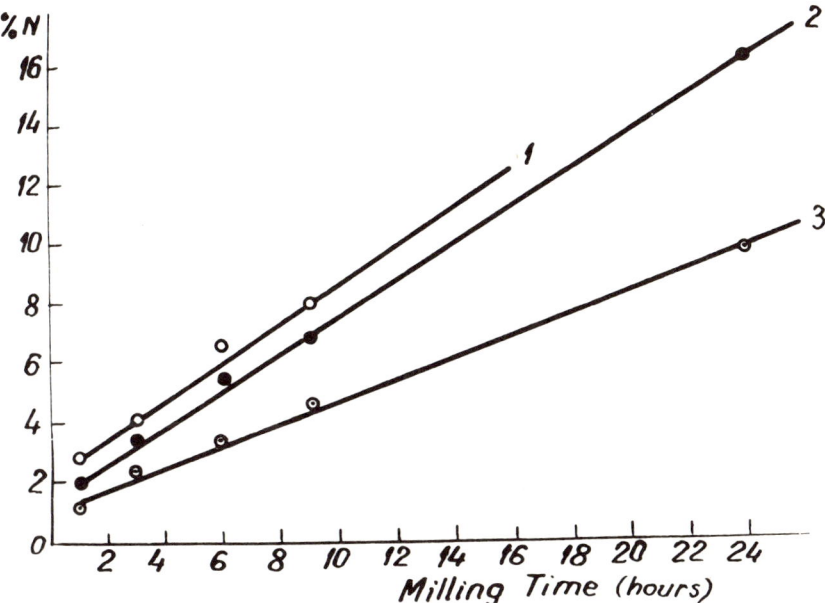

Figure 11. Influence of filling ratio and milling duration on mechanochemical polycondensation reaction: 1, $\eta=0.2\%$; 2, $\eta=0.8\%$; 3, $\eta=3.2\%$

The formation of polychelate compounds was determined by the percentage of nitrogen linked to the polyester backbone. This nitrogen percentage was compared with the nitrogen percentage of the products obtained without metallic salts. The results are presented in Figures 14 and 15 for complexations achieved with chloride salts (Fe^{3+} and Mn^{2+}) and in Figure 16 for manganese acetate. The nitrogen percentages of chelate products are smaller than those of the polycondensation product. Figures 15 and 16 show that in less than 20 hours of reaction, the percentage of chemically linked nitrogen changes with time in an ascending linear fashion. After that, the reaction proceeds more slowly and, at the maximum of 96 hours, the nitrogen percentage reaches a limit.

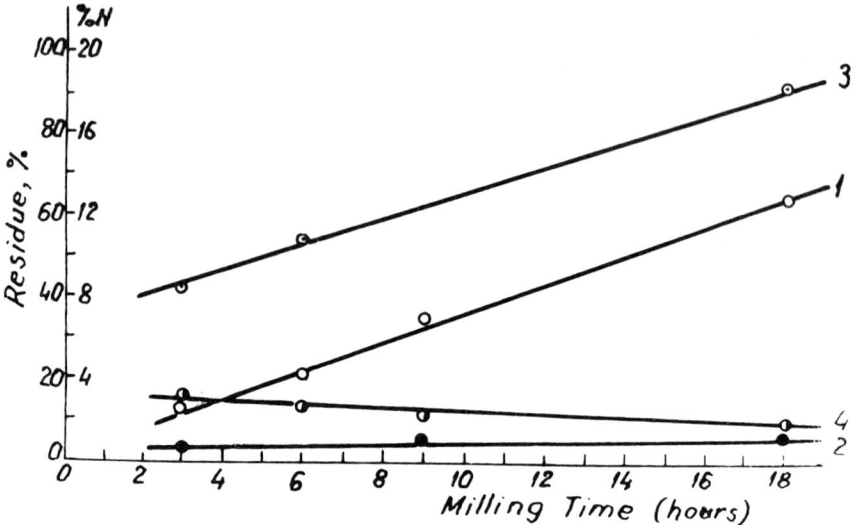

Figure 12. Influence of the nature of building material of equipment on percentage of chemically linked nitrogen; 1, percentage of nitrogen for stainless-steel equipment; 2, percentage of residue for stainless-steel equipment; 3, percentage of residue for porcelain equipment; 4, percentage of nitrogen for porcelain equipment

Diamine Quantity. The proportion of the reactants introduced into the working medium represents another significant parameter in the development of the process since it regulates the ratio between the mechanochemical polycondensation (ligand synthesis) and the complexation. This proportion was modified by introducing some variable

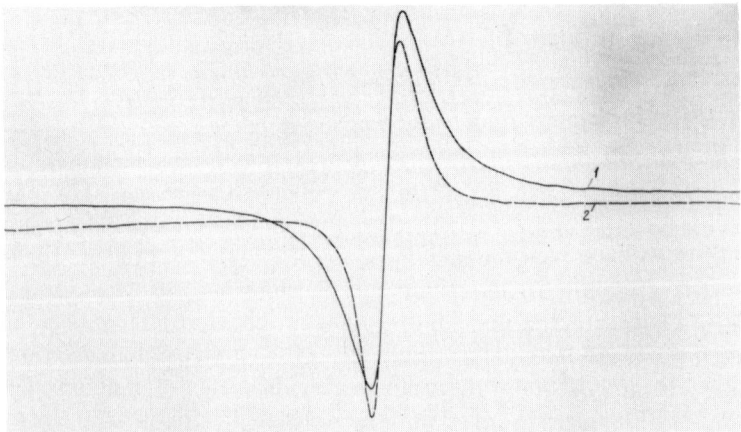

Figure 13. ESR spectrum of soluble fraction obtained after 96 hours; 1, soluble fraction; 2, standard (DPPH)

quantities of ethylenediamine (between 10 and 50 ml) and keeping constant the quantities of polyethylene terephthalate and metallic salt. The syntheses were carried out for a milling duration of nine hours. The data obtained in the presence of the two salts are presented in Figure 17. In both cases, the curves are ascending and have a flattening tendency for large diamine quantities (more than 50 ml).

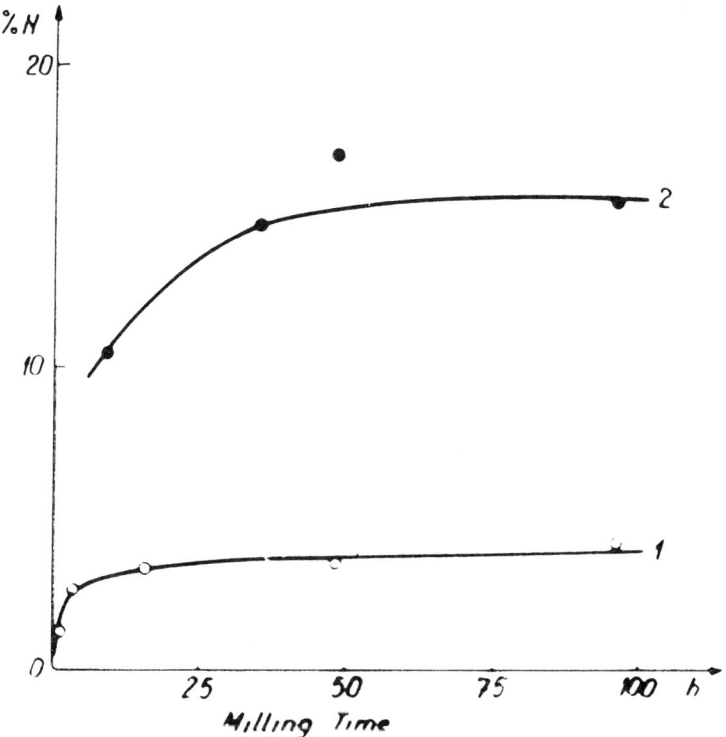

Figure 14. Influence of milling duration on mechanochemical complexation for the system polyethylene terephthalate, ethylenediamine and ferric chloride

Nature of the Metallic Salt. The data in Figures 15, 16, and 17 also show the influence of another parameter—the nature of the metallic salt. The two kinds of salts were chosen because the chlorides are salts of a strong acid, and the acetate is that of a weak acid. The data obtained from both the duration of the reaction and the diamine quantity show that the kind of salt used is important to mechanochemical complexation. Use of manganese acetate gives scission of a larger number of complexing centers than does use of choride salts.

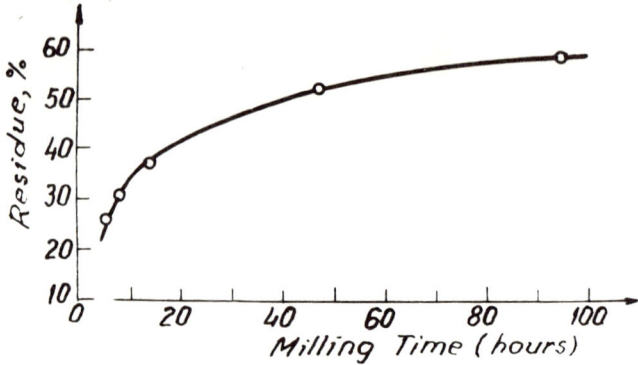

Figure 15. Influence of milling duration on mechanochemical complexation for the system polyethylene terephthalate, ethylenediamine, and manganese chloride

Characterization of the Products of Mechanochemical Complexation

As with the products of polycondensation, the products of mechanochemical complexation were characterized by analysis of chemically linked nitrogen by ligand synthesis (Kjeldahl method), elementary analysis, and IR spectroscopy. The nitrogen variance for different working conditions is discussed elsewhere in this paper; results of the elementary analysis are in Table III. IR spectra confirm that the ligand has the same structure as the polycondensation products, obeying the rule that for short durations, the band for ester carbonyl remains unaltered. For longer times, this band disappears.

The thermal decomposition of mechanochemically synthesized products was achieved in a combustion furnace at 700°C in a nitrogen stream. The data show that chemical composition of the polymers depends directly on the conditions of mechanochemical synthesis. Carbon content of the original polymer unmilled polyethylene terephthalate) was 63%; in polymer milled in the presence of metallic salt (manganese acetate and 96 hours) without diamine, it amounted to 59.17%. Carbon content drops from 34.04% (for 15 hours) to 19.84% as milling time increases to 96 hours. An increase in the amount of linked nitrogen by polycondensation agrees with the percentage obtained by the Kjeldahl method. Finally, the percentage of residue also increases with milling time—from 36.25 to 57.40% when the milling was done with manganese acetate.

The residue cannot be identified as one of the manganese oxides and used in calculating manganese content by the usual methods; such attempts gave exaggerated results. In fact, the residue is a macromolecular compound resulting from the thermal transformation of the polychelate; it is not transformed any farther at high temperatures (about 1000°C).

Another conclusion arising from the data in Table III is that the largest carbon loss takes place during the first reaction hours (less than 15), when the synthesis is probably intensely competitive because of the release of volatile derivatives. (Some gas chromatographic data indicate decarboxylation of the polyester). The polymer becomes more and more thermostable as reaction time increases.

Properties of Mechanochemically Synthesized Polychelates

Thermal Stability. The thermal behavior of the complexes was determined two ways. The first was thermogravimetric and differential thermal analysis. The second was to keep the polychelates in a furnace at up to 1000°C in air.

The first method used a Pauling-Erdeli-Pauling thermobalance at temperature intervals between 20° and 700°C in an inert medium. The temperature was raised 12.4°C per minute, the results being compared with those of the standard polymer. With a Fe^{3+} complex synthesized for 96 hours, the polychelate reached a maximum loss of only 28% at 440°C and often showed a constant thermal stability. Unmodified polyethylene terephthalate continually lost weight showing a loss of 95.9% at 577°C.

For the second method, samples obtained from reactions run under different conditions (milling duration, ratio of reactants and inorganic salt, etc.) were used (see Figures 18 through 22). The results show that the fraction of thermostable polymer from both Fe^{3+} and Mn^{2+} complexes increased with an increase in chemically linked nitrogen quantity which, in turn, was determined by the established parameters.

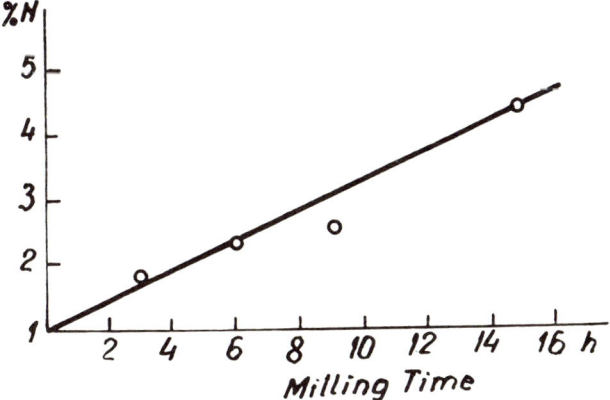

Figure 16. Influence of milling duration on mechanochemical complexation for the system polyethylene terephthalate, ethylenediamine and manganese acetate

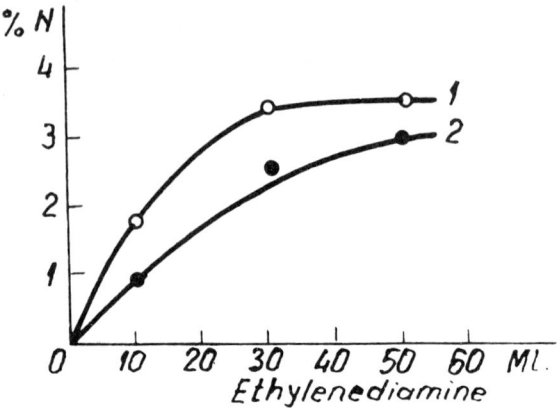

Figure 17. Influence of diamine quantity on mechanochemical complexation for the system polyethylene terephthalate, ethylenediamine and metallic salt; 1, manganese acetate; 2, manganese chloride

By using the most stable polychelate—the one synthesized using ferric chloride and a milling time of 96 hours—weight loss was only about 30% even at 990°C. The residue obtained showed some interesting properties, such as electrical conductivity and chemical resistance.

Electrical Conductivity. The existence of metal in the structural unit of a polychelate involves, besides the properties already discussed, perturbation of the electric moment of the macromolecule. Electrical con-

Table III. Elementary Analysis of Mechanochemically Synthesized Polychelates

Polyethylene Terephthalate	Milling Duration (hours)	Elementary Analysis, %				N Determined by Kjeldahl Method (%)
		C	H	N	Residue	
	0	63.60	4.81	0.00	0.00	0.00
Milled with $FeCl_3 \cdot 6H_2O$	48	47.33	4.00	0.00	27.67	—
		47.47	3.70	0.00	28.93	—
Milled with $FeCl_3 \cdot 6H_2O$	96	—	—	—	20.50	—
Complexed with $FeCl_3 \cdot 6H_2O$ and ethylenediamine	96	14.98	2.41	3.80	9.51	—
Milled with $(CH_3COO)_2Mn \cdot 4H_2O$	96	59.17	0.0	2.39	0.0	—
Complexed with manganese acetate and ethylenediamine	15	34.04	3.64	4.54	36.25	4.8
	48	21.98	3.42	5.35	51.65	5.9
	96	19.84	3.53	5.98	57.40	6.0

ductivity measurements were made for samples containing different amounts of nitrogen selected either from the series concerning the composition dependence vs. milling duration, or from the series concerning the same dependence vs. quantity of condensing agent. The measurements were made over temperatures from 12° to 100°C.

The values obtained obey the fundamental rule of semiconductors, with activation energies varying between 1.0 and 1.6 ev for manganese

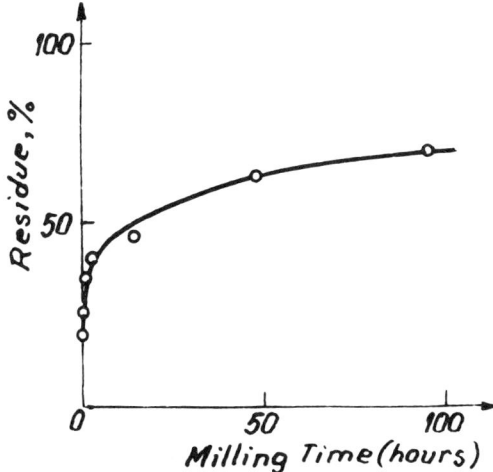

Figure 18. Variation of amount of thermostable polymer from the complex of polyethylene terephthalate, ethylenediamine, and Fe^{3+} with milling time

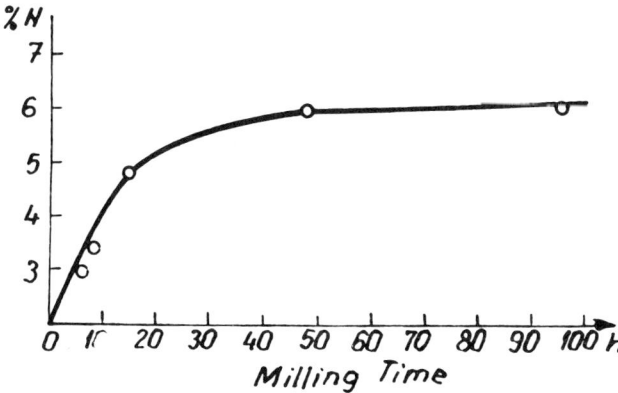

Figure 19. Variation of amount of thermostable polymer from the complex of polyethylene terephthalate, ethylenediamine, and Mn^{2+} (from manganese acetate) with milling time

polychelates, and between 2.09 and 1.03 ev for ferric polychelates. The data prove that the activation energy of electrical conductivity depends on the conditions of the mechanochemical synthesis (milling duration,

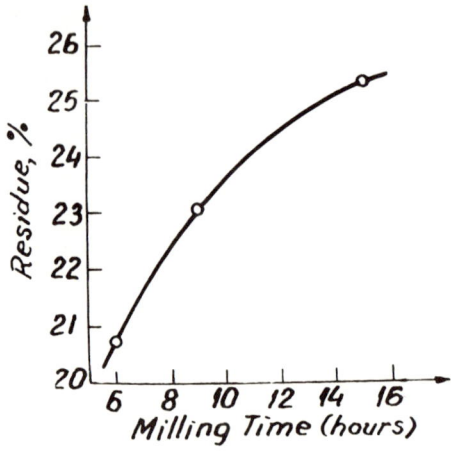

Figure 20. *Variation of amount of thermostable polymer from the complex with Mn^{2+} (from manganese chloride) with milling time*

Table IV. **Dependence of Electrical Conductivity on Parameters of the Mechanochemical Synthesis**

Polymer	Parameter of Mechanochemical Synthesis	Electric Conductivity σ ohm^{-1} cm^{-1}
Polychelate based on Fe^{3+}	Milling Time (hours)	
	0.25	$(2.66 \times 10^{-13})_{30°C} - (1.03 \times 10^{-9})_{106°C}$
	15.00	$(2.24 \times 10^{-10})_{30°C} - (5.6 \times 10^{-7})_{108°C}$
	96.00	$(3.85 \times 10^{-10})_{30°C} - (1.6 \times 10^{-6})_{106°C}$
	Ethylenediamine (grams)	
	15.00	$(1.59 \times 10^{-10})_{30°C} - (2.6 \times 10^{-7})_{121°C}$
	30.00	$(2.24 \times 10^{-10})_{30°C} - (5.6 \times 10^{-7})_{108°C}$
Polychelate based on Mn^{2+} (polyethylene terephthalate + manganese acetate + ethylenediamine)	Milling Time (hours)	
	96.00	$(1.15 \times 10^{-10})_{12°C} - (5.41 \times 10^{-:})_{96°C}$
Polyethylene terephthalate + manganese acetate	Milling Time (hours)	
	96.00	$(1.15 \times 10^{-14})_{134°C} - (1.5 \times 10^{-15})_{119°C}$

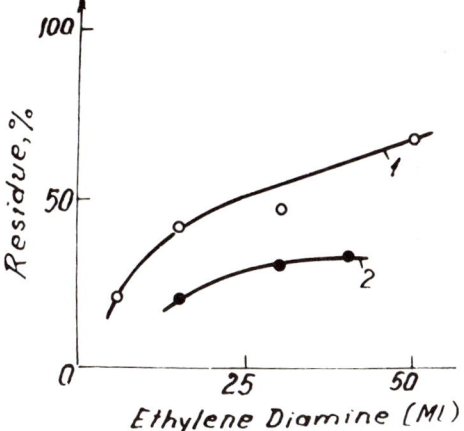

Figure 21. Variation of amount of thermostable polymer from the complex with Fe^{3+} with diamine quantity

amount of condensing agent, etc.) and thus on polychelate composition and structure. (See Figures 23 through 27).

Table IV contains some comparative data regarding the electrical conductivity of some polychelates based on Fe^{3+} and Mn^{2+}. The data dealing with electrical conductivity of polychelates, the starting polymers (for polyethylene terephthalate, $\sigma = 10^{-15}$ ohm^{-1}cm^{-1}), and polyethylene terephthalate milled with metallic salt but without diamine show essential differences. However, only the polychelates are characterized by electrical conductivity values and activation energies that justify placing them in the semiconducting class.

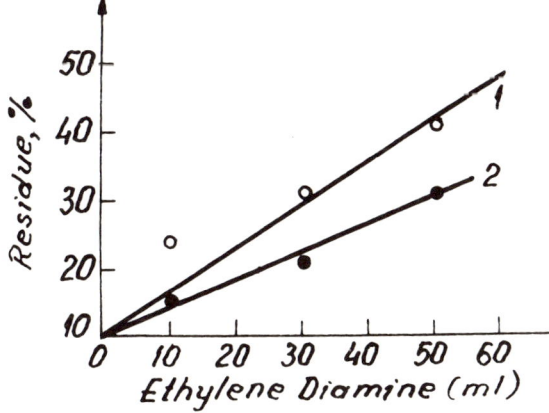

Figure 22. Variation of amount of thermostable polymer from the complex with Mn^{2+} with diamine quantity; 1, manganese acetate; 2, manganese chloride

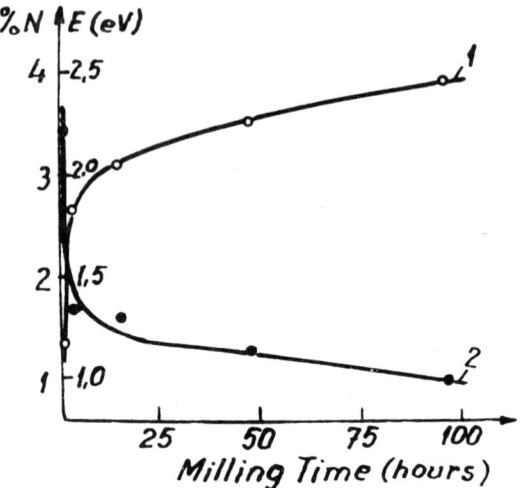

Figure 23. Variation of nitrogen percentage (1) and of activation energy of electrical conductivity (2) with milling time for ferric polychelates

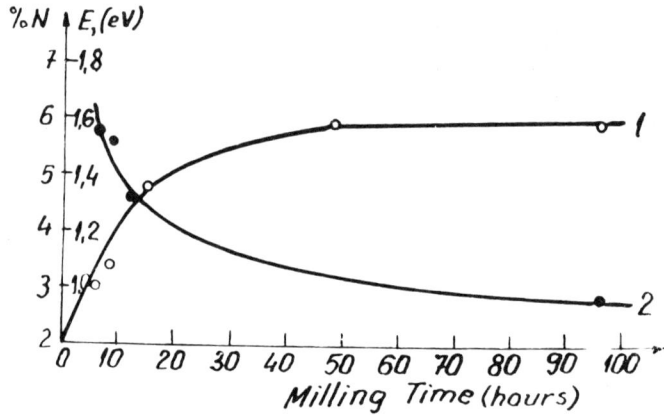

Figure 24. Variation of nitrogen percentage (1) and of activation energy of electrical conductivity (2) with milling time for manganese polychelates

Dielectric Constant. Insertion of electropositive elements such as Fe^{3+} and Mn^{2+} into molecular backbones that contain exclusively electronegative elements such as nitrogen and oxygen significantly modifies dielectric constant values. The initial material passes by complexation from being an electrical insulator to being a semiconductor, as shown in Figures 28 and 29. The data in those figures were obtained by measuring dielectric constants and tangents of loss-angles; these were plotted vs. frequency and temperature.

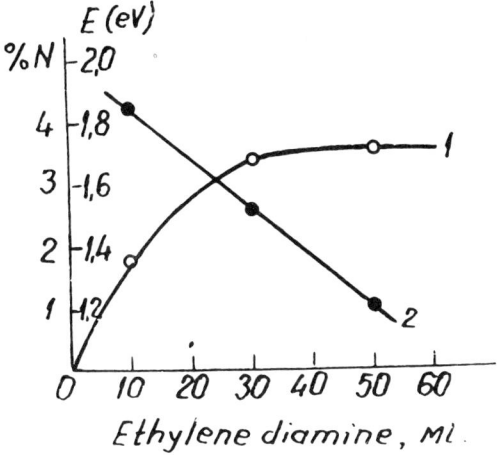

Figure 25. Variation of nitrogen percentage (1) and of activation energy (2) with ethylenediamine quantity for manganese polychelates

The change in the value of tangent of loss-angle vs. frequency was measured at two constant temperatures, 20° and 40°C. At 40°C, this magnitude reaches a maximum over the frequency interval of 10^3 and 10^4 Hz. This maximum does not appear at 20°C; it was probably shifted toward lower frequencies that were not measured (Figure 29).

Magnetic Properties. Insertion of the two metals into polyesters modified by mechanochemical polycondensation with ethylenediamine confers magnetic properties on the products. To prove this, ESR spectra

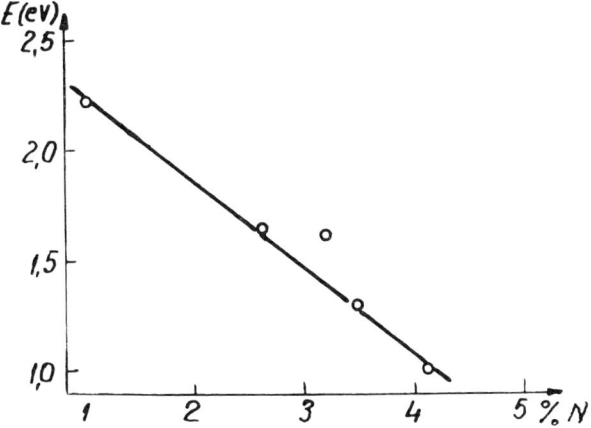

Figure 26. Variation of activation energy with nitrogen percentage for samples obtained with different milling times

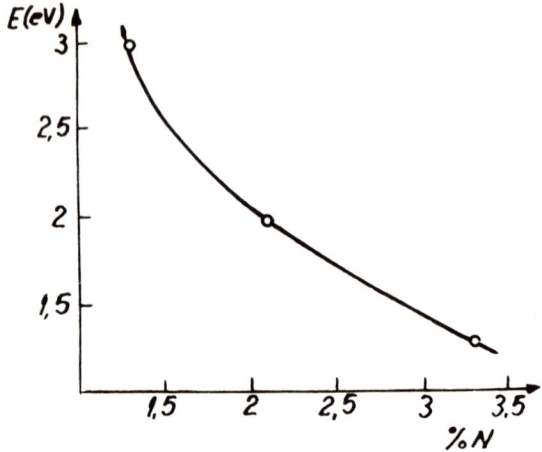

Figure 27. Variation of activation energy with nitrogen percentage for samples obtained with different diamine quantities

were recorded, using a sample of polychelate synthesized from polyethylene terephthalate, ethylenediamine, and manganese acetate, and a milling time of 12 hours.

The usual ESR spectrum of manganese is characterized by the existence of six maxima. With our complexes, however, owing either to the screening produced by spins of nonparticipating electrons along the

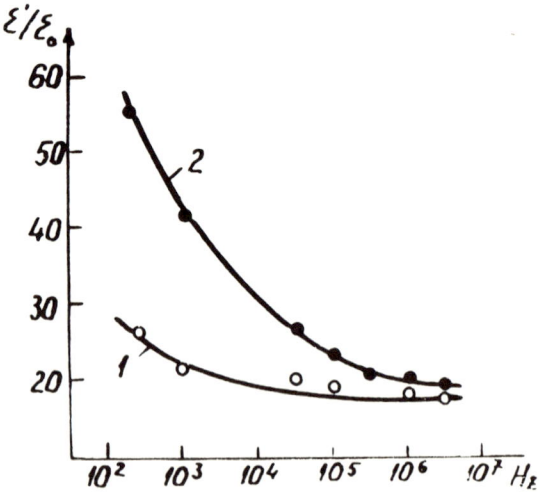

Figure 28. Dielectric constant of a ferric polychelate obtained for a milling time of 96 hours vs. frequency; 1, 20°C; 2, 40°C

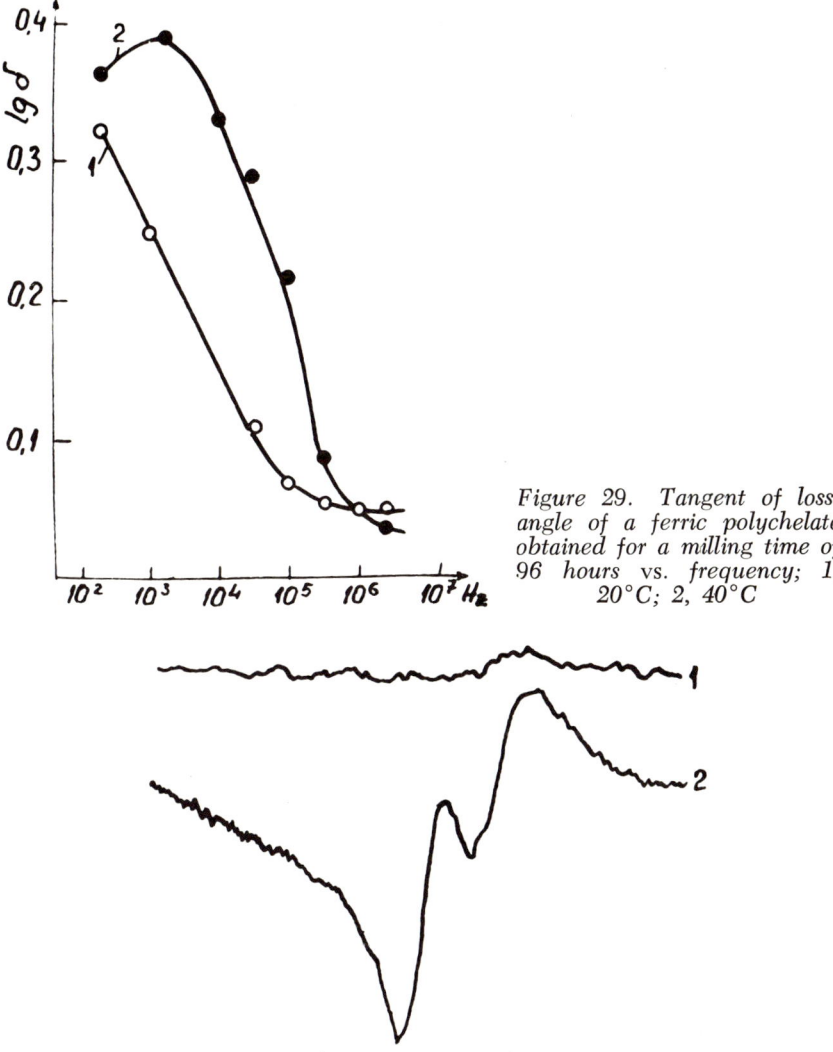

Figure 29. Tangent of loss-angle of a ferric polychelate obtained for a milling time of 96 hours vs. frequency; 1, 20°C; 2, 40°C

Figure 30. ESR spectra of polyethylene terephthalate (1) and mechanochemically synthesized polychelate (2)

ligand backbone or to too large a concentration of manganese electrons, only two maxima appear (Figure 30).

Discussion of Results

The mechanochemical synthesis of polychelates in the system polyethylene terephthalate/ethylenediamine/metallic salts confirms the idea set forth previously (22) concerning the ability of mechanical energy to activate mechanochemical polycondensation between the active destructive fragments of polyethylene terephthalate and ethylenediamine. The polymer supplies chains to complex with metallic centers because of the presence of nonparticipating electron pairs along the backbone of some atoms (oxygen and nitrogen). Data obtained by chemical analysis, IR spectroscopy, ESR, etc., together with the properties of the new polymers, point to formation of some polychelates.

Starting with the electronic structures of the complexing metal and the ligand, the structural unit of mechanochemically synthesized polychelates can be represented this way:

$$\left[\begin{array}{c} \cdots\text{—CH}_2\text{—CH}_2\text{—O—C}\underset{\underset{O}{\|}}{}\!\!\left\langle\!\!\bigcirc\!\!\right\rangle\!\!\underset{\underset{O\cdots}{\|}}{\text{—C—NH—R—NH—C}}\!\!\left\langle\!\!\bigcirc\!\!\right\rangle\!\!\underset{\underset{O}{\|}}{\text{—C—O—CH}_2\text{—CH}_2}\text{—}\cdots \\ \text{—Me—} \\ \cdots\text{—CH}_2\text{—CH}_2\text{—O—C}\underset{\underset{O}{\|}}{}\!\!\left\langle\!\!\bigcirc\!\!\right\rangle\!\!\underset{\underset{O}{\|}}{\text{—C—NH—R—NH—C}}\!\!\left\langle\!\!\bigcirc\!\!\right\rangle\!\!\underset{\underset{O}{\|}}{\text{—C—O—CH}_2\text{—CH}_2}\text{—}\cdots \end{array} \right]^{n+}$$

where R = alkyl or aryl, and $Me^{n+} = Fe^{3+}$, Mn^{2+}, etc.

The IR spectra prove that by increasing milling time, the ligand tautomerizes, as shown earlier, to give a form that is more thermally and chemically stable:

$$\left[\begin{array}{c} \cdots\text{—CH}_2\text{—CH}_2\text{—O—C}\!\!\left\langle\!\!\bigcirc\!\!\right\rangle\!\!\text{—C=N—R—N=C}\!\!\left\langle\!\!\bigcirc\!\!\right\rangle\!\!\text{—C—O—CH}_2\text{—CH}_2\text{—}\cdots \\ \text{HO} \quad \text{OH} \\ \text{—Me—} \\ \text{HO} \quad \text{OH} \\ \cdots\text{—CH}_2\text{—CH}_2\text{—O—C}\!\!\left\langle\!\!\bigcirc\!\!\right\rangle\!\!\text{—C=N—R—N=C}\!\!\left\langle\!\!\bigcirc\!\!\right\rangle\!\!\text{—C—O—CH}_2\text{—CH}_2\text{—}\cdots \end{array} \right]^{n+}$$

Diamine plays a prominent role in the complexing reaction because the polymers obtained by vibratory milling of both polyethylene terephthalate and inorganic salts with diamine do not differ essentially in their structures and properties from those of the original polymer.

In addition, such a method of synthesis can be generated by making use of another destructive means for ligand synthesis. In that sense, the method was verified by achieving some polycondensation products of polyethylene terephthalate and ethylenediamine and their complexation with different metals by irradiating the system with cobalt-60 (gamma rays). These results also make up the subject of other studies.

We consider use of the destructive methods to be helpful in synthesizing new polymers with properties differing from those of the support polymers. Polymers made that way enhanced thermal stability, electrical conductivity, magnetic properties, and chemical resistance.

Literature Cited

1. Butyagin, P. Yu., *Dokl. Akad. Nauk. SSSR* (1961) **140**, 145.
2. Baramaboim, N. K., "Mechano-chemistry of Polymers," Moscow, 1971.
3. Jurkov, S. N., Safirova, J. P., *J. Tech. Fiziki* (1958) **28**, 1720.
4. Regel, V. R., Tchornyi, N. N., *Vysokomol. Soedin.* (1963) **5**, 925.
5. Regel, V. R., *Vysokomol. Soedin.* (1964) **6**, 1092.
6. Kuksenko, V. S., Stutsker, A. I., *Mechan. Polym. (SSSR)* (1970) **1**, 43.
7. Kasatkin, B. S., Grinyuk, V. D., *Mechan. Polym. (SSSR)* (1968) **1**, 3.
8. Kuvinski, E. V., *U.F.N.* (1959) **67**, 177.
9. Lebedev, G. A., Kuvschinski, E. V., *Dokl. Akad. Nauk SSSR* (1956) **108**, 1096; *Fiz. Tverd. Tela* (1965) **3**, 2672.
10. Simionescu, Cr., Vasiliu-Oprea, C., *Mat. Plast.* (1970) **8**, 385; *Plaste Kautschuk* (1971) **18**, 484.
11. Kuzminski, A. S., Maisels, M. G., Lejnev, N. N., *Dokl. Akad. Nauk SSSR* (1950) **71**, 319.
12. Dogatkin, B. A., *Kaut Gumni, Kunstst.* (1959) **12**, 5.
13. Kuzminski, A. S., Lyubtehauskaya, L. J., "The Aging and Fatigue of Natural and Vulcanized Rubber and the Increase of their Stability," Moscow, 1955.
14. Grohn, H. R., Paudert, R., *J. Prakt. Chem.* (1960) **11**, 64
15. Grohn, H., and Bishof, K., *Plaste Kautschuk* (1959) **6**, 361.
16. Baramboim, N. K., "Mechanochemistry of Polymers," Nauk. Techn. Liter., Moscow, 1961.
17. Grohn, H., Vasiliu-Oprea, Cl., *Rev. Roum. Chim.* (1964) **9**, 757.
18. Vasiliu-Oprea, Cl., Neguleanu, Cl., Simionescu, Cr., *Europ. Polym. J.* (1970) **6**, 181.
19. Kargin, V. A., Plate, N. A., *Vysokomol. Soedin.* (1959) **1**, 330.
20. Plate, N. A., Prokopenko, V. V., Kargin, V. A., *Vysokomol. Soedin.* (1959) **1**, 713.
21. Kargin, V. A., Kabanov, I. N., *Vysokomol. Soedin.* (1961) **3**, 787.
22. Vasiliu-Oprea, Cl., Neguleanu, Cl., Simionescu, Cr., *Makromol. Chem.* (1969) **126**, 217.

RECEIVED May 5, 1972.

6

Branching and Crosslinking in Styrene-Butadiene Polymerizations

G. M. BURNETT and G. G. CAMERON
University of Aberdeen, Old Aberdeen, Scotland

> *A method based on partial conversion properties for studying crosslinking during polymerization of dienes is described. With the method, the competition between crosslinking and processes that produce new polymer molecules can be studied with considerable sensitivity. This competition depends upon polymerization variables such as temperature, and modifier type and concentration. The problem of controlling crosslinking during styrene-butadiene copolymerization is considered, and it is shown that incremental addition of modifier is more satisfactory for this purpose than is a single initial charge. Mechanisms of branch and crosslink formation during copolymerization of styrene and butadiene are discussed.*

Active free radicals will almost inevitably react in some way with molecular species with which they come in contact. The interaction of radicals in polymerizing systems with either monomer or solvent is well recognized as a transfer step that brings about the formation of inactive polymeric molecules. At the same time, though, new radicals are generated; the kinetic chain reaction is usually not seriously impeded although, in some instances, this is not true.

As the amount of polymer in the system increases, the probability of interaction of growing free radicals with polymeric (inactive) species either by transfer or, sometimes, by copolymerization through residual double bonds must increase. Because such reactions automatically give rise to branched and crosslinked species, reliable experimentation aimed at studying these processes is difficult to achieve. This follows from the fact that the region of the reaction in which study is essential is precisely that region in which there are rapid changes in the characteristics of the polymer produced, particularly in solubility. This is of significance in the methods described here in that these depend on rapid, reliable, and precise molecular-weight measurements.

Although fundamental studies of these phenomena have been slow to develop, the deleterious effects of branching and crosslinking in commercially produced polymers have been recognized for many years. In particular, when significant crosslinking occurs during polymer manufacture, the processing characteristics of the resulting polymer are suddenly, and catastrophically, altered in many instances. With styrene-butadiene copolymers, the usual indicator of good processability of the polymer is the delta Mooney index, a negative value being characteristic of a processible rubber. The change from negative to positive values takes place over a very narrow conversion range.

The objective of this paper is to outline methods by which control of the deleterious reactions can be achieved, bearing in mind that any solution to the problem should not interfere with those reactions that lead to a normally acceptable product.

Because of the solubility difficulties, methods that can get as close to the insoluble situation as possible must be used. Although the processability index undergoes a very sharp change, there is no doubt that the branching and crosslinking reactions are possible from the time that polymer forms; these reactions grow in importance as the reaction proceeds. What is required, therefore, is a reliable method of assessing the balance between the normal "productive" reactions and those leading to crosslinks at all stages of the reaction.

About 25 years ago, F. T. Wall proposed such a method, which depended on the measurement of the so-called "partial conversion molecular weight" (1). If N is the number of polymeric molecules and W the weight of polymer, then the number average molecular weight $\overline{M}_n$ is given by W/N, from which it is easy to show

$$dN/dW = \overline{M}_n{}^{-2}(\overline{M}_n - W\, d\overline{M}_n/dW) \qquad (1)$$
$$= \overline{\overline{M}}_n{}^{-1}$$

where $\overline{\overline{M}}_n$ is the partial conversion molecular weight. During a reaction, dN/dW is the measure of the rate at which the population of polymeric molecules varies with the extent of reaction, as defined by W. Normally, this quantity will be positive. However, each intermolecular crosslinkage formed reduces the number of polymeric molecules by one. Consequently, when crosslinking becomes of importance, dN/dW will tend toward negative values. When $dN/dW = 0$, the production rates of new macromolecules and of reduction caused by crosslinking are equal.

The application of this technique demands accurate determinations of a large number of molecular weights. This is no longer a formidable task with the advent of automated osmometers. Figure 1 shows the sort of results that have been obtained using this technique (2). Number-average molecular weights are determined at various conversions, and from plots of these against conversion, dN/dW can be calculated from Equation 1. The correlation of the results of this approach with the conventional delta Mooney procedure can be assessed from Figures 2 and 3 (3).

In Figure 2, dN/dW is plotted as a function of W for two series of commercial samples withdrawn from an experimental continuous reactor line in which butadiene and styrene were copolymerized in emulsion using a 1500-type recipe at 5°C. The conversion marking the onset of predominant crosslinking is given by the point at which the curves cut the W axis—that is, at 53 and 57% conversions.

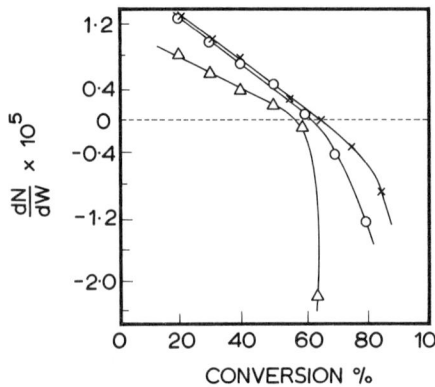

Figure 1. Rate of change in macromolecular population dN/dw as a function of conversion in emulsion copolymerization of styrene and butadiene (styrene: butadiene = 29:71 by wt); concentration of tert-dodecyl mercaptan (TDM) modifier in phm: △ — run 4, 0.10; ○ — run 6, 0.20; X — run 7, 0.23

Figure 3 shows that the delta Mooney indexes show an abrupt change from negative to positive values in the same region. Thus, there is a clear relationship between the onset of dominant crosslinking and processability. The deterioration in physical characteristics is not coincidental with the formation of gel since, in both of these examples, the gel point occurs at conversions greater than 60%.

Although the serious effects of crosslinking on the properties of commercial polymers have long been recognized, systematic study of the phenomenon has been somewhat neglected. It is true that Flory (4) outlined the basic kinetics of the crosslinking process for the homopolymerization of dienes which gave the proportion of crosslinks v as a function of the fraction conversion α as:

$$dv/d\alpha = 2KN_0[\alpha/(1-\alpha)] \qquad (2)$$

where the density of crosslinks, ρ, defined as $v/\alpha N_0$ is given by

$$\rho = -2K[1+\alpha^{-1}\ln(1-\alpha)] \qquad (3)$$

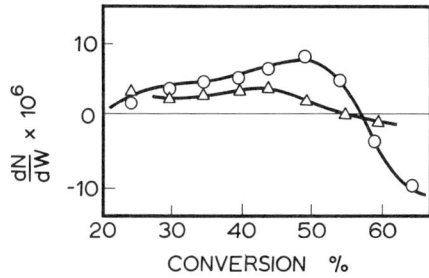

Figure 2. dN/dw as a function of conversion for series A (△) and B (○) (3)

where N_0 is the number of monomer molecules originally present.

In this derivation, it is assumed that crosslinks are formed by the addition of propagating radicals with either internal or pendent double bonds on polymer molecules. K is the ratio of the rate constant k_x for this essentially copolymerization process and the propagation rate constant k_p. While this is logical enough, there must be some doubt as to its ultimate validity in that the internal double bonds are essentially 1,2-disubstituted ethylenes where reactivity is notoriously low. There is, indeed, little evidence to suggest that these bonds disappear in the later stages of the reaction. So it may be that crosslinking (even in diene systems) is preceded by branching that results from transfer to the polymeric molecules followed by coupling of branched radicals.

For the SBR case, the numerator α in Equation 2 should be replaced by α_B, the fraction of consumed monomer, which is butadiene. The analysis of this situation is complex but may be simplified by the experimental observation that

$$\alpha_B/(1 - \alpha) = A + B\alpha^2 \qquad (4)$$

where A and B are constants at a fixed temperature. (5).

Flory has also shown that at the gel point

$$\nu/\alpha = \rho = \overline{Y}_{wg}^{-1} \qquad (5)$$

where $\overline{Y}_{wg}$ is the weight-average degree of polymerization of the primary chains.

Substituting Equation 4 in Equation 2 modified as indicated and integrating between $\alpha = 0$ and $\alpha = \alpha_g$ (the gel point) gives

$$\overline{Y}_{wg}^{-1} = 2K(A + (B/3)\alpha_g^2).$$

When $\overline{Y}_{wg}$ and α_g are known, K can be found and thus its dependence on temperature. In this analysis, however, the parameter $\alpha_B/(1-\alpha)$ must vary with conversion, initial composition, and temperature so that a limitation is imposed on the range of application of the data.

The variations in K are reflected in the partial conversion properties. Figure 4 shows the variation in dN/dW with conversion for a number

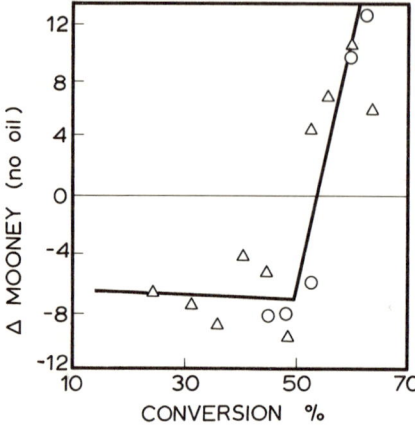

Chemistry and Industry (London)

Figure 3. Delta Mooney indexes (no oil) as a function of conversion for series A (△) and series B (○) (3)

of runs carried out at 5°, 15°, and 25°C (5). In each case, concentration of the initial modifier, *tert*-dodecyl mercaptan (TDM), was the same (0.22 phm), but the concentration of initiator was adjusted to give the same polymerization rate. The macromolecular population tends to increase up to about 50% conversion. Beyond this stage, the curve falls, cutting the conversion axis at a point somewhat above 70% conversion. As the temperature increases, the crossover point occurs at lower conversions, as expected from the fact that K increases with temperature. A further contributory factor to the effects shown in Figure 4 is that the mercaptan regulating index decreases with increasing temperature.

The method that we have sought to exploit provides a very sensitive indication of the balance between processes leading to production of new polymer molecules and those producing crosslinks. Since this indicator operates under conditions where no gel has appeared, it makes it much more reliable than any technique that depends on the actual detection of gel. In all the instances studied, the gel point lay 6-8% above the crossover points. Thus, we would appear to be in a position to study more precisely and objectively the effects of varying the experimental parameters.

The effect of temperature has already been demonstrated. By increasing the concentration of modifier, the incidence of crosslinking should be delayed. This delay, though small, is clearly discernible in Figure 1, which shows the results of increasing the initial TDM concentration from 0.10 through 0.20 to 0.23 phm— runs 4, 6, and 7, respectively (2). However, increasing the modifier charge must also have the less desirable effect of reducing the molecular weight of the polymer, particularly that formed in the early stages of the process. This could mean that the resulting polymer would exhibit poorer physical characteristics.

Essentially, the problem is how to achieve simultaneously a delay in the formation of gel and a reasonably high molecular weight in the elastomer produced. It is, therefore, necessary to ensure that there is not too much modifier present in the early stages of the reaction to cause deterioration of mechanical properties of the product, and that there is not too little present in the later stage so as to encourage crosslinking and gel formation. The general practice of adding the modifier as a single, initial charge is unlikely to meet satisfactorily the two requirements. This is particularly true in the case of highly reactive modifiers (for example, Dixie, diisopropyl xanthogen disulfide), which are substantially removed from the system at conversions around 50%. Their effectiveness in producing control later in the reaction is thus negligible. There appear to be two methods by which the necessary conditions can be met.

The concentration of modifier will tend to diminish during the reaction, and the more reactive the modifier, the more serious is the effect likely to be. An extension of the Flory treatment of crosslinking to take into account the presence of modifier demonstrates the point. When branches arise by transfer to polymer and crosslinks by the subsequent coupling of branched radicals, it is easy enough to show that the rate of formation of crosslinks, X, as a function of conversion is:

$$\frac{dX}{d\alpha} = \frac{K_1\alpha^2}{[K_2 + K_3\alpha + kS_0(1-\alpha)^K]^2(1-\alpha)} \qquad (6)$$

K_1, K_2, and K_3 are composite constants involving a variety of rate constants, and K is the ratio of the rate constants for transfer to modifier, k, and for propagation (6). When K is large, $(1-\alpha)^K$ diminishes rapidly as α increases. Figure 5 shows plots of a number of curves derived from Equation 6. Apart from K, all other parameters are

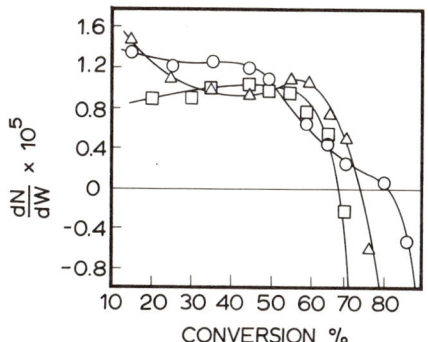

Journal of Polymer Science

Figure 4. dN/dw as a function of conversion at varying temperatures for emulsion copolymerization of styrene and butadiene (styrene:butadiene = 29:71 by wt); modifier TDM (0.22 phm); ○, 5°; △, 15°; □, 25°C (5)

constant in Figure 5. The higher K is, the less effective is the modifier in controlling crosslinking in the later stages of the reaction. Because of this, addition of the total modifier charge incrementally should provide a solution to the problem.

Figure 6 shows the result of a simple experiment in which the modifier TDM (0.23 phm) was added as a single initial charge (run 7) and in two parts: two-thirds initially and the rest at 30% conversion (run 16) (2). The effectiveness of this procedure is obvious. No attempt has yet been made to optimize the conditions for incremental addition, but there is no doubt that the situation could be very much improved.

A second possible approach is to use a mixture of modifiers of very different activity. Initial control of molecular weight would be achieved this way mainly by the more active component, and later control of crosslinking by the less active. Experiments thus far are much less conclusive than those for incremental addition, but mixtures of Dixie and *tert*-hexadecyl mercaptan (active and relatively inactive, respectively) show that the general hypothesis is borne out. The mixed modifier procedure has the advantage that the whole charge is added initially.

As far as the mechanisms of branching and crosslinking are concerned, there appear to us to be certain weaknesses in those commonly accepted. With ethylenic monomers, there can be little doubt that if branching were to occur at all, it will arise from radical attack upon the polymer already formed. It would be immaterial whether this transfer takes place on backbone carbon atoms or *via* side chains, as is almost certainly true for, say, vinyl acetate. When dienes are present, it has been generally accepted that the residual double bonds are the main seat of reaction, thereby creating the immediate possibility of crosslinking. However, the internal residual double bonds—that is, those

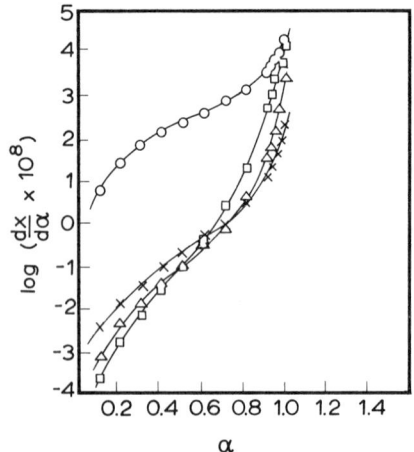

Figure 5. Rate of formation of crosslinks $dX/d\alpha$ *in relation to fractional conversion* α *according to equation (6); no modifier,* ○; *modifier at initial concentration 0.1 phm,* X, $K=0.5$; △, $K=1.0$; □, $K=2.0$

resulting from a 1,4-addition of butadiene—are also 1,2-substituted ethylenes in nature. These are notoriously difficult to polymerize or copolymerize.

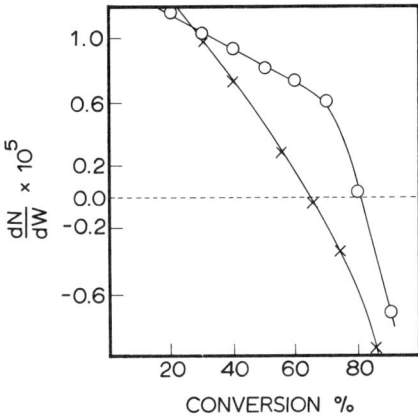

Figure 6. Effect of incremental addition of TDM modifier (0.23 phm) on copolymerization of styrene and butadiene; single initial charge, X; incremental addition as in text, O

Pendent double bonds—that is, from 1,2-addition—will be more likely to copolymerize but will be likely to be much fewer in number and, therefore, contribute less effectively. Polymerization of styrene in the presence of poly(cis-1,4-butadiene) shows no detectable branching unless benzoyl peroxide is used as initiator. This would tend to suggest that a vital role in the reaction is played by the initiator, probably by an abstraction reaction involving α-hydrogen atoms. It may be, therefore, that these sites are the most important ones when considering the branching mechanism; and that even in the presence of residual double bonds, the major part of the branches formed results from transfer reactions involving polymer with growing or initiator radicals or both. The kinetic analysis of these possible mechanisms is rather complex but should be completed relatively soon.

Literature Cited

1. Wall, F. T., *J. Amer. Chem. Soc.* (1945) **67**, 1929.
2. Burnett, G. M., Cameron, G. G., Kale, V. D., unpublished work.
3. Burnett, G. M., Cameron, G. G., Dukiet, A., Pearson, J. M., *Chem. Ind.* (London) (1968) 1518.
4. Flory, P. J., *J. Amer. Chem. Soc.* (1947) **69**, 2893.
5. Burnett, G. M., Cameron, G. G., Thorat, P. L., *J. Polym. Sci.*, A–1 (1970) **8**, 3443.
6. Burnett, G. M., unpublished work.

RECEIVED May 26, 1972.

7

Popcorn Polymers

J. W. BREITENBACH

University of Vienna, Austria

H. AXMANN

Isovolta, Österreichische Isolierstoffwerke, AG.

> *This paper reviews some of the more basic contributions over the past 20 years or so to the study of "popcorn" polymers. Some results from the authors' laboratories are then presented in an attempt to characterize further the nature of these polymers and their growth mechanisms. These more recent results include those obtained by studies of the "popcorn" polymers using the polarizing and electron-scanning microscopes and ESR measurements. The importance of crosslinking and entanglements is discussed, and industrial applications—present and potential— are considered.*

The first observations of popcorn (PC) polymer formation go back to the beginning of this century when Kondakow (*1*) studied the polymerization of dimethylbutadiene. The name "popcorn polymers," however, is of much more recent origin. It was first used in the U.S. in 1940, where the similarity of appearance of these materials to a sponge or to cauliflower was replaced by the obviously more popular popcorn. Today, it is possible to prepare polymers of this type having no similarity at all to popcorn, but the name is retained.

Some of the more remarkable contributions to a better understanding of popcorn polymer formation and growth are summarized here. Staudinger and Husemann (*2*) showed that in the styrene-divinylbenzene system, an optimum divinylbenzene concentration exists for popcorn polymer formation varying with polymerization conditions. Whitby and Zomlefer (*3*) found proliferous growth of popcorn polymers in pure monovinyl compounds. Welch, Swaney, Gleason, Beckwith, and Howe (*4*) made a quantitative study of the growth of butadiene popcorn polymers, finding the rate of growth proportional to the amount of popcorn polymer present; they also observed a mechanical scission of the popcorn polymer by the growth process.

Breitenbach and Frank (5) showed that with styrene-divinylbenzene, no further additives (such as peroxides) are necessary for popcorn polymer formation. Breitenbach and Fally (6) found, in methyl acrylate polymerization, the possibility of crosslinking in the polymerization of a monovinyl compound. Miller and coworkers (7) developed the kinetics of the process; Pravednikow and Medvedev (8) studied the chain scission, and assumed radical formation by that process as an important step.

The occurrence of straining in a microscopic range was shown by Breitenbach, Preisinger, and Tomschik (9), and further microscopic studies of popcorn polymer formation and growth were carried out by Breitenbach and coworkers (10).

This paper presents some results obtained with the polarizing microscope and the scanning electron microscope. The importance of crosslinking and entanglements is discussed, and results of ESR measurements on popcorn polymer systems are presented.

Proliferous Growth

A necessary condition for the appearance of popcorn polymers is that growing chains initiated at different sites of the polymerizing system are fixed in their place and not able to interact. They form a great number of independently growing centers. The growth of spontaneously formed globular particles in the system styrene-divinylbenzene (0.6 weight %) at 60°C is shown in Figures 1-4.

Monomer is taken up by the particle by swelling and is transformed to polymer in the interior of the particle. Figures 3 and 4 show the proliferation; new, small globules are growing from the mother globule. The strain induced in the growing particle can be seen in the polarizing microscope photographs.

Particles in different growth stages in the n-butyl acrylate-glycol dimethacrylate system appears in Figure 5. There, no optical anisotropy appears because of the low straining birefringence of acrylates.

At later growth stages, strains are also induced in the surrounding polymer gel. Figure 6 may show the first step in the transformation of a preformed glassy polymer into a popcorn polymer.

The primary popcorn particles give a variety of different shapes in the final product. Electron microphotographs are given in Figures 7-9. The morphological elements of the popcorn polymer have dimensions of several μm.

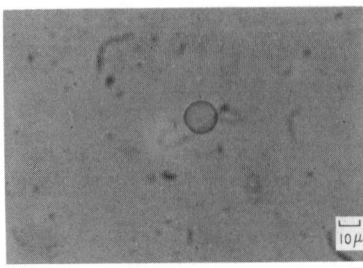

Figure 1. Formation of popcorn polymer particle; styrene with 0.6 wt % p-DVB; 17 hours at 60°C.

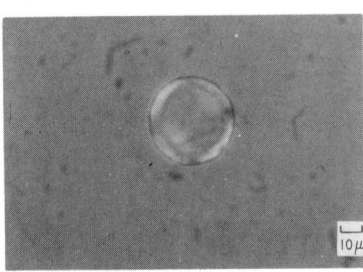

Figure 2. Growth of particle for 4 hours at 60°C; V/V_o 17.

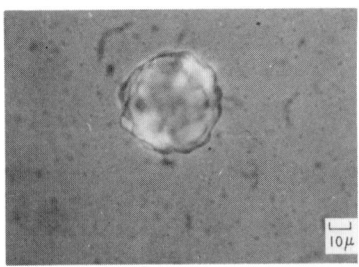

Figure 3. Growth of a popcorn polymer particle; styrene with 0.6 wt % p-DVB at 60°C.

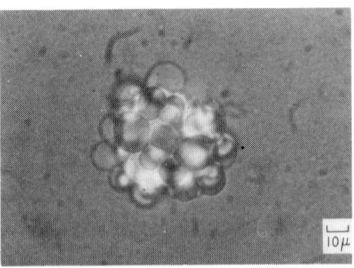

Figure 4. Growth of particle for 8 hours at 60°C.

Chemical Crosslinking

Popcorn polymer only forms in those polymerizing systems where a certain amount of crosslinking takes place. In all systems investigated, an optimum crosslinking range for popcorn polymer formation exists (*11*); In thermal polymerization of a styrene-*p*-divinylbenzene mixture

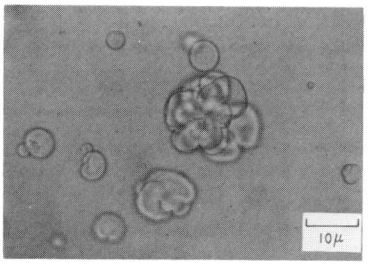

Figure 5. Formation of popcorn polymer particles; n-butylacrylate with 3.6 mole % glycol dimethacrylate, 0.1 mole % AIBN at 20°C; 3 hours after first particles formed.

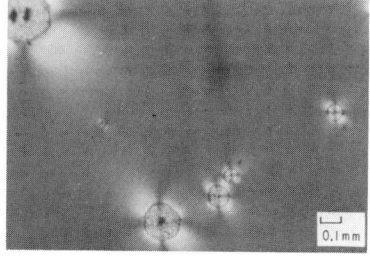

Figure 6. Formation of popcorn polymer particles; m-bromostyrene with 1.9 wt % p-DVB; 144 hours at 70°C.

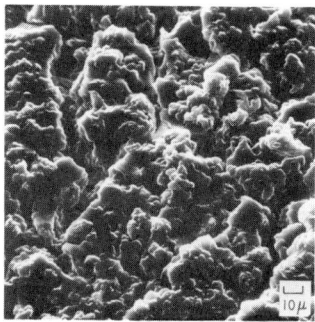

Figure 7. Surface of a popcorn polymer proliferated in styrene

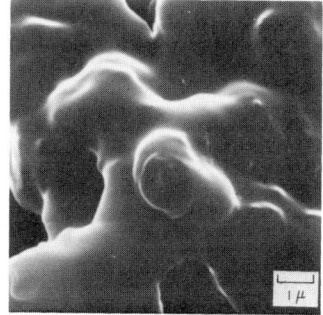

Figure 8. Central part of Figure 7

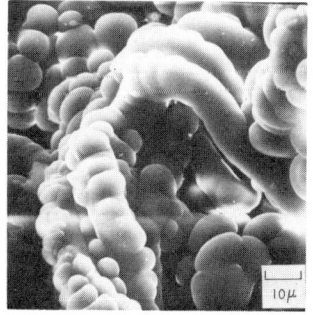

Figure 9. Surface of an acrylonitrile–styrene–glycol dimethacrylate popcorn polymer

at 70°C, the optimum is at p-divinylbenzene concentration of 0.35 mole %; see Figure 10. Between 0.038 and 0.47 mole %, the first tiny popcorn particles are free to move in the liquid monomer-polymer system. Only at higher divinylbenzene concentrations is the system transformed into a gel before the popcorn polymer appears. As important as the content of divinyl compound is the primary mean chain length of the polymer formed under the same condition. It is possible to vary the chain length by adding a chain-transfer agent (*11*) Table I. With 0.60 mole % *p*-divinylbenzene the optimum chain length is at 3300, compared with 6500 when 0.35 mole % *p*-divinylbenzene is used. The PC-polymer formation is further influenced by solvent properties of the medium (*12*); see Table II.

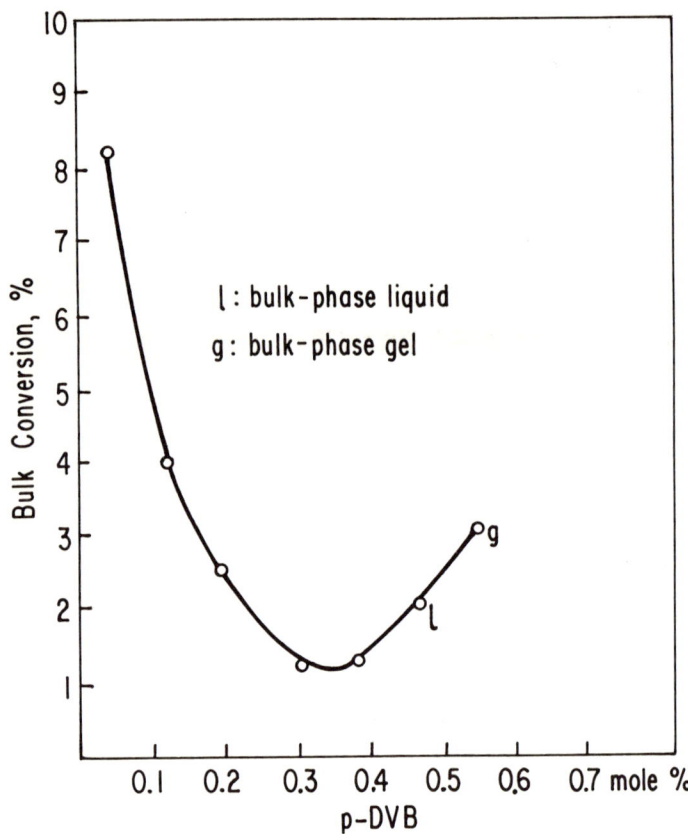

Figure 10. Thermal polymerization of styrene-p-divinyl benzene at 70°C; optimum for popcorn polymer formation

Table I. Influence of Primary Chain length on PC Formation

p-DVB mole %	CCl_4 mole %	First Visible PC Particles After hours	Consistency of Bulk Phase	Primary Polymer $\bar{P}$
0.60	0	15	Gel	6500
0.60	1.5	11	Liquid	3300
0.56	9.1	100	Liquid	860

Table II. Influence of Medium on PC Formation Styrene with 0.35 mole % p-DVB at 70°C

Volume % Diluent	Diluent	
	Benzene	Methanol
	First Visible PC Particles at Bulk Conversion %	
0	1.1	
20	2.2	0.9
30	2.5	0.7

Under the most favorable conditions, the formation of a popcorn polymer particle is a highly improbable process; for 10^{16} molecules of polymer produced in the system, only one popcorn polymer particle has been grown, possibly from a single macromolecular precursor. By contrast, growth under optimum conditions is a fast process and approximately obeys the relation—given by Welch et al. (4)—for butadiene popcorn polymer:

$$kt = \ln \alpha$$

where α is the degree of proliferation P/P_0, P being the amount of polymer after proliferation time t, and P_0 the amount of polymer at the beginning of proliferation. For k, values up to $8h^{-1}$ have been measured in the vinyl acetate-glycol-dimethacrylate system (13), where growth from macromolecular dimensions to visible size is possible in a few hours.

There are many similar systems consisting of mono- and divinyl compounds in which optimum conditions for PC-polymer formation have been found (14) Table III. The general existence of an optimum concentration of a divinyl compound for popcorn polymer formation indicates that at least two opposite effects are working. One of the essential features of popcorn polymer growth—that is proliferous polymerization—is free-radical production by chain cleavage via the combined action of polymerization and swelling. On the one hand, higher content of divinyl compound gives a higher crosslinking density and, therefore, a higher strength of the gel. On the other hand it gives a higher content of pendant double bonds on the polymer network and, therefore, a greater number of growing chains in the polymer network and a higher driving force for chain splitting. At low concentrations of divinyl compound, the rate of polymer formation in the network is too low, and at high concentrations, the strength of the network is too high.

If this interpretation is correct, it should also be possible to prepare crosslinked glassy polymers that proliferate in contact with a monovinyl compound. Such materials have been obtained by polymerizing styrene with a high concentration of divinylbenzene in the presence of a diluent

(15); see Figure 11. A polymer prepared with 33 volume % isooctane in the feed has the optimum capacity for proliferation. This polymer has a permanent porosity of 22 volume % and contains 2.6 mole % of pendant double bonds. Its proliferation leads to growth (which can be observed microscopically) of opaque popcorn polymer out of the glassy material.

Chemical Cleavage of Crosslinks

A chemical crosslinking by copolymerization of mono- and divinyl compounds takes place in these systems. Therefore, the swelling capa-

Table III. Optimum Concentration of Divinyl Compound for PC Polymer Formation

Monovinyl Compound	Polymerization Conditions	p-Divinylbenzene	m-Divinylbenzene	Glycol Dimethacrylate
Styrene	70°C without initiator	0.35 mole %; PC in solution at 1.1% conversion	0.6 mole %; PC in solution at 1.7% conversion	2 mole %; PC in gel at 1.8% conversion
p-Chlorstyrene	50°C without initiator	1 mole %; PC in gel and from gas phase		
Methyl methacrylate	70°C with 1×10^{-4} mole % AIBN	0.5–1 mole %; PC from gas phase and in gel		2 mole %; PC from gas phase and in gel
Vinyl acetate	70°C with 1×10^{-4} mole % AIBN			5 mole %; PC in solution at 1% conversion
n-Butylacrylate	40°C with 0.1 mole % AIBN			4 mole %; PC from gas phase and in gel

city of the PC-polymers should be related to the divinyl content of the monomer feed. A fairly general result of such swelling measurements is that the degree of swelling of PC-polymer is much lower than can be accounted for by the divinyl compound. The nature of popcorn crosslinking has been investigated by using a dimethacrylate Schiff base as crosslinking agent (16). The crosslinks formed have this structure:

$$\}-\underset{\underset{O}{\overset{CH_2}{\underset{\|}{C}}}}{H_3C-C}-C-O-O-\bigcirc-CH=N-CH_2-CH_2-N=CH-\bigcirc-O-\underset{\underset{O}{\overset{CH_2}{\underset{\|}{C}}}}{C}-C-CH_3-\{$$

The crosslinks can be easily destroyed by acid hydrolysis of the –CH=N– bonds. A styrene PC-polymer prepared with 1.5 mole % of the Schiff base swells only to a very low degree in benzene and shows the typical anisotropy in the polarizing microscope. When a few drops of a 0.1M dichloracetic acid solution in tetrahydrofuran are added to the swollen polymer, most of the optical anisotropy of the polymer disappears, but the polymer still remains insoluble; only the degree of swelling increases somewhat. The remaining crosslinks, at least to a large degree, are formed by chain entanglements and are therefore more movable than chemical bonds.

Chain Entanglements

The question of chain entanglements has been studied more extensively in acrylonitrile styrene copolymerization (17) (Table IV). At the first appearance of popcorn particles in this system, the copolymer has a 1:1 molar ratio of acrylonitrile to styrene units. Its mean chain length is 20,000.

The popcorn formation tendency is enhanced by addition of suitable liquids. In a system with 20 volume % butanone, the first popcorn particles are visible after 16 hours at a bulk conversion of 8.5% in a mobile liquid. The optical anisotropy of these polymers disappears in dimethylformamide, where they swell but do not dissolve. On the contrary, when a popcorn polymer is formed from a similar feed but with additional glycol dimethacrylate, this material—which contains a high amount of chemical crosslinks—retains its optical anisotropy also after swelling in dimethylformamide for a long time.

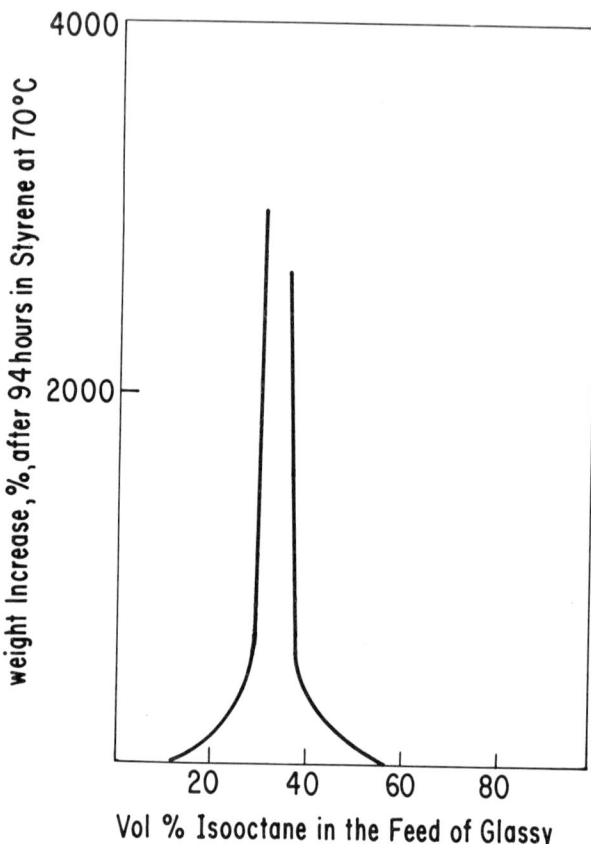

Figure 11. Proliferous growth of glassy polymers prepared from a monomer feed with 80 wt % styrene and 20% divinylbenzene

Table IV. Acrylonitrile–Styrene Copolymerization at 70°C; 70 mole % Acrylonitrile in Feed

Time (hours)	Conversion (%)	PC Particles in 1.7 g Feed
16.7	6.9	No
23.6	11.1	No
33.4	15.1	No
43.6	23.6	6 Particles
50.0	29.0	9 Particles
62.0	38.2	14 Particles and 1 Proliferated
71.0	44.5	9 Particles and 2 Proliferated
83.0	55.5	
120.0	73.8	

The association phenomena in polyacrylonitrile solution caused by the nitrile dipoles are well known (18). Also, gelation in concentrated polyacrylonitrile solution has been found (19). Accordingly, crosslinking by dipolar interaction of nitrile groups may be assumed between acrylonitrile-styrene copolymer chains. There is an alteration of phenyl and nitrile groups along the chain and a greater number of nitrile groups may thus participate in the formation of one crosslink. In dimethylformamide, the dipolar crosslinking disappears; the remaining chain entanglement allows a removal of internal strains by the enhanced segment mobility of polymer chains, but it inhibits dissolution of the polymer.

When chemical crosslinking by glycol-dimethacrylate is present, the chain fixation also remains in dimethylformamide, and optical anisotropy is retained. All these results show that chain entanglement plays a predominant role in the structure of PC-polymers.

A chemical crosslinking also seems possible in the acrylonitrile polymerization by a copolymerization activity of the nitrile group (20). In fact, we found a certain extent of acid hydrolysis in acrylonitrile-styrene popcorn copolymers, which could be an indication for crosslinking groups such as:

$$CH-C=N- \quad \text{or} \quad C=C-N- $$
$$\phantom{CH-C=N- \quad \text{or} \quad C=C-N}|$$
$$\phantom{CH-C=N- \quad \text{or} \quad C=C-N}H$$

Whether entanglement alone can give an insoluble polymer with equilibrium swelling properties, or whether a small part of chemical crosslinks is necessary cannot be determined with certainty. Entanglements formed by mutual penetration of polymer molecules in solution should also be dissolved by thermal motion of polymer segments. However, when entanglements are formed by polymerization of a monomer in a polymer matrix, a geometrical arrangement that resists a dissolution by thermal motion may be visualized.

Popcorn polymers are also formed by some pure monovinyl compounds—for example, methyl acrylate, ethyl acrylate, and n-butyl methacrylate. In these cases, chemical crosslinking by combined polymer chain transfer and combination termination also seems to take place.

Free Radicals and ESR Signals

One reaction of popcorn polymers is their very rapid, proliferous growth in appropriate monomers. The rapid growth reaction corresponds to a relatively high content of the growing material on radical chain ends. It is possible to measure the growth rate directly by observing growing popcorn materials under the microscope. Monomer radical concentrations in the monomer-swollen popcorn particles are obtained from the propagation rate constant. The concentrations range up to 10^{-5} mole/l. The ESR signals of the growing polymer radicals can easily be observed; see Figures 12, 13, 14, and 15.

The best resolved is the methyl methacrylate spectrum; it is identical with the spectrum observed in nonpopcorn polymerization of methyl methacrylate in highly crosslinked gel systems (21). According to Fischer (22), in the nine visible lines are hidden 16 hyperfine structure lines obtained in diluted, noncrosslinked systems.

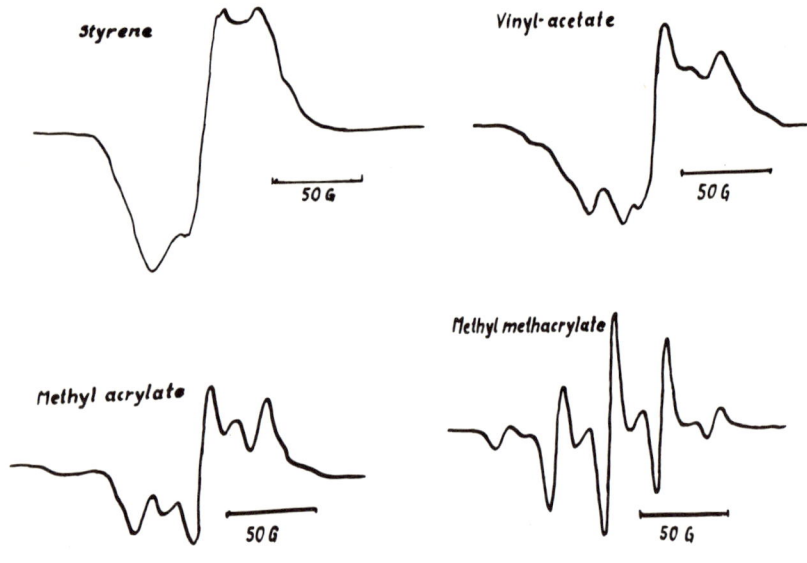

Figure 12.   ESR spectra of growing popcorn polymer chains

Similarily, Yoshida and Rånby (23) found a well-resolved 24-line spectrum for the growing vinyl acetate chain in dilute aqueous solution; five nonresolved lines may be seen for the vinyl acetate popcorn system. Least structured is the styrene spectrum, with three nonresolved lines. In principle, the same features appear in the substituted styrenes represented in Figures 13 and 14, and the vinylpyridines (Figure 15). Only the form of the shoulders changes. With better resolution, these spectra should consist of at least six lines.

The intensity of the ESR signal remains nearly unchanged after complete consumption of the monomer so long as the popcorn polymers are prevented from coming in contact with air. ESR signals have been observed in methyl methacrylate popcorns after 14 months at room temperature, and in styrene popcorns after five years of oxygen-free storage. These materials are well-suited for studying the reactions of the corresponding radicals.

The thermal decay process of the radicals is a bimolecular reaction. The decay rate increases with increasing temperature. At 100°C, a half-life of 12 minutes has been observed for methyl methacrylate popcorn radicals. A fast decay rate takes place when the dry popcorn is swollen in a liquid such as benzene. That means that during the proliferous growth process, when the polymer is swollen by the monomer, a decay process also occurs, and a stationary radical concentration in the growing polymer popcorn results. A liquid that does not swell the polymer (for example, methanol for polystyrene) does not influence the decay rate. A much higher rate of radical decay is obtained with a benzene solution of diphenylpicrylhydrazil. The reaction rate between the polymer radical and inhibitor radical may be measured.

A solution of carbon tetrobromide in benzene also gives a high rate of radical decay. The reaction occurs between polymer radicals and

$$\mathrm{\sim\!\!\!\!-C \cdot \ + \ CBr_4 \ \longrightarrow \ \sim\!\!\!\!-C-Br \ + \ \cdot CBr_3}$$

carbon tetrabromide at a high rate. The freely movable $\cdot CBr_3$ radicals give a fast reaction with other radicals and a fast decay is thus obtained. This shows that the fixation of radicals on the polymer is the reason for their longevity.

Some Possible Uses

The properties of PC-polymers have not been systematically evaluated. Their mechanical stability depends on the void content, which may be varied by a wide range of conditions (*13*). Under normal pressure, void contents as low as 20 volume % are easily reached; this material seems to have good mechanical qualities. Polymerization at higher pressure (about 100 atmospheres) gives a popcorn material practically without pores.

It is possible to obtain popcorn polymers in the form of flakes (*24*). For this purpose, a pulverized popcorn polymer is brought into the monomer feed in a stirred reactor. The proliferous growth of the seed material in the stirred reactor leads to a finely distributed material

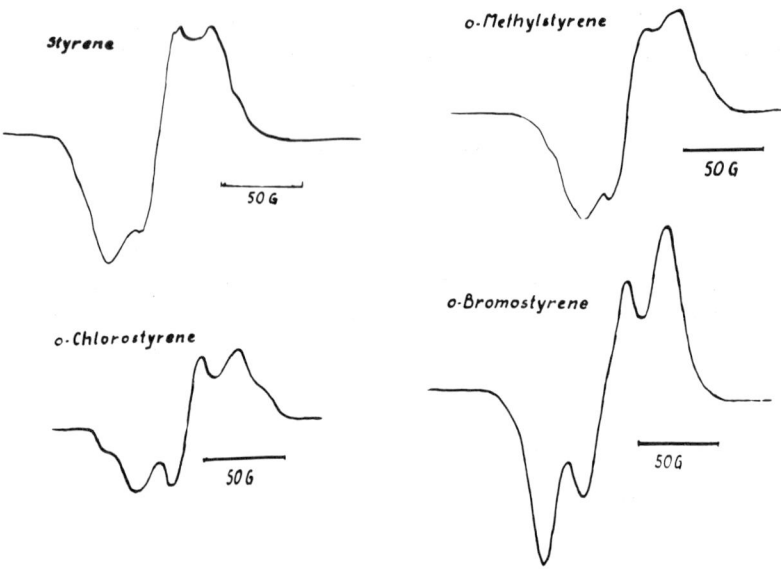

Figure 13. ESR spectra of ortho-substituted styrene popcorn polymers

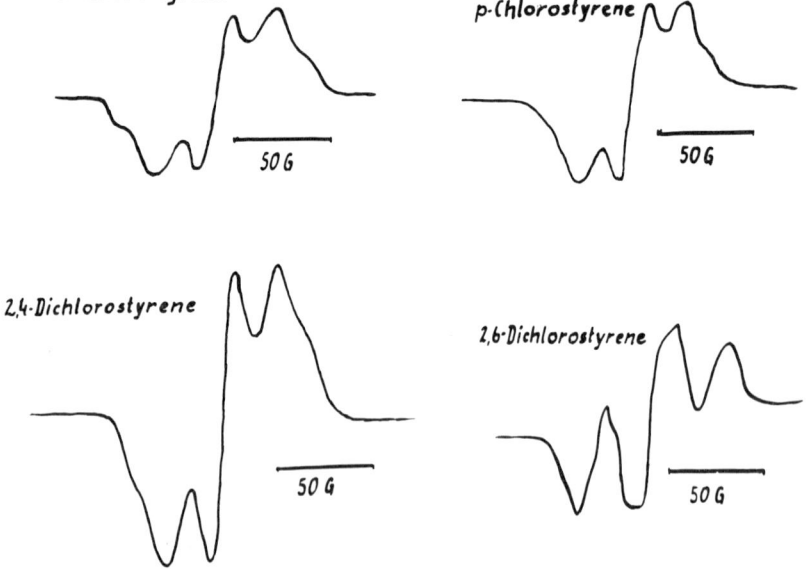

Figure 14. ESR spectra of halogen-substituted styrene popcorn polymers

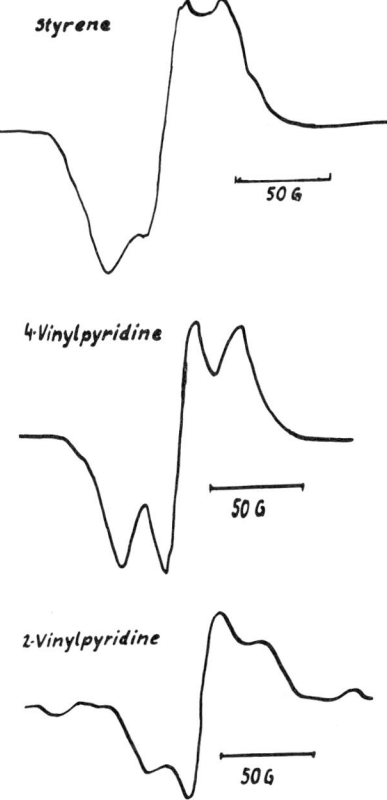

Figure 15. ESR spectra of vinylpyridine popcorn polymers

rather than a coherent mass of polymer. After a conversion of 10 to 30%, the greater part of the popcorn polymer flakes is replaced by monomer feed, and the operation repeated. In a relatively short time, a conversion of 10 to 30% is again reached. Five kilograms of popcorn polymer have been obtained per hour in a 5-liter reactor this way. So far, styrene and vinyl acetate have been polymerized this way with divinylbenzene, monoallyl maleate, allyl methacrylate, glycol acrylate methacrylate, and glycol dimethacrylate as crosslinking agents.

The popcorn polymer flakes may be used as supporting materials for ion exchangers. Depending on the monomers units, the polymers have functional groups of different kinds and may be used for synthetic purposes. The materials also adsorb and absorb a variety of organic vapors and thus may be used as filter materials.

Acknowledgment

We thank Fonds zur Förderung der wissenschaftlichen Forschung, Wien, for supporting this work, and for purchasing the ESR spectrometer. Figures 1-4 are from a series of time-lapse motion pictures made by H. Dworak and H. Sulek. The microphotographs in Figures 5 and 6 were made by H. Burgmann. The stereo-scan electron micrographs in Figures 7-9 were made by H. Hendus of Badische Anilin & Soda Fabrik, Ludwigshafen. The ESR spectra (Figures 12-15) were determined by O. F. Olaj and H. Goldenberg.

Literature Cited

1. Kondakow, J., *J. Prakt. Chemie* (1901) **64**, 109.
2. Staudinger H., Husemann, E., *Berichte* (1935) **68**, 1618.
3. Whitby, G. S., Zomlefer, J., Technical Reports, Office of Rubber Reserve (1944-1946).
4. Welch, L. M., Swaney, M. W., Gleason, A. H., Beckwith, R. K. Howe, R. F., *Ind. Eng. Chem.* (1947) **39**, 826.
5. Breitenbach, J. W., Frank, H. P., *Mh. Chem.* (1948) **78**, 923.
6. Breitenbach, J. W., Fally, A., *Mh. Chem.* (1951) **82**, 1118.
7. Miller, G. H., Alumbaugh, R. L., Brotherton, R. J., *J. Polym. Sci.* (1952) **IX**, 452.
    Miller, G. H., *J. Polym. Sci.* (1953) **XI**, 269.
8. Pravednikov, A. N., Medvedev, S .S., *Dokl. Akad. Nauk SSSR* (1955) **103**, 461; (1956) **109**, 579.
9. Breitenbach, J. W., Preisinger, A., Tomschik, E., *Mh. Chem.* (1963) **94**, 807.
10. Breitenbach, J. W., *Mh. Chem.* (1964) **95**, 1225.
11. Breitenbach, J. W., Sulek, H., *Makromol Chem.* (1967) **108**, 255.
12. Breitenbach, J. W., Sulek, H., *Mh. Chem.* (1968) **99**, 1130.
13. Breitenbach, J. W., Hermann, B., *Angew. Makromol. Chem.* (1970) **10**, 197.
14. Breitenbach, J. W., *IUPAC, Int. Symp. Macromol. Chem.*, Budapest (1969), 529.
15. Breitenbach, J. W., Fucik, I., *Monatsh. Chem.* (1968) **99**, 2436.
16. Breitenbach, J. W., Dworak, H., *J. Polym. Sci.* (1966) **4**, 1328.
17. Breitenbach, J. W., Axmann, H., *Bul. Inst. Politeh. Iasi, Sect. II.* (1970) **XVI**, 53.
18. Beevers, R. B., *Polymer* (1967) **8**, 463.
19. Labudzinska, A., Wasiak, A., Ziabicki, A., *J. Polym. Sci. Part C*, (1967) **16/5**, 2835.
20. Brandrup J., Kirby, J. R., Peebles,, Jr., L. H., *Macromolecules*, (1968) **1**, 53, 59.
21. Fraenkel, G. K., Hirshon, J. M., Walling, C., *J. Amer. Chem. Soc.* (1954) **76**, 3606.
22. Fischer H., *Z. Naturforsch.* (1964) **19a**. 866.
23. Yoshida, H., Rånby, B., *J. Polym. Science, Part C*, (1967) **16**, 1333.
24. Isovolta, Österreichisches Isolierstoffwerk K. G., Austrian patent application.

RECEIVED April 13, 1972.

8

Polyether Modifiers for Polyvinyl Chloride and Chlorinated Polyvinyl Chloride

P. DREYFUSS [1], M. P. DREYFUSS, and H. A. TUCKER

B. F. Goodrich Research Center, Brecksville, Ohio 44141

> *Rather remarkable effects on the properties of polyvinyl chloride and chlorinated polyvinyl chloride were found when rubbery polyethers were added to standard recipies. The effects were related to the chemical structure of the polyether. For example, addition of polytetrahydrofuran of sufficiently high molecular weight led to resins with high Izod impact strength and good resistance to discoloration. The necessary molecular weight could be achieved either by polymerization or by crosslinking a lower-molecular-weight resin. Most other polyethers led to incompatible mixtures that were not successfully milled to a smooth sheet. Low levels of addition of certain linear and especially branched polyepoxides to modified chlorinated polyvinyl chloride led to enhanced Izod impact strength and faster flow rates through a constant load rheometer without sacrificing other properties significantly.*

Apart from a few isolated hints in the literature (1, 2), no systematic investigation of the effects of polyether additives on the properties of polyvinyl chloride (PVC) and chlorinated polyvinyl chloride (CPVC) has been reported before this study. The early hints dealt only with low-molecular-weight materials. This work covers a wide molecular-weight range for one polyether, polytetrahydrofuran, and examines the effects of limited changes in chemical structure. The polyethers included in the study are listed in Table I; most of the polyethers examined are rubbers.

---

[1] Present Address: Department of Chemistry, Case Western Reserve University, Cleveland, Ohio 44106.

Table I. Polyether Added to Polyvinyl Chloride, Chlorinated Polyvinyl Chloride, or Both

Polyether	Chemical Structure	Symbol
Polytetrahydrofuran	$[-CH_2CH_2CH_2CH_2O-]_n$	PTHF
Polyethylene Oxide	$[-CH_2CH_2O-]_n$	PEO
Polypropylene Oxide	$[-CHCH_2O-]_n$ $\quad\vert$ $\;\;CH_3$	PPO
Polybutene-1 Oxide	$[-CHCH_2O-]_n$ $\quad\vert$ $\;\;C_2H_5$	PBO
Polyhexene-1 Oxide	$[-CHCH_2O-]_n$ $\quad\vert$ $\;\;C_4H_9$	PHO
Polyphenylglycidyl Ether	$[-CHCH_2O-]_n$ $\quad\vert$ $\;\;CH_2O\phi$	PPGO

Experimental

Preparation of Blends. The blends were prepared by milling the desired quantity of polyether additive with a portion of a masterbatch chosen from those described in Table II. Both the temperature and procedure used for the milling operation were critical. As is discussed in a subsequent section, temperatures that can be attained on a steam-heated mill were generally too low to attain good properties. Consequently, most milling was carried out on a 4-inch oil-heated mill at the desired temperature. The temperatures given are the mill roll temperatures. However, when the stock temperature was measured, it was about the same as the mill roll temperature.

When the mill was closed tightly at the time the polyethers were added, excessively long banding times, 10 to 15 minutes, were encountered during the early stages of the milling operation. The banding time could be reduced to about five minutes when the mill was opened somewhat during the banding stage and tightened down before measurement of milling time was begun. Once the initial band was achieved, no further difficulties were encountered. When desired, the blends could also be remilled at a later time without any further evidence of banding problems.

Before testing, the samples were compression molded, usually at 10°F above the milling temperature, using a cycle with a five-minute preheat and three minutes at a pressure of 40,000 pounds ram force. Specimens were then transferred to a cooling press and cooled under pressure.

Testing of Blends. Notched Izod impact strengths were determined for four specimens broken according to ASTM B648-56, Method A. Heat distortion temperatures were run according to ASTM D648-56 at 264 psi applied pressure. Flow rates were determined in a constant-load extrusion rheometer. Tensile strengths were determined on an Instron, using 1/8-inch dumbells and a strain rate of 2 inches per minute.

Results and Discussion

Three different kinds of behavior were observed on addition of the polyethers to PVC and CPVC. First, high-impact resins were sometimes produced without the addition of any other rubber (3, 4). Second, incompatible mixtures with poor properties were obtained. Third, certain polyethers, which in themselves did not lead to PVC's of high impact strengths, seemed to enhance the properties of an already modified CPVC (5).

Table II. Recipes for Masterbatches [a]

A. *Masterbatch for Chlorinated Polyvinyl Chloride with Modifier*

Material	Parts by Weight
Chlorinated Polyvinyl Chloride (Geon 603X560)	100.0
Stabilizer [b]	2.5
TiO_2 (Rutile)	5.0
Lubricant [c]	1.0
Chlorinated Polyethylene	8.75

B. *Masterbatch for Chlorinated Polyvinyl Chloride Alone*

Material	Parts by Weight
Chlorinated Polyvinyl Chloride (Geon 603X560)	100.0
Stabilizer [b]	2.5
TiO_2 (Rutile)	5.0
Lubricant [c]	1.0

C. *Masterbatch for Polyvinyl Chloride*

Material	Parts by Weight
Polyvinyl Chloride (Geon 103)	100.0
Processing Aid [d]	3.0
Calcium Stearate	2.0
TiO_2 (Rutile)	5.0
Stabilizer [b]	3.0

[a] The masterbatches were prepared by mixing in a Henschel mixer. Typically, the ingredients were mixed four to six minutes at 2600 rpm, and a final mixing temperature of 150°–60° C was reached.
[b] Dibutyltin thioglycollate
[c] Low-molecular-weight polyethylene, "Ac629A," made by Allied Chemical Corp.
[d] Styrene/acrylonitrile copolymer

High-Impact Resins from Polytetrahydrofuran. The only polyether successfully used alone to produce high-impact resins was PTHF. (Brittle resins were obtained from all the other polyethers tried.) Both PVC (Figure 1) and CPVC (Figure 2) could be improved in Izod impact strength by addition of PTHF. As usual, high impact strength did not develop below a certain critical concentration, and did not change significantly as the concentration was increased further. In the PVC blends, about 7 phr was needed for good impact strength. In CPVC blends 5 phr seemed to be enough.

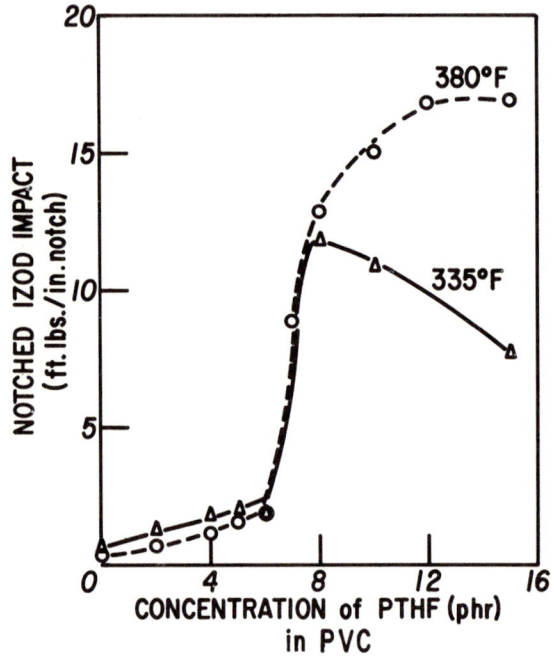

Figure 1. Effect of concentration of PTHF on notched Izod impact strength of PVC–PTHF blends. Intrinsic viscosity of PTHF is 3.2 dl/gram (C_6H_6 at 25°C)

Figure 2 also illustrates the poor properties obtained when milling temperature is too low. When milling was carried out at 335°F (steam-heated mill), the impact strength that developed was lower than the maximum obtained after milling at 380°F and *decreased* as the concentration of PTHF was increased above 8 phr.

The PTHF resins used in the blends shown in Figures 1 and 2 had intrinsic viscosities in C_6H_6 at 25°C of 2 to 3 dl/g. This viscosity corresponds to a molecular weight in the range of 100,000. As shown in Figure 3, these are not the only molecular weights that will lead to high-impact resins. The PTHF apparently can have any molecular weight in this range or above. The highest viscosities shown correspond to molecular weights of a million or more. Again, the importance of using a high enough temperature is evident. Most of the PTHF's led to brittle resins when blended at 335°F.

Lower-molecular-weight PTHF's did not lead to high impact resins, as shown by the results in Figure 3. However, if these PTHF's were first crosslinked (6), high-impact resins could be obtained. Some typical results are shown in Table III.

As might be predicted, the degree of crosslinking required to produce a high-impact resin is inversely proportional to the molecular weight of the parent PTHF. However, when the crosslink density of the PTHF is too high, brittle resins are again produced. This is reasonable because at very high crosslink densities, the PTHF loses its rubbery character.

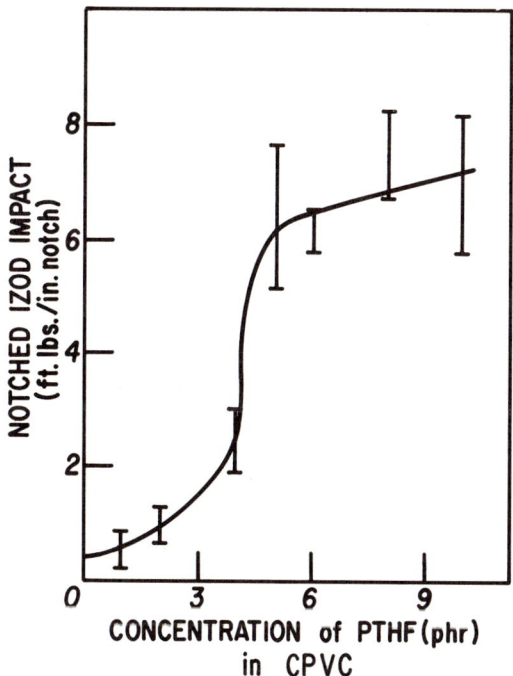

Figure 2. Effect of concentration of PTHF on notched Izod impact strength of CPVC–PTHF blends. Intrinsic viscosity of PTHF is 1.82 dl/gram (C_6H_6 at 25°C)

Table IV shows that crosslinking of the PTHF leads to somewhat increased stability of the blends during protracted milling experiments. A blend containing an uncrosslinked resin loses a substantial proportion of its strength after milling for 30 minutes, whereas the blend containing crosslinked PTHF seems to be essentially unchanged in impact strength after the same time. Nevertheless, there is a price to be paid since initial banding of a crosslinked PTHF blend requires a longer time.

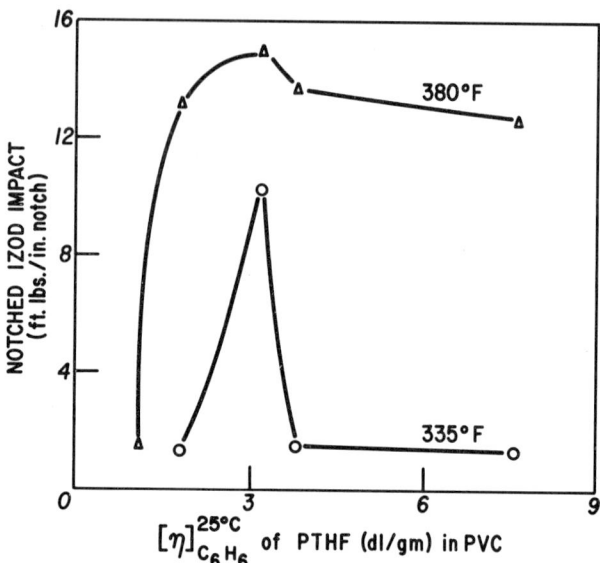

Figure 3. Effect of intrinsic viscosity of PTHF on the notched Izod impact strength of PVC-PTHF blends

Further evidence of the heat stability of the PTHF blends with PVC and CPVC was obtained in an oven aging test at 350°F. Samples were removed from the oven at 10-minute intervals and compared with similarly treated blends containing the impact modifier, a styrene/acrylonitrile/butadiene terpolymer. The blends containing PTHF were outstanding in their resistance to discoloration at elevated temperature. The blends containing the terpolymer showed some darkening at 90 minutes and were brown at 370 minutes. The PTHF blends did not discolor at all until 190 minutes, and 920 minutes were required for the

Table III. Polyvinyl Chloride Blends Containing Crosslinked Polytetrahydrofuran [a]

PTHF $[\eta]$ Before Crosslinking (dl/g) [b]	% Gel After Crosslinking	Notched Izod Impact Strength (ft lbs/in Notch)
1.1	0	1.5
	79.6	10.0
	86.6	1.1
3.8	0	13.9
	58.9	11.0
	80.8	0.5

[a] All blends contained 10 phr PTHF and were milled five minutes after banding at 380°F.
[b] Determined in C_6H_6 at 25°C

development of the brown color obtained in the terpolymer blend after 370 minutes.

Since PTHF is a crystallizable polymer, there is some concern that on aging at room temperature under ordinary conditions, the blends might lose their strength. In a comparison of Izod impact strengths obtained on blends from a single batch tested at intervals from one hour to 28 days after milling, the Izod impact strength values obtained were unchanged. In the blends, the PTHF has a melting temperature of about 31°C, compared with 52°C observed for the original PTHF. Apparently, the PTHF is sufficiently compatible with the CPVC or with one of the compounding ingredients so that its melting temperature is significantly lowered.

Table IV. Effect of Milling Time on Impact Strength of PVC-PTHF Blends [a]

% Gel	Notched Izod Impact Strength (ft lb/in Notch)			
	5 Minutes	10 Minutes	20 Minutes	30 Minutes
0	13.9	12.1	13.0	4.5
58.9	11.0	10.6	13.0	10.7

[a] All blends contain 10 phr of PTHF with an intrinsic viscosity before crosslinking of 3.8 dl/g in C_6H_6 at 25°C and were milled at 380°F.

The particular recipe used for the masterbatches given in Table II is not the only one that will lead to high-impact resins. As illustrated in Table V, changes in the lubricant and processing aid can be made without loss of impact strength. Changes in the recipe were not extensively explored. In particular, the effect of milling temperature on the ease of blending with different recipes was not studied. It is possible that appropriate changes in recipe could increase the blending case enough to permit use of lower milling temperatures.

Incompatible Mixtures. Even at very low levels, many of the polyether additives led to "incompatible mixtures." These blends were not successfully milled to a smooth sheet under any conditions tried. Instead, a mass of crumbs was obtained. These crumbs could be molded into a coherent mass, but the physical properties were poor. For example, addition of 8.75 parts of polybutene-1 oxide to Masterbatch B for CPVC alone gave a brittle, free-flowing material with these properties; notched Izod impact strength, 0.7 lb/in notch, flow rate 452 g/10 min. This is a particularly interesting result, since PBO has the same chemical formula as PTHF but structurally is a substituted ethylene oxide polymer rather than a linear homopolymer. No further studies were made of such blends.

Table V. Effect of Varying Lubricant/Processing Aid in PVC/PTHF Blends [a]

Lubricant/Processing Aid	phr	Notched Izod Impact Strength (ft lbs/in Notch)
Ca Stearate/—	2/0	1.6
" /Sty-ACN [b]	2/3	12.3
" /Polyethylene	2/1	15.6
" /Polypropylene	2/1	16.3
Polyethylene/Sty-ACN [b]	1/3	4.8
Polypropylene/Sty-ACN [b]	1/3	9.1

[a] All blends contain 10 phr PTHF of intrinsic viscosity 3.22 dl/g in C_6H_6 at 25°C.
[b] Styrene/acrylonitrile copolymer

Polyethers that Enhance the Properties of Modified CPVC. The third type of behavior observed on the addition of polyethers can be illustrated by the addition of PBO to Masterbatch A. When a small amount of PBO was added to CPVC already modified with chlorinated polyethylene, the results shown in Figure 4 were obtained (5). Both the notched Izod impact strength and the flow rate measured in a constant load rheometer increased. Again, high milling temperatures, about 400°F, and high shear rates were necessary during blending if good properties were to be obtained. The effect of PBO concentration on several other properties of the modified resin are shown in Table VI. The heat distortion temperature was hardly changed and the tensile strength was slightly decreased. Similar effects on heat distortion temperature and tensile strength were also observed for the high impact strength blends where PTHF was used as the sole modifier.

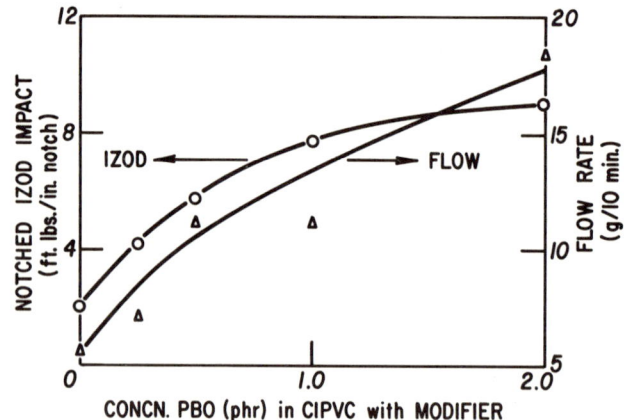

Figure 4. Effect of concentration of PBO on flow rate and notched Izod impact strength of CPVC containing 8.75 phr chlorinated polyethylene as modifier

The effects of longer and shorter side chains on the epoxide compared to PBO are shown in Table VII. Polyethylene oxide led to a small improvement in impact strength and melt flow rate, but the heat distortion temperature was decreased. Polypropylene oxide and polyhexene-1 oxide had enhancing effects similar to and even a bit greater than those of PBO. Polyphenylglycidyl ether appeared to be inert when added to modified CPVC. Finally, in this application, the linear PTHF was harmful to properties.

Contrasted with the resins modified with chlorinated polyethylene, addition of the substituted epoxides to resins modified with PTHF led to blends with very poor properties.

Table VI. Effect of Concentration of PBO on Other Properties of Modified CPVC

phr	0	0.25	0.5	1.0	2.0
HDT, °C	105	103	102	105	105
Tensile Strength (psi)	7680	7380	7470	7350	7210

Table VII. Addition of Epoxides with Other Side Chains to Modified CPVC [a]

Epoxide	phr	Izod ft lbs/in Notch	Melt Flow Rate (g/min)	HDT (°C)
—	—	3.9	6.3	101
PEO	1	7.9	8.8	97
PPO	1	8.8	13.7	105
PBO	1	7.8	10.9	105
PHO	1	8.8	12.9	105
PPGO	1	3.9	7.4	101
PTHF	1	1.0	5.1	—

[a] PBO and PTHF are included for comparison.

Conclusions

Polyether additives to PVC and CPVC sometimes have rather remarkable effects on the properties of the resulting blends. Polytetrahydrofurans of appropriate molecular weight lead to resins of high Izod impact strength and good resistance to thermal discoloration. Low levels of addition of certain linear and especially branched polyepoxides to modified CPVC lead to enhanced Izod impact strength and faster flow rates through the constant load rheometer.

Literature Cited

1. Farthing, A. C., "High Polymers," Vol. XIII, p. 309, Interscience, New York, 1963.
2. Kliner, G. J., Attridge, J. F., U.S. Patent **3,155,745** (1964).
3. Dreyfuss, P., Dreyfuss, M. P., U.S. Patent **3,548,037** (1970).
4. Dreyfuss, P., U.S. Patent **3,463,834** (1969).
5. Dreyfuss, P., Tucker, H. A., U.S. Patent **3,453,347** (1969).
6. Bak, K., Elefante, G., Mark, J. E., *J. Phys. Chem.* (1967), **71,** 4007.

RECEIVED May 26, 1972.

9

Developments in Vinyl Chloride Graft Copolymers

F. WOLLRAB, J. DUMOULIN, F. DECLERCK, P. GEORLETTE, and M. OBSOMER

Solvay et Cie S.A., Centre Recherche, rue de Ransbeek, 310, 1120 Brussels, Belgium

> *To obtain vinyl chloride graft copolymers with a high backbone-polymer content of the desired homogeneous morphology, all of the liquid vinyl chloride must be absorbed by the backbone polymer during polymerization. This condition is realized when, at polymerization temperature, the partial pressure of vinyl chloride in the autoclave is lower than the vapor pressure of pure vinyl chloride. On this basis, a graft polymerization process for the production of homogeneous vinyl chloride/polyethylene graft copolymers with high backbone-polymer content has been developed. One of these graft copolymers has already found industrial application as a peroxide-crosslinkable material in cable insulation. The graft polymerization process also applies to rubbery ethylene-propylene copolymers and amorphous epichlorohydrin polymers.*

Grafting of vinyl chloride (VC) onto various polymer backbones is a useful way of preparing modified polyvinyl chloride (PVC) type resins. This paper deals with the grafting of vinyl chloride onto polyethylene (PE), onto rubbery ethylene-propylene copolymers (EPR), and onto amorphous epichlorohydrin polymers. This kind of PVC modification should lead to resins with improved processability or impact strength, or to crosslinkable polymers.

Among the various techniques of grafting vinyl chloride described in the literature (1), we chose the one based on the polymerization, in aqueous suspension, of vinyl chloride in the presence of the backbone polymer with an organic peroxide. Some important improvements in this technique are the main subject of this paper.

With a polyethylene backbone, this grafting technique will not yield pure poly(ethylene-g-vinyl chloride)—"pure graft copolymer"—but a mixture of this compound with PVC homopolymer and unmodified polyethylene. We call this raw graft polymerization product "VC/PE graft copolymer." Sometimes, we add its gross composition between parentheses—for example, VC/PE (50-50) graft copolymer. The same statement applies to the other backbone polymers.

The grafting of vinyl chloride onto polyethylene by following these principles has already been studied in several industrial laboratories (2-5). In the case of the EPR backbone, earlier work was essentially done by Natta (6) and Severini (7), and processes are described in the patents literature (8, 9). As far as we know, that work did not lead to commercial applications.

Graft Polymerization Processes

Our own work in this field followed two lines: preparation of graft copolymers with high backbone-polymer content, such as VC/EPR (50-50) graft copolymers; and preparation of products with low backbone-polymer content, such as VC/EPR (95-5) graft copolymers.

Graft Copolymers with High Backbone-Polymer Content. Work on VC/PE graft copolymers of this type led to a process (10) now being tested on pilot-plant scale. Its industrial application is being considered. The process is also applicable to other backbone polymers that are sufficiently swelled by this monomer. Most grades of ethylene-propylene

Table I. Characteristics of the Polyethylene Backbone Polymers

Type	*Microthene* N 710	*Alkathene* WRM 19	*Eltex 6037*
Process	High Pressure		Low Pressure
Particle Form	Powder	Pellets	Fluff
Producer	USI	ICI	Solvay
Density g/cm³ (ASTM D 1505)	0.916	0.916	0.960
Melt index g/10 min (ASTM D 1238)	22	20	3.7
Crystallinity, % (differential thermal analysis)	26	28	73
Double bonds/1000 C atoms			
trans	0.09	0.09	<0.05
vinyl	0.04	0.02	1.55
vinylidene	0.64	0.53	<0.05
CH₃/1000 C atoms	15	16	0

rubber (*11*) and the amorphous epichlorohydrin polymers and copolymers (*12*) are convenient for this procedure, which also applies to mixtures of polyethylene and all types of ethylene propylene rubber (*11*).

The backbone polymer is used in powdered form or as pellets or chips. The end product is generally of the VC/backbone (50-50) type. The process described here is for the VC/PE graft copolymers.

Vinyl Chloride/Polyethylene Absorption Isotherms. The absorption isotherms have been determined by adding increasing quantities of vinyl chloride to an aqueous suspension of polyethylene. After saturation of the polyethylene, there is no further increase in the pressure with increasing quantities of vinyl chloride. The principal characteristics of the various polyethylenes examined are shown in Table I.

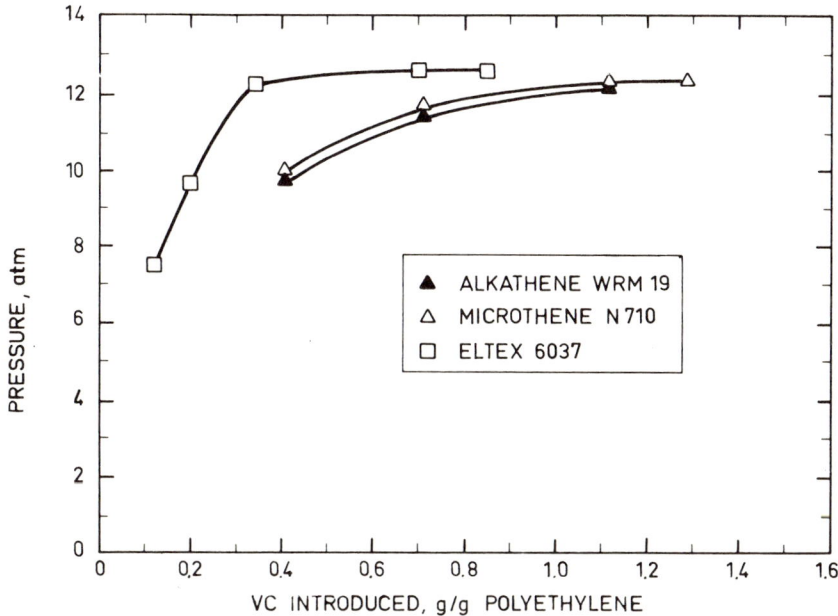

Figure 1. Vinyl chloride/polyethylene absorption isotherms at 68°C

Figure 1 shows that two polyethylenes of nearly identical chemical nature (Alkathene WRM 19 and Microthene N 710) have nearly the same absorption isotherms in spite of their very different particle sizes. This proves that the vinyl chloride is homogeneously absorbed in the mass, not only adsorbed on the surface. Figure 1 also shows that, at 68°C, low-density polyethylenes (Alkathene WRM 19 and Microthene N 710) absorb much more vinyl chloride than does high-density polytheylene (Eltex 6037).

Figure 2 shows that this absorption increases considerably with temperature. The maximum pressure reached by these absorption isotherms is higher than the value that could be explained by the vapor pressures of vinyl chloride and water. This increment is caused by residual inert gases present in the autoclave.

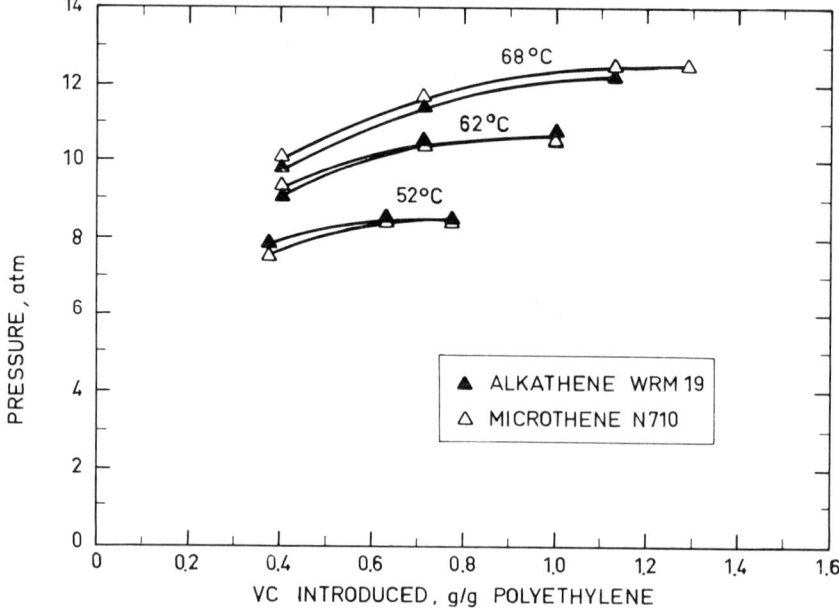

Figure 2. Vinyl chloride/polyethylene absorption isotherms at different temperatures

Characteristics of the Graft Polymerization Process. Under certain polymerization conditions such as when the vinyl chloride/polyethylene ratio is high, heterogeneous products are obtained in the sense that, beside the polyethylene grains gorged with PVC, independent suspension PVC particles are formed. Since the optimum conditions for production of PVC and graft copolymers are not necessarily identical, it is advantageous to avoid the formation of these independent PVC particles.

Figures 3 through 6 show interferential photomicrographs of cuts of the different polymer particles. The structural differences are considerable between the particles of the initial polyethylene (Figure 3), of supension PVC (Solvic 229) (Figure 4), of a heterogeneous VC/PE (90-10) graft copolymer (Figure 5), and of a homogeneous VC/PE (50-50) graft copolymer (Figure 6).

To obtain a homogeneous VC/PE graft copolymer, all the liquid vinyl chloride must be absorbed in the polyethylene. This condition is realized when, at the polymerization temperature, the partial pressure

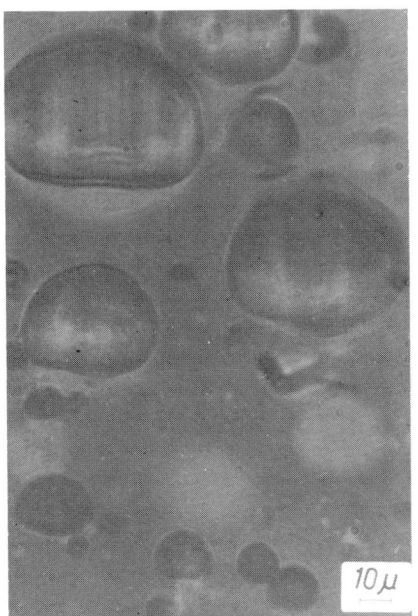

Figure 3. Polyethylene powder (Microthene FN 500)

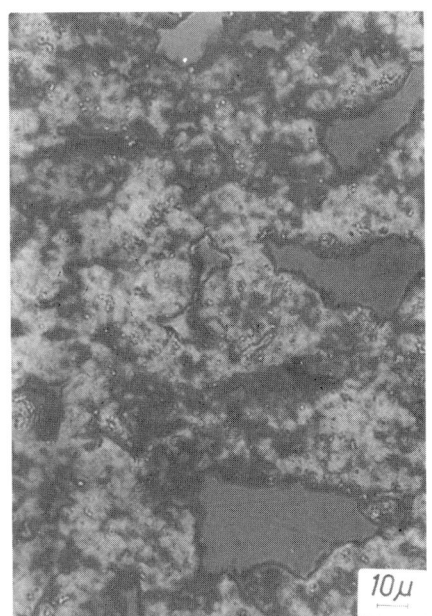

Figure 4. Suspension PVC (Solvic 229)

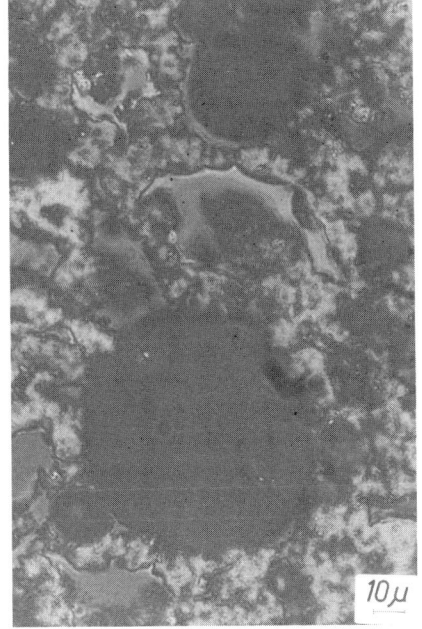

Figure 5. VC/PE (90-10) graft copolymer

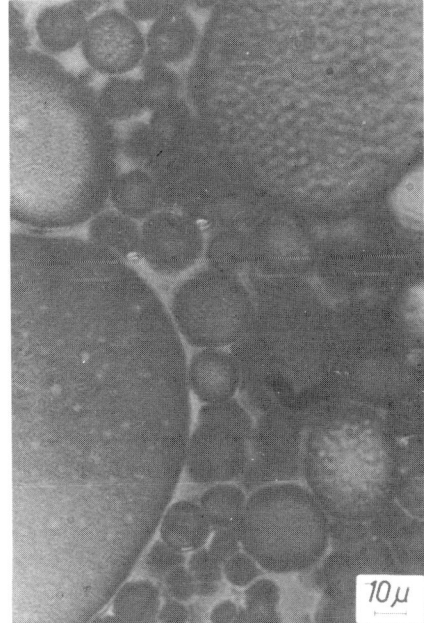

Figure 6. VC/PE (50-50) graft copolymer

of vinyl chloride in the autoclave is lower than the vapor pressure of pure vinyl chloride. This is an essential characteristic of the graft process.

Experimental

Into a 5-liter autoclave, continuously stirred at a speed of 450 rpm, water, a dispersing agent, and polyethylene (powder or pellets) are introduced in the proportions set forth in Table II. After introduction of the initiator, the vinyl chloride is injected in such a quantity that, at the polymerization temperature, the vinyl chloride's partial pressure is lower than the vapor pressure of pure vinyl chloride at the same temperature. The vinyl chloride quantities compatible with this condition are easily determined by the absorption isotherms. When the pressure has dropped to at least half of its maximum value, the nontransformed vinyl chloride is removed. After filtering, washing, and drying, the product is collected.

In this procedure, stirring speeds and polyvinyl alcohol concentrations have been chosen to give a good dispersion of the polyethylene in water. Initiator concentration is adjusted to give a reaction time of

Table II. Polymerization Conditions

Nature of Backbone Polymer	Microthene N 710		Alkathene WRM 19
Vinyl chloride introduced, g		680	
Polymerization recipe, g/kg VC			
Total vinyl chloride	1000	1000	1000
Initial vinyl chloride	1000	600	600
VC per supplementary fraction	—	100	100
polyethylene	1000	1000	1000
demineralized water	3500	3500	3500
polyvinyl alcohol	1.8	1.8	1.8
lauroyl peroxide	4.2	4.2	3
epoxidized soybean oil	—	—	5
Polymerization temperature, °C	62	62	62
Initial pressure (maximum), atm	10.5	10.3	10.3
Pressure drop before addition of supplementary fraction, atm	—	2	3
Final pressure, atm	3.9	4	4.1
Polymerization time, min	6.00	6.00	7.00
Vinyl chloride conversion, %	84	89	93

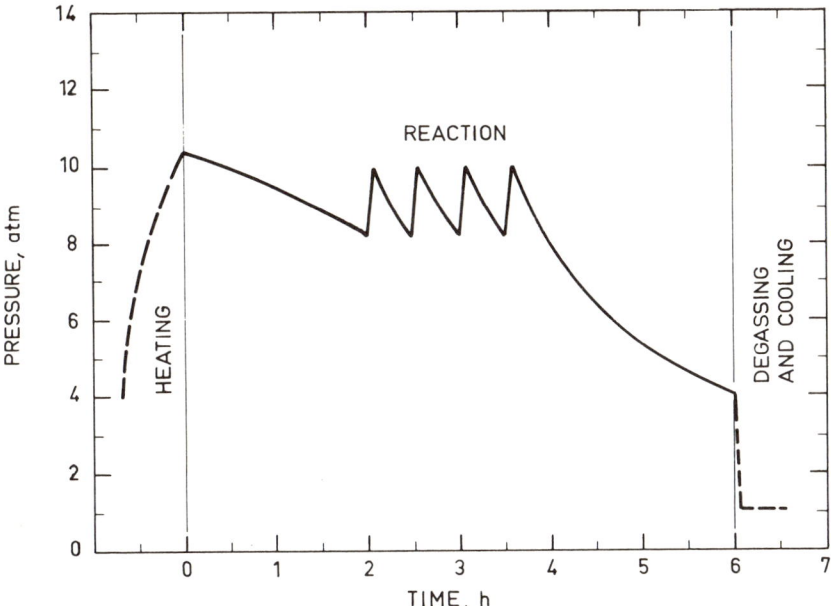

Figure 7. Pressure curve of a representative laboratory polymerization (62° C)

about six hours. Products in pellet form were prestabilized with epoxidized soybean oil to maintain good color during the drying operation.

Vinyl chloride absorption is, in general, very rapid, and takes place during the heating of the autoclave. Immediately after reaching the reaction temperature, the pressure in the autoclave begins to drop. This absence of a period of constant pressure shows that the system is effectively below saturation.

When, in the case of low-density polyethylene, a vinyl chloride/polyethylene ratio near 1 is used, the system is close to saturation and a few per cent of a fine powder containing principally PVC may form. This is avoided when a vinyl chloride/polyethylene ratio of 0.6 is used, with injection of the rest of the vinyl chloride during the polymerization. This way, the system is always definitely below saturation. Moreover, this technique improves the grafting efficiency (defined by the ratio: grafted VC x 100/total polymerized VC), which increases with a decreasing vinyl chloride/polyethylene ratio during polymerization.

Figure 7 shows the pressure curve of a representative laboratory polymerization. In the pilot plant, the addition of supplementary fractions of vinyl chloride during the polymerization is replaced by a continuous injection of monomer into the autoclave.

By polymerization at different temperatures, it is possible to vary the length of the grafts and, at the same time, the K value of the PVC homopolymer formed as a by-product.

Degree of Grafting of the Backbone Polymer. With graft copolymers based on low-density polyethylene, the degree of grafting (defined by: grafted PE x 100/total PE) is determined by selective extraction of the homopolymers. With high-density polyethylenes, it is determined by a liquid-solid absorption technique which will be described elsewhere (*13*).

The degree of grafting is often between 40 and 60%. The degree is mainly influenced by the nature of the backbone polymer in the sense that an increase in its unsaturation or its CH_3 content increases the grafting degree. The choice of a convenient polyethylene is therefore another essential characteristic of the process. Use of a low vinyl chloride/polyethylene ratio and the addition of a large proportion of the vinyl chloride during polymerization also favors a higher grafting degree. Examination of the ungrafted polyethylene present in these raw graft copolymers shows that grafting occurs preferentially onto the high molecular weight fractions.

Graft Copolymers with Low Backbone-Polymer Content. The procedure for preparing this kind of graft copolymer is based on the dissolution of the backbone polymer in the monomer, dispersion of this solution in water, and polymerization by means of an organic peroxide. It applies only to soluble backbone polymers, such as most EPR's. As the handling of a too-viscous vinyl chloride/backbone polymer solution is impractical, this procedure is normally used for preparing end products of the type VC/backbone polymer (95-5) or (90-10).

Our work has shown rapidly that a considerable amount of know-how is necessary to obtain a product having satisfactory processing characteristics. More particularly it is necessary to have a material that gives no matrix-rubber separation during fabrication (extrusion, injection molding, etc.).

Properties of the Graft Copolymers

VC/PE Graft Copolymers. The most useful VC/PE graft copolymers are those with a high backbone-polymer content.

RAW GRAFT COPOLYMERS. The main characteristic of the raw VC/PE (50-50) graft copolymers is their brittleness. For this reason, no detailed studies of their mechanical properties have been made.

The compatibilizing effect of poly(ethylene-g-vinyl chloride), the characteristic constituent of these VC/PE graft copolymers, may be shown by electron photomicrographs of films consisting of a PVC-PE/50-50 mixture on the one hand, and of a VC/PE (50-50) graft copolymer containing about 50% of poly(ethylene-g-vinyl chloride), 25% of polyethylene and 25% of polyvinyl chloride on the other hand (*see* Figure 8). The films were obtained from a solution in *o*-dichlorobenzene. In the first case, the polyvinyl chloride and polyethylene phases are clearly separated; in the second case, they interpenetrate closely.

POLY(ETHYLENE-G-VINYL CHLORIDE): "PURE GRAFT COPOLYMER." Small quantities of a poly(ethylene-g-vinyl chloride) containing 46%

vinyl chloride have been isolated by selective extraction of the homopolymers. This pure graft copolymer is a good impact modifier for PVC.

To increase the knowledge of these pure graft copolymers, we determined the temperature dependence of the dynamic mechanical properties of such a product. These measurements were made with the Rheovibron apparatus at a frequency of 110 Hz.

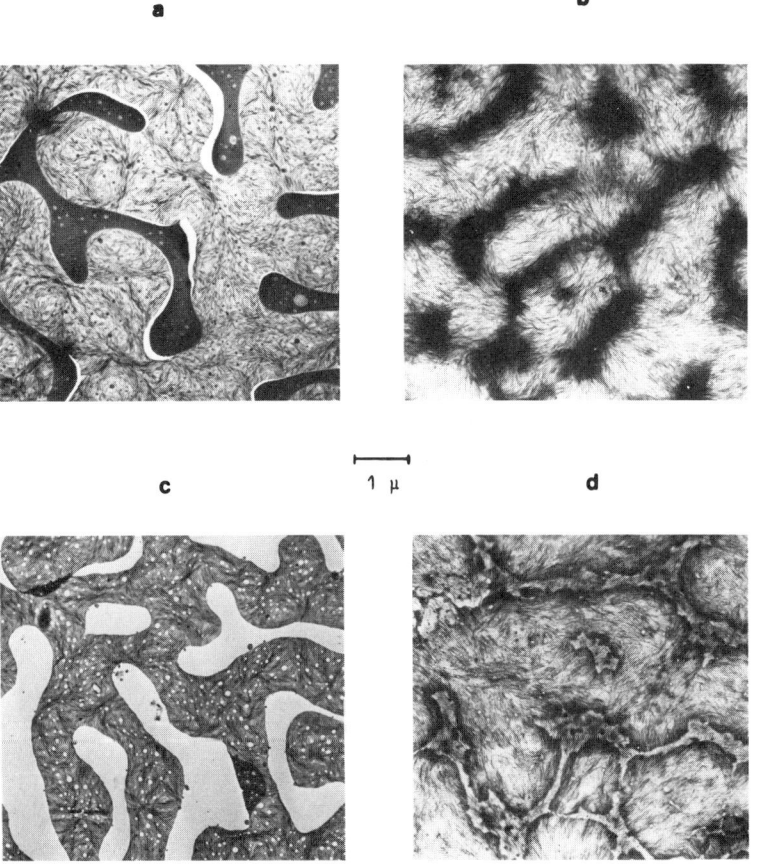

Figure 8. Electron photomicrographs showing the compatibilizing effect of poly(ethylene-g-vinyl chloride)

(a) PVC-PE/50-50 mixture
(b) VC/PE (50-50) graft copolymer
(c) Mixture after treatment with tetrahydrofuran
(d) Graft copolymer after treatment with tetrahydrofuran

Figure 9 shows the mechanical loss curves of a low-density polyethylene, a PVC, a raw VC/PE (50-50) graft copolymer, and a poly(ethylene-g-vinyl chloride) containing 46 weight % of vinyl chloride.

The curves of the raw graft copolymer and of the poly(ethylene-g-vinyl chloride) are rather close to that of the low-density polyethylene. The outstanding fact is the absence of the PVC transition peak (between 60° and 100°C) in the mechanical loss curves of these two products. This means that they contain no rigid PVC phase in spite of the presence of about 25 weight % of ungrafted polyvinyl chloride in the raw graft copolymer. This PVC seems thus to be strongly compatibilized with the other constituents by the poly(ethylene-g-vinyl chloride).

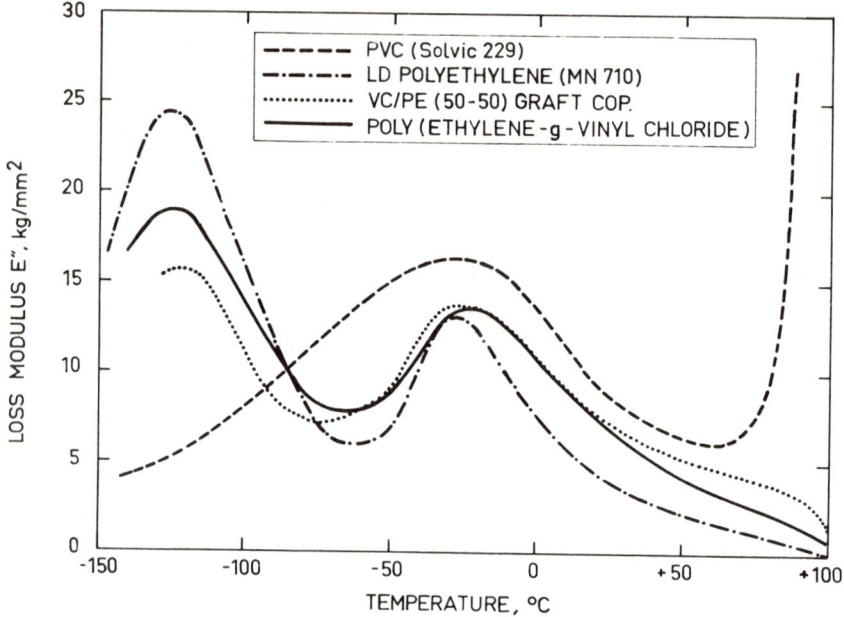

Figure 9. Temperature dependence of the loss moduli E"(Rheovibron, 110 Hz)

Figure 10 compares the moduli of elasticity of the same four products: low-density polyethylene, PVC, raw VC/PE (50-50) graft copolymer, and poly(ethylene-g-vinyl chloride).

In spite of their high contents of grafted and ungrafted PVC, the behavior of the raw graft copolymer and of the poly(ethylene-g-vinyl chloride) is not very different from that of low-density polyethylene. The stiffening action of this PVC is thus rather low.

MIXTURES OF PVC AND RAW GRAFT COPOLYMER. *State of Dispersion of Graft Copolymer in a PVC Matrix.* We tried to follow the dispersion of a raw VC/PE (50-50) graft copolymer in a PVC matrix during the gelling operation. This was done by gelling the PVC Solvic 223 on a roll mill at 170°C in a transparent formulation, and by incorporating into the crepe the quantity of graft copolymer necessary to get a PVC-graft copolymer/85-15 mixture. At different gelling times, we took samples of the crepe and examined them with an interferential microscope (thickness of the cut = 10μ) (Figures 11 through 15). The dark zones in Figures 12 and 13 correspond to the graft copolymer (refractive index 1.528, compared with 1.543 for the PVC and 1.515 for the backbone polyethylene).

The initial graft copolymer particles (diameter about 200μ) disappear rapidly, and the distinct graft copolymer phase existing at the end of its incorporation is easily dispersed. The state of dispersion remains practically unchanged after two minutes of gelling.

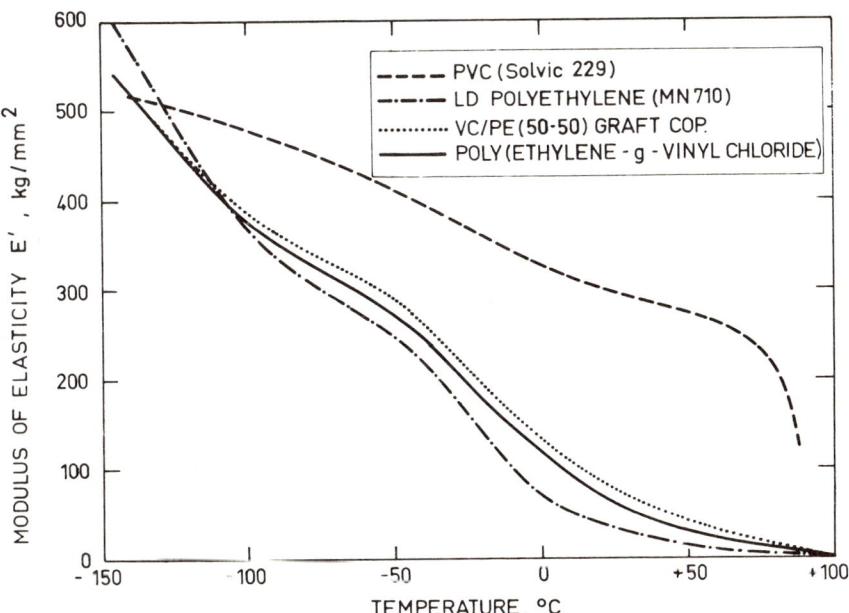

Figure 10. Temperature dependence of the moduli of elasticity E'(Rheovibron, 110 Hz)

Melt Viscosity of PVC-Graft Copolymer Mixtures. Table III shows that the presence of 15 phr of a VC/PE (50-50) graft copolymer in a rigid PVC compound lowers its melt viscosity considerably. This effect is more pronounced for the graft copolymers with a low-molecular-weight PVC fraction.

Temperature Dependence of Stiffness of PVC-VC/PE Graft Copolymer Alloys. Figure 16 shows that the presence of a VC/PE graft copolymer in a rigid PVC compound changes its stiffness only slightly. At 20°C, the introduction of 13 weight % of a VC/LD·PE (50-50) graft

 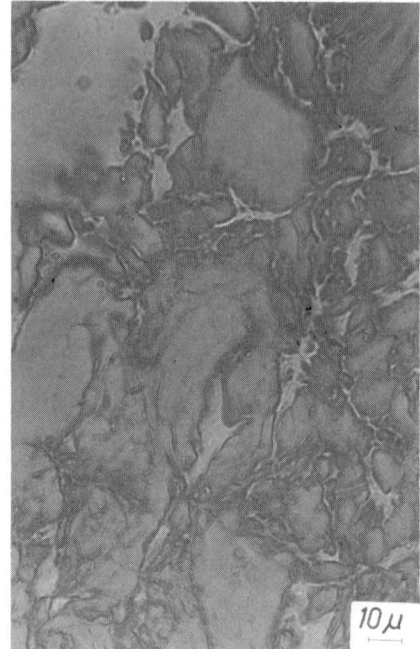

Figure 11. PVC   Figure 12. PVC + graft copolymer, immediately after incorporation

copolymer or a VC/HD·PE (40-60) graft copolymer lowers the value of the shear modulus from 9.000 kg/cm^2 to 8.600 or 8.400 kg/cm^2.

Impact Strength of PVC-VC/PE Graft Copolymer Alloys. Figure 17 shows the influence on Izod impact strength of the graft copolymer content of a PVC-VC/LD·PE graft copolymer alloy. The maximum impact strength is obtained with a total polyethylene content of about 7.5%. The nature of the backbone polymer influences the Izod value in the sense that the impact strength decreases as the crystallinity of the backbone polymer increases.

Figure 18 shows that the impact strength of such an alloy depends strongly on the working conditions and especially on the gelling time. A high proportion of unbroken specimens is obtained only after a rather long gelling time.

Other Mechanical and Thermal Properties. Table IV shows that the tensile properties, the Shore D hardness, and the deflection temperature of a PVC graft copolymer alloy containing 13% of a VC/LD·PE (50-50) graft copolymer are only slightly lower than the corresponding properties of the pure PVC (Solvic 239). There is no difference between the heat stabilities of the PVC and of the alloy.

Transparency. Alloys of PVC and VC/PE graft copolymer are more transparent than PVC-polyethylene mixtures having the same total polyethylene content. Graft copolymers based on high-density polyethylene give better transparency than do similar graft copolymers based on low-density polyethylene. Perfect transparency, however, cannot be obtained.

VC/EPR Graft Copolymers. For most practical applications, compositions with low EPR content should be used. Specimens of this type may be obtained by dilution with PVC of a VC/EPR graft copolymer with high EPR content, or directly from a VC/EPR graft copolymer with low EPR content.

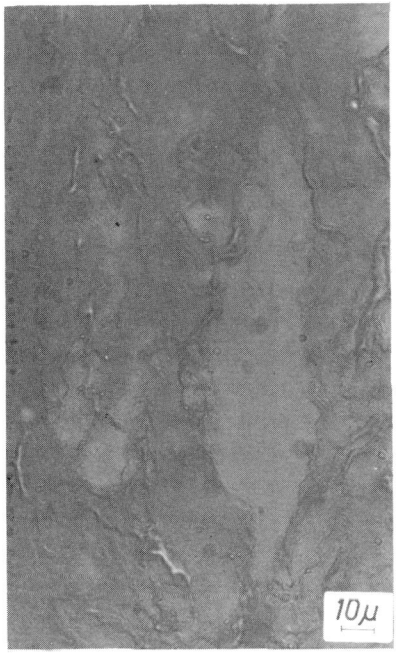

Figure 13. PVC + graft copolymer, 1 minute of gelling

Figure 14. PVC + graft copolymer, 2 minutes of gelling

STATE OF DISPERSION OF THE EPR. Specimens obtained from a VC/EPR (95-5) graft copolymer and from a PVC-EPR/95-5 mechanical mixture have been examined with an interferential microscope (Figures 19 and 20).

In the mechanical mixture, the dispersion of the rubbery phase is coarse; in the graft copolymer, by contrast, it is so fine that it is difficult to detect the elastomer by this optical technique of limited resolving power. However, measurements of the temperature dependence of the mechanical loss show that the elastomer is present as a distinct phase.

OPACITY. Specimens prepared from the VC/EPR graft copolymers are opaque. This seems to be because of the two-phase structure of the system, the large difference between the refractive indexes of the two phases, and the fact that the elastomer domains are large enough.

Figure 15. PVC + graft copolymer, 18 minutes of gelling

MECHANICAL PROPERTIES. The properties of a composition containing 5% of EPR are shown in Table V. The tensile properties and the shear modulus of this composition are slightly lower than those of a rigid PVC of the same K value. Shore hardness and deflection temperature are comparable in the two cases. In comparison with rigid PVC, the grafted compositions have an excellent impact strength that increases with the molecular weight.

VC/Polyepichlorohydrin and VC/Poly(epichlorohydrin-co-ethylene oxide) Graft Copolymers. Graft copolymers of this type with high backbone-polymer content should be useful PVC additives.

TRANSPARENCY. Alloys of PVC with up to 20% of this kind of graft copolymer show excellent transparency.

MECHANICAL PROPERTIES. The tensile properties of these alloys are slightly lower than those of unmodified PVC, but their impact strength is improved.

Table III. Melt Viscosity of PVC-VC/PE Graft Copolymer Mixtures

Formulation, phr	1	2	3
Solvic 223	100	100	100
VC/PE (50-50) graft copolymer (medium K value)	—	15	—
VC/PE (50-50) graft copolymer (low K value)	—	—	15
dibasic lead phthalate	3	3	3
calcium stearate	0.5	0.5	0.5
lead stearate	0.4	0.4	0.5
Melt viscosity at 180° C, 10^{-4} poise			
Shear rate $\gamma = 100$ s^{-1}	62	42	34.5
$\gamma = 500$ s^{-1}	17	13	11

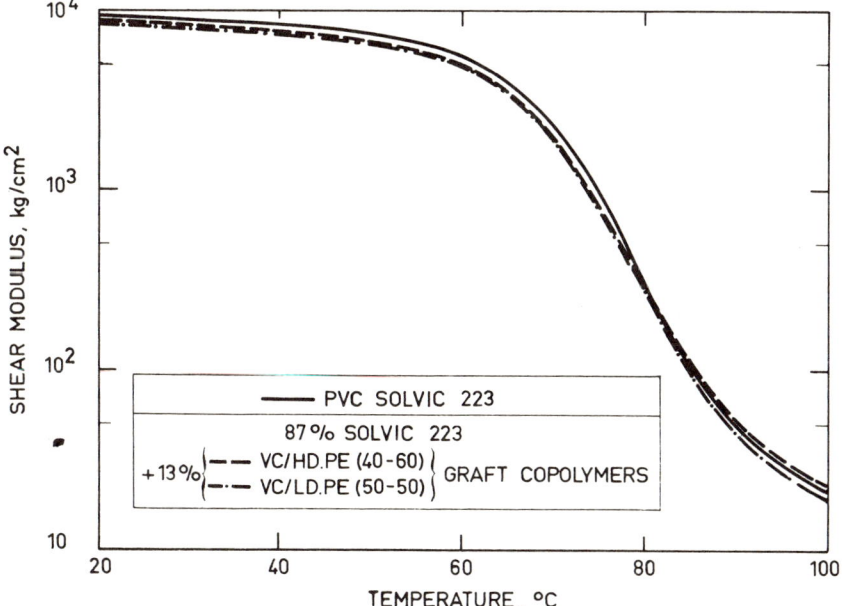

Figure 16. Temperature dependence of the stiffness of PVC-VC/PE graft copolymer alloys. The three formulations are stabilized with 1% Advastab 17 MO.

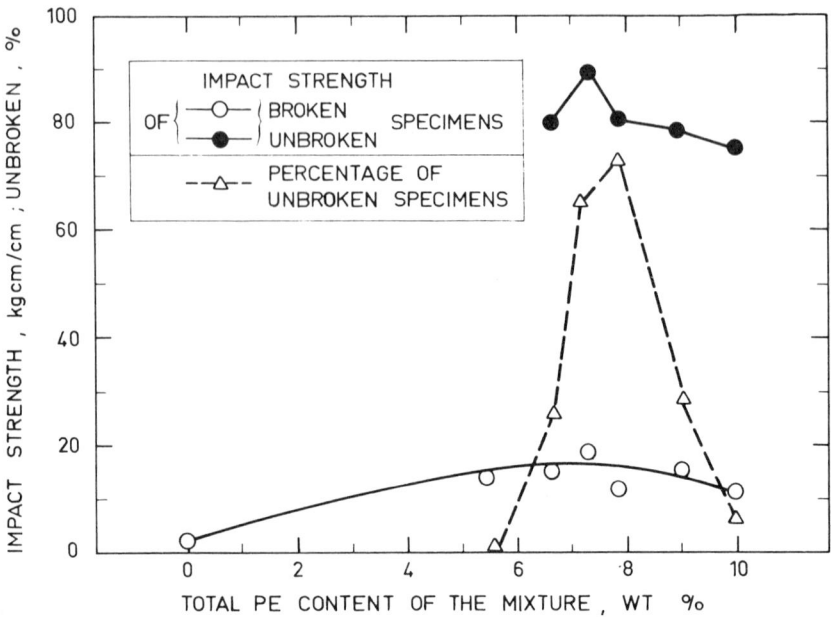

Figure 17. Notched Izod impact strength of PVC-VC/PE graft copolymer alloys and proportion of unbroken specimens; hammer: 10 ft-lb

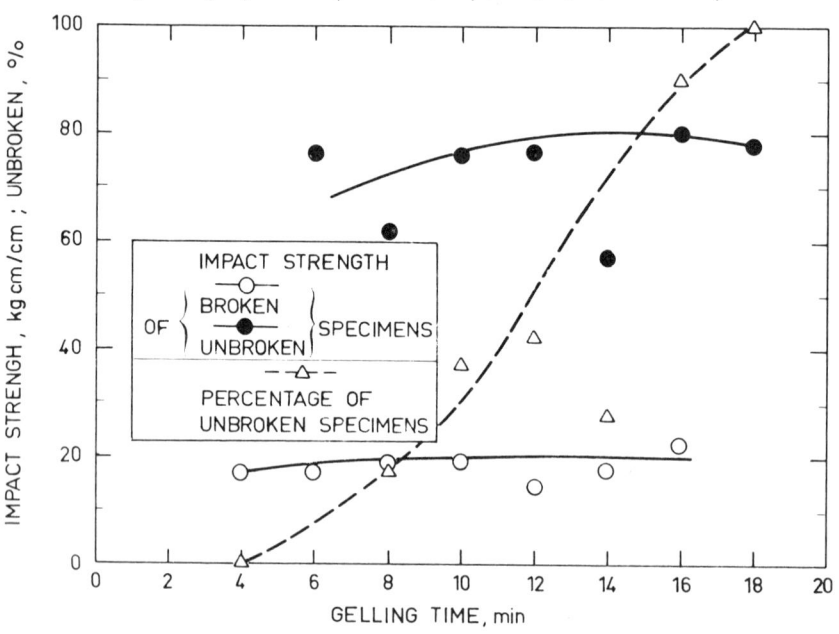

Figure 18. Influence of the gelling time on the Izod impact strength (notched) of a PVC/PE graft copolymer alloy (87% Solvic 239 + 13% VC/LD·PE (50-50) graft copolymer); hammers: 3 ft-lb, 10 ft-lb

Table IV. Properties of Solvic 239 Modified by VC/LD-PE (50-50) Graft Copolymer

	Solvic 239	Alloy of Solvic 239 and Graft Copolymer (13% Graft Copolymer)
Tensile yield strength, kg/cm^2	550	430
Tensile break strength, kg/cm^2	500	415
Ultimate elongation, %	100	150
Shore D hardness	82	80
Deflection temperature, °C (264 psi)	75	73

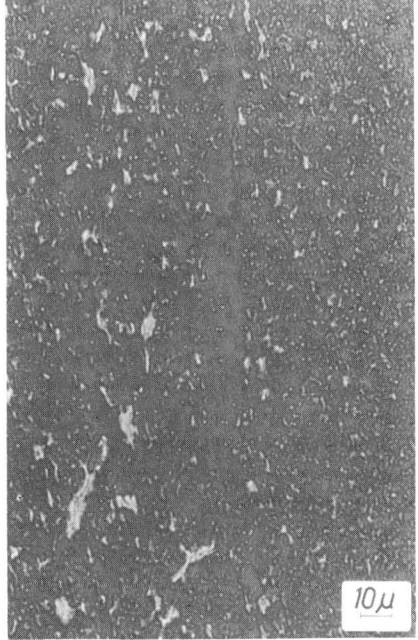
Figure 19. PVC/EPR/(95-5) mechanical mixture

Figure 20. VC/EPR (95-5) graft copolymer

Applications of the Graft Copolymers

VC/PE Graft Copolymers. These applications have already been studied in some detail:
- PVC additive for increasing impact strength or improving the ease of injection molding.
- Processing-aid and antisticking agent for PVC compounds (*14*).
- Crosslinking (*15*).

USE OF CROSSLINKED VC/PE GRAFT COPOLYMERS FOR CABLE INSULATION. At high temperatures, plasticized PVC cable insulations show a rather high flow. This should be improved by crosslinking of the thermoplastic insulation material. Unfortunately, PVC is not easily crosslinkable by chemical agents. The presence of about 50% of grafted or

Table V. Properties of VC/EPR (95-5) Graft Copolymer

	PVC Solvic 229	Graft Copolymer
Tensile yield strength, kg/cm^2	533	454
Tensile break strength, kg/cm^2	496	432
Ultimate elongation, %	175	175
Izod impact strength (notched), kg cm/cm	3.6	15
Shore D hardness	81	82
Shear modulus, kg/cm^2		
at 20° C	10,600	8,600
60° C	6,700	5,900
70° C	3,500	3,600
80° C	250	520
100° C	36	36
Deflection temperature (264 psi), °C	73.5	73
Melt viscosity at 170° C, 10^{-4} poise	8.4	4.5

Table VI. Crosslinking Ability of VC/PE Graft Copolymer

Composition, phr			
PVC Solvic 229	100	—	—
VC/LD.PE (50-50) graft copolymer	—	100	100
dioctylphthalate	30	—	—
dibasic lead phthalate	7	7	7
lead stearate	1	1	1
Agerite resin D [a]	0.1	0.1	0.1
Varox [b]	—	—	4
triallyl cyanurate	—	—	2
Crosslinking time at 180°C, min	—	—	10
Tensile break strength, kg/cm^2	240	155	228
Ultimate elongation, %	250	34	145
Shore D hardness	67	59	61
Brittle point, °C	> −20	< −60	< −60
Flow at 200° C			
deformation under load, %	98	97	12
permanent deformation, %	96	91	0

[a] Polymerized trimethyldihydroquinoline (antioxidant)
[b] 2,5-Bis(tert-butylperoxy)-2,5-dimethylhexane (50% act. mat.)
[c] Williams plastometer

ungrafted polyethylene in our raw VC/PE graft copolymers should allow crosslinking by organic peroxides.

Table VI shows that crosslinking ability of the VC/PE (50-50) graft copolymers is good (little flow at 200°C). To work with products of similar hardness, the PVC compound mentioned in that table for comparison was plasticized with 30 phr of dioctylphthalate. Another outstanding property of the graft copolymer compositions is their very low brittle point. Crosslinking of these compositions is rapid.

Table VII. Crosslinking of Plasticized VC/PE Graft Copolymer Compositions

Composition, phr				
VC/LD.PE (50-50) graft copolymer	100	100	100	100
dioctyl phthalate	—	10	20	30
dibasic lead phthalate	7	7	7	7
lead stearate	1	1	1	1
Agerite resin D [a]	0.1	0.1	0.1	0.1
Varox [b]	4	4	4	4
triallyl cyanurate	2	2	2	2
Crosslinking time at 180°C, min	5	10	10	15
Tensile break strength, kg/cm^2	213	201	158	154
Ultimate elongation, %	140	190	215	265
Shore D hardness	60	57	44	37
Shear modulus at 20° C, kg/cm^2	1030	790	300	146
Flow at 200° C				
deformation under load, %	12	12	15	25
permanent deformation, %	0	0	0	0

[a] Polymerized trimethyldihydroquinoline (antioxidant)
[b] 2,5-Bis(*tert*-butylperoxy)-2,5-dimethylhexane (50% act. mat.)
[c] Williams plastometer

Table VII shows that incorporation of a plasticizer in the graft copolymer compositions gives vulcanizates of lower hardness. Their flow resistance remains good. Elongation of the plasticized vulcanizates is improved.

The graft copolymer compositions are easily extrudable at temperatures low enough to avoid scorching.

VC/EPR Graft Copolymers. Possible applications of the VC/EPR graft copolymers all concern high-impact PVC: extrusion of pipes and profiles, calendering, and injection molding.

VC/Polyepichlorohydrin and VC/Poly(epichlorohydrin-co-ethylene oxide) Graft Copolymers. Alloys of PVC and graft copolymers of this type with high backbone-polymer content give interesting results in the field of bottle blowing (12). The combination of transparency and impact strength that can be realized with these compositions should enable them to penetrate into the field covered at present by PVC-MBS mixtures.

Conclusions

Our work on vinyl chloride graft copolymers has led to the development of a graft polymerization process for the production of homogeneous VC/PE graft copolymers with high backbone-polymer content. This process is now being tested on a pilot-plant scale, and industrial applications are being considered.

The brittleness of the VC/PE graft copolymers excludes their use without modification. Nevertheless, they should be useful PVC additives for improving impact strength and lowering melt viscosity.

The fact that these graft copolymers contain polyethylene chains makes them crosslinkable by organic peroxides. One of them has already found industrial application as a peroxide crosslinkable material in cable insulation.

The graft polymerization process also applies to other backbone polymers, such as rubbery ethylene-propylene copolymers and amorphous epichlorohydrin polymers. Further work is necessary to bring the resulting products to industrial maturity and to develop their applications.

Acknowledgment

This work was sponsored in part by I.R.S.I.A., Institut pour l'Encouragement de la Recherche Scientifique dans l'Industrie et l'Agriculture, Brussels, whose support is gratefully acknowledged. The authors thank P. Walch and Y. Gobillon for the determination of the dynamic mechanical properties, the preparation of the interferential photomicrographs, and the examinations with the electron microscope. They are grateful to J. Franklin for his help in the preparation of the English manuscript. They are also indebted to Solvay et Cie S.A. for permission to publish this work.

Literature Cited

1. Battaerd, H. A., Tregear, G. W., "Graft Copolymers," Interscience, New York 1967.
2. U.S. Patent **2,947,719** (Sept. 9, 1954).
3. British Patent **783,790** (Dec. 9, 1954).

4. Belgian Patent **652,069** (Aug. 20, 1964); **657,762** (Dec. 30, 1964; U.S. Patent **3,347,956** (Aug. 27, 1964).
5. Japanese Patent **14228/68** (Nov. 30, 1964).
6. Natta, G., et al, *Chim. Ind.* (1965) **47/9** 960.
7. Severini, F., *Polymer Preprints* (1970) **11/1** 371.
8. Italian Patent **565.419** (June 28, 1956); French Patent **1,401,601** (June 12, 1964).
9. German Patent Application **1,203,959** (May 31, 1963).
10. Belgian Patent **769,043** (June 25, 1971).
11. Belgian Patent **772,261** (Sept. 7, 1971).
12. Belgian Patent **773,224** (Sept. 29, 1971).
13. Heyns, H., unpublished data.
14. Belgian Patent **771,684** (Aug. 24, 1971).
15. French Patent Application **71,43,015** (Nov. 30, 1971).

RECEIVED April 1, 1972.

10

The Copolymerization of Tetrachloroethylene and Ethylene

H. HOPFF and N. BALINT

Swiss Federal Institute of Technology, Zurich, Switzerland

> *Tetrachloroethylene can be copolymerized with ethylene in the ratio 7–14/86–93 mole %. Higher ratios of tetrachloroethylene lead to telomers and mixtures of telomers and copolymers. With increasing chlorine content, the molecular weight decreases from 7860 to 720. The products contain trichlorovinyl and chlorine end groups and vary from liquids to waxes and solids.*

The homopolymerization of tetrachloroethylene has been studied by several scientists (1-8) but has always failed. Perchlorinated paraffins have low heat stability and an unusual tendency to cyclization with growing chain length (9, 10) because of sterical hindrance. Our own experiments have confirmed this result. Neither radical-forming nor ionic catalysts showed any tendency to form polymers. Ziegler catalysts removed the chlorine completely to give polyethylene. Benzoyl peroxide gave an oily addition product of the monomer, from which trichlorovinylbenzene was isolated. Little is known of the copolymerization of tetrachloroethylene. It shows the least tendency of all monomers studied to enter copolymer chains. According to a Dupont patent (11) telomers are obtained when tetrachloroethylene is heated with ethylene in the presence of benzoyl peroxide and water at 115°C. A telomer of the formula $Cl(CH_2)_4CCl=CCl_2$ (63.42% chlorine) was isolated.

Our own studies of the copolymerization of tetrachloroethylene with ethylene showed that besides telomerization, copolymerization occurs simultaneously. The influence of the experimental variables (pressure, ratio of the monomers, structure and concentration of the initiator, temperature, and time) were studied. As catalysts azoisobutyrodinitrile, *tert*-butylperoxyisopropyl carbonate, benzoyl peroxide, cyclohexylperoxy carbonate and *tert*-butyl peroctoate were used. The reactive ratio r_1

Table I. Chlorine Content, Yield, Softening Point, Density, and Molecular Weight of Copolymers And Telomers

M_1 C_2H_4 Mole %	M_2 C_2Cl_4 Mole %	Chlorine, %	Yield, %	Softening Point, °C	Density	Molecular Weight
92.55	7.45	2.77	25.25	120.5	0.91	7860
91.41	8.59	5.24	24.92	118	0.915	6640
89.87	10.13	7.75	24.45	116.5	0.918	5780
89.22	10.78	9.15	23.94	115	0.920	5170
87.65	12.35	13.26	23.55	114	0.922	3730
87.65	12.35	14.33	23.23	113	0.923	3360
86.17	13.87	19.26	22.76	111	0.94	2030
85.53	14.47	19.28	22.74	111	0.945	2030
84.18	15.82	19.37	22.53	111.5	0.95	2070
84.18	15.82	22.94	22.04	109	0.97	1640
83.24	16.76	24.90	20.12	105	0.99	1460
82.55	17.45	23.38	20.65	106	0.98	1510
81.60	18.40	28.58	18.24	100.5	1.12	1120
80.98	19.02	29.61	17.44	98	1.19	980
78.01	21.99	28.75	17.36	101	1.16	1140
73.94	26.06	32.08	13.84	94	1.26	930
72.88	27.12	38.90	12.76	88.0	1.35	720

Table II. Reaction Conditions for the Polymerization of Tetrachloroethylene and Ethylene
Initiator: *tert*-Butylperoctoate

Monomer C_2H_4		Comonomer C_2Cl_4		Initiator, %	Polymerization Conditions			Analysis		
grams	mole %	grams	mole %		(max) P atm	(max) T °C	min.	C %	H %	Cl %
105	92.55	50	7.45	1.1	725	101	117	83.83	13.75	2.77
90	91.41	50	8.59	1.0	530	98	107	81.68	13.55	5.24
75	89.87	50	10.13	0.9	498	92	129	78.47	12.67	7.75
60	87.65	50	12.35	0.8	462	89	143	73.53	12.27	14.33
105	86.17	100	13.87	0.8	865	101	138	69.65	11.52	19.26
90	84.18	100	15.82	0.9	655	88	117	66.71	10.77	22.94
75	81.60	100	18.40	1.0	612	99	112	61.15	9.57	28.58
60	84.18	100	15.82	1.1	565	91	128	69.36	11.44	19.37
90	87.65	75	12.35	1.0	648	93	68	83.22	13.56	3.38

for ethylene was 1.67 and $r_2 = 0$ for tetrachloroethylene. There is a definite influence of the ratio of the monomers in the reaction mixture on the nature of the product: high ethylene and low tetrachloroethylene ratios (86-93/7-14 mole %) lead to copolymers with chlorine contents

Table III. Solubility[a] of Copolymers

Chlorine, %	Methanol °C 25	Methanol °C 50	Acetone °C 25	Acetone °C 50	Ether °C 25	Tetrahydrofuran °C 25	Tetrahydrofuran °C 50	Tetrahydrofuran °C 75	Heptane °C 25	Heptane °C 50
2.77	—	—	—	—	—	—	sl	sl	—	sl
5.24	—	—	—	—	—	—	sl	sl	—	sl
7.75	—	—	—	—	—	—	sl	sl	—	sl
9.15	—	—	—	—	—	—	sl	sl	—	sl
13.26	—	—	—	—	—	—	sl	sl	—	sl
14.33	—	—	—	—	—	—	sl	sl	—	sl
19.26	sl	sl	sl	sl	sl	sl	sl	sl	—	sl
19.28	sl	sl	sl	sl	sl	sl	sl	sl	—	sl
19.37	sl	sl	sl	sl	sl	sl	sl	sl	—	sl
22.84	sl	sl	sl	sl	sl	sl	sl	s	sl	sl
24.90	sl	sl	sl	sl	sl	sl	sl	s	sl	sl
23.38	sl	sl	sl	sl	sl	sl	sl	s	sl	sl
28.58	sl	sl	sl	s	s	sl	sl	s	sl	sl
29.61	sl	sl	sl	s	s	sl	sl	s	sl	s
28.75	sl	sl	sl	s	s	sl	sl	s	sl	s
32.08	sl	sl	sl	s	s	sl	sl	s	sl	s
38.90	sl	sl	sl	s	s	sl	sl	s	sl	s

[a] Bak, K., Elefante, G., Mark, J. E., *J. Phys. Chem.* (1967) **71**, 4007.

between 1.5 and 18.4%. Table I shows the results, the yield, density, softening point, and molecular weight of the copolymers and telomers; Table II shows the influence of the reaction conditions, and Table III gives the solubility.

With increasing amounts of tetrachloroethylene in the reaction mixture (14-18 mole %) the yield and softening points are decreased.

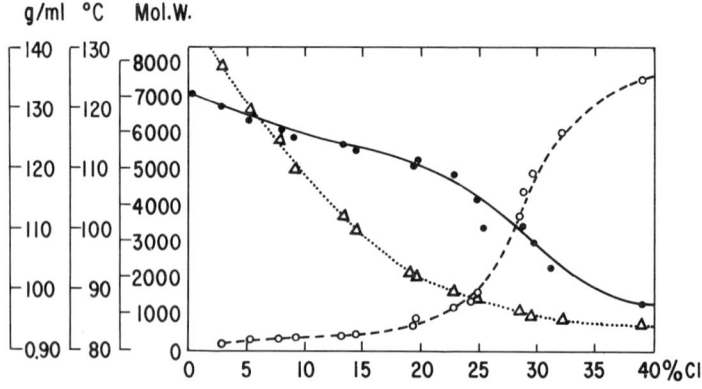

Figure 1. Density (○), softening point (●), and molecular weight (△) of the copolymers in relation to chlorine content

of Tetrachloroethylene and Ethylene

Benzene °C			Toluene °C			Xylene °C			CCl_4 °C		C_2Cl_4 °C		$CHCl_3$
25	50	75	25	50	75	25	50	75	25	50	25	50	25
—	sl	sl	—	sl	s	—	sl	s	sl	sl	sl	sl	sl
—	sl	sl	—	sl	s	—	sl	s	sl	sl	sl	sl	sl
—	sl	sl	—	sl	s	—	sl	s	sl	sl	sl	sl	sl
—	sl	sl	—	sl	s	—	sl	s	sl	sl	sl	sl	sl
—	sl	sl	—	sl	s	—	sl	s	sl	sl	sl	sl	sl
—	sl	s	—	sl	s	—	sl	s	sl	sl	sl	sl	sl
sl	s	s	sl	s	s	sl	s	s	sl	s	sl	s	sl
sl	s	s	sl	s	s	sl	s	s	sl	s	sl	s	sl
sl	s	s	sl	s	s	sl	s	s	sl	s	sl	s	sl
sl	s	s	sl	s	s	sl	s	s	sl	s	sl	s	sl
sl	s	s	sl	s	s	sl	s	s	s	s	s	s	sl
sl	s	s	sl	s	s	sl	s	s·	s	s	s	s	sl
sl	s	s	sl	s	s	sl	s	s	s	s	s	s	s
sl	s	s	sl	s	s	sl	s	s	s	s	s	s	s
sl	s	s	sl	s	s	sl	s	s	s	s	s	s	s
sl	s	s	sl	s	s	sl	s	s	s	s	s	s	s
sl	s	s	sl	s	s	sl	s	s	s	s	s	s	s

The chlorine content increases to 19-24%. The product is a mixture of a telomer and copolymer. By further increasing the amount of tetrachloroethylene, only waxlike telomers are formed, containing trichlorovinyl and chlorine end groups with chlorine content between 25 and 39% (Table I) and molecular weights as low as 720. By fractionated vacuum distillation and preparative gas chromatography, the following products could be isolated:

> 1,1,2,6-tetrachloro-1-hexene
> 1,1,2,8-tetrachloro-1-octene
> 1,1,2,10-tetrachloro-1-decene
> 1,1,2,12-tetrachloro-1-dodecene
> 1,1,2,14-tetrachloro-1-tetradecene
> 1,1,2,16-tetrachloro-1-hexadecene
> 1,1,2,18-tetrachloro-1-octadecene
> 1,1,2,20-tetrachloro-1-eicosene

The waxlike products are successive members of these homologous series.

The properties of the products depend on the speed at which the reaction mixture was heated. Rapid heating favors the formation of telomers and homopolymerization of ethylene. The relationship of chlorine content to density, softening point, and molecular weight of the products is shown in Figure 1; the influence of the heating rate is shown in Figure 2.

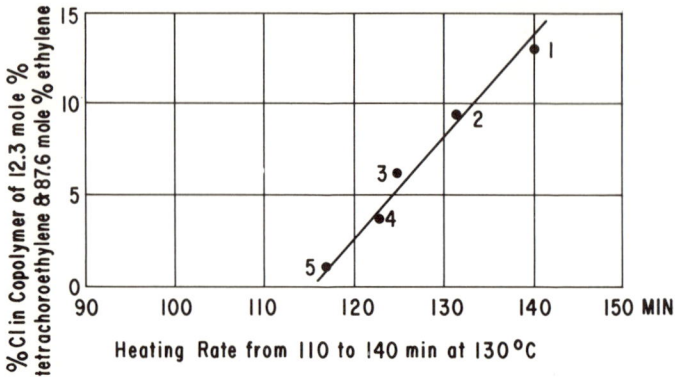

Figure 2. *Influence of heating rate, 90-150 minutes*

Copolymerization in aqueous emulsion at pH 9–11 with ammonium persulfate as catalyst gave only a low melting wax (mp., 86°C) with a chlorine content of 22.5%.

All products were characterized by DTA, DTG analysis, infrared and x-ray spectra. The pressure was varied between 50 and 1,000 atm. Copolymerization with other monomers, vinyl methyl ether, propylene, and hexavinylbenzene gave products with chlorine contents of 2-15% Vinyl chloride and vinylidene chloride gave only homopolymers.

Acknowledgment

Support of this work by Pittsburgh Glass Co., Barberton, Ohio, is gratefully acknowledged.

Literature Cited

1. Korshak V. V., Samplanskaya, K. K., *Dokl. Akad. Nauk. USSR* (1948) **59**, 497; *Chem. Abstr.* (1948) **42**, 6204a.
2. Breitenbach J. W., Schindler, A., Pflug, Ch., *Monatsh. Chem.* (1950) **81**, 21.
3. Korshak V. V., Mateeva, N. G., *Dokl. Akad. Nauk. USSR* (1952) **85**, 797; *Chem. Abstr.* (1953) **47**, 281.
4. Goerrig D., Jonas, H., German patent **937,919** (1956).
5. Solvic S. A., Belgian patent **562,433** (1958).
6. Klaassens K., Gisolf, J., *J. Polymer Sci.* (1953) **10**, 149.
7. Gonikberg M. G., Zhulin, V. M., *Bull. Acad. Sci. USSR, Div. Chem. Sci.* (1957).
8. Marcel C. J., Price M., *AEC Access. No. 37185, Rept.* **BNL 874**, 24 (1964).
9. Roedig A., *Ann. Chem.* (1951) **574**, 128.
10. Roedig, A., Voss, G., Kuchinke, E., *Ann. Chem.* (1953) **580**. 27.
11. Roland, J. R., British Patent **589,065** (1947).

RECEIVED April 1, 1972.

11

Radiation-Induced Chlorination of Polyisobutene

CHRISTEL SCHNEIDER and PETR LOPOUR [1]

Institut für Physikalische Chemie der Universität zu Köln, West Germany

Radiation-induced chlorination of polyisobutene in carbon tetrachloride was studied at various temperatures. The process is a chain reaction with a G value of about 10^4 to 10^5, depending on the reaction conditions. At very low dose rates (0.1 to 0.2 rad/sec), the chlorination rate is directly proportional to the dose rate. At higher dose rates, the rate approaches a square-root dependence on the dose rate. The termination reaction and the influence of oxygen are discussed. The reaction is first order with respect to chlorine concentration. An activitation energy of about 4 kcal/mole was obtained. In connection with the chlorination reaction, degradation of the polyisobutene takes place. This degradation was followed by osmometric measurements. The structure of the chlorinated product was briefly investigated by IR spectroscopy.

Modification of polymers by irradiation has been little studied. Chlorination is one of the most interesting processes for polymer modification, and is usually carried out by means of catalysts or by UV irradiation. Since 1960, radiation-induced chlorination of polyethylene and polypropylene has been studied, especially by Soviet workers (*1-3*). As the polymers used in that work are insoluble in the usual solvents at normal temperature, chlorination was done in the heterogeneous phase—for example, by leading continuously a stream of chlorine over the finely ground polymer or through an aqueous suspension of the polyolefin. It is, therefore, difficult to compare the results obtained under the different conditions used.

[1] Present address. Institute of Macromolecular Chemistry, Czechoslovak Academy of Sciences, Prague.

To get a better insight into the chlorination reaction, we wanted to avoid a heterogeneous process. Instead of polyethylene or polypropylene, we used polyisobutene, which is soluble in carbon tetrachloride, as are its chlorination products. In addition, we were interested in the structure and properties of the chlorinated products, especially in comparison with polyvinyl chloride (PVC) and vinyl chloride/isobutene (VC/IB) copolymers.

Experimental

Polyisobutene (Oppanol 15 B, BASF) was purified by precipitation from n-hexane with acetone; the number-average molecular weight of the polymer used was $\overline{M}_n = 77,300$. CCl_4 was purified with H_2SO_4 and distilled from CaH_2 immediately before use. The solutions of polyisobutene (PIB) in CCl_4 and Cl_2 in CCl_4 were carefully prepared separately at a high vacuum line, and special precautions were taken to remove impurities, especially oxygen.

Figure 1 shows the mixing and irradiation ampules. The solution of Cl_2 in CCl_4 was condensed from the high-vacuum line into the lower part (A), which was then sealed off at 2. The ampules were connected at point 3 to a stock solution of PIB in CCl_4, a fixed volume of which was allowed to pass into B. Before the solutions of the compounds (separated by break seal 1) were mixed and filled into the irradiation

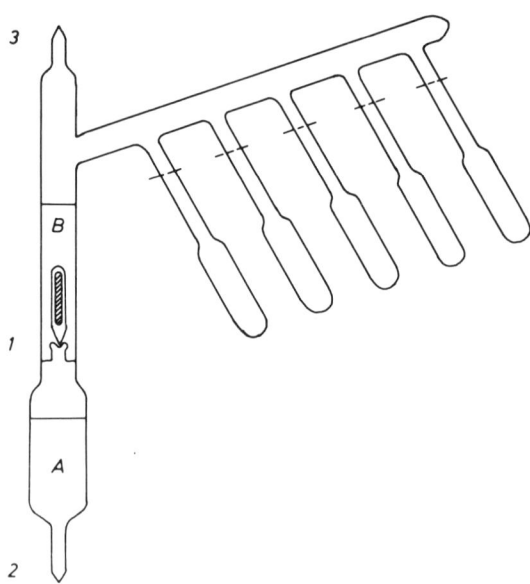

Figure 1. Apparatus for preparation of reaction mixtures; A, Solution of Cl_2 in CCl_4; B, Solution of PIB in CCl_4

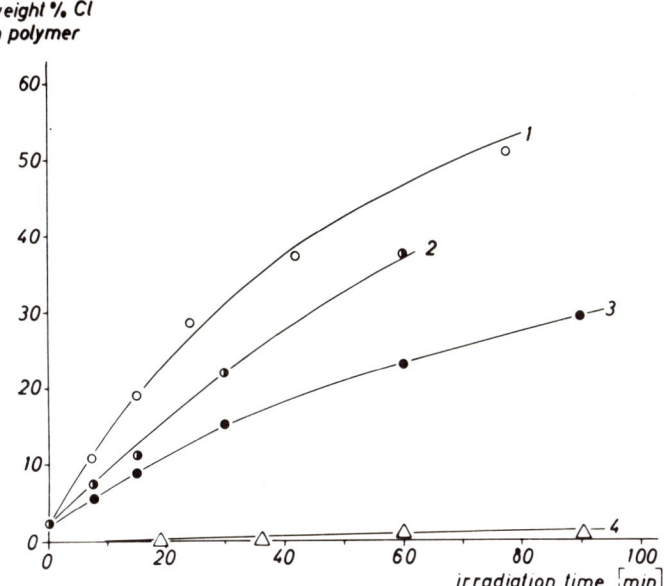

Figure 2. Conversion-time curves for different temperatures: (1) 25°C; (2) 5°C; (3) −15°C; (4) 25°C in the presence of air; [PIB]= 3.4 g/l; [Cl₂]=32.7 g/l; dose rate=0.128 rad/sec

ampules, all glassware was painted black to avoid a photochemical reaction. The chlorine content of the feed was determined iodometrically. The sealed ampules were irradiated at between −15°C and 25°C using Co-60-γ-rays, with dose rates varying from 0.13 to 1.3 rad/sec. After irradiation, the ampules were opened in the dark, and the excess chlorine was removed by shaking the solution with potassium iodide, which subsequently was reduced by alkaline sodium thiosulfate. The solvent was evaporated and the chlorine content of the polymer determined. The molecular weight of the chlorinated product was measured either by vapor-phase osmometry or by membrane osmometry.

Results and Discussion

Figure 2 shows the percentage of chlorine in the polymer as a function of the irradiation time at various temperatures. Only rather low radiation doses are necessary to obtain highly chlorinated products. The radiation chemical G value (substituted hydrogen atoms per 100 eV absorbed energy) for the results shown in Figure 2 are about 10^4 to 10^5, meaning that the chlorination proceeds *via* a chain reaction with a high kinetic chain length as expected in comparison with other chlorination processes. This is confirmed by the strong inhibiting effect of oxygen traces on the reaction rate, as shown in curve 4 at 25°C; all other conditions were kept constant.

The time-conversion curves do not pass through the origin; in spite of all precautions taken, there was always a small "dark reaction." This dark reaction is obviously caused by traces of light penetrating into the ampule during the preparation and is not caused by a thermal reaction. The conversion of the dark reaction in experimental times is always so small that it does not have any implications for the size of the radiation-induced reaction. Irradiation of a solution of polyisobutene in carbon tetrachloride in the absence of chlorine showed that only negligible amounts of chlorine (about 2 weight %) were taken up from the solvent.

In analogy to classical chlorination reactions, the mechanism of the radiation induced process is presumed to be:

$$Cl_2 \xrightarrow{\gamma} 2\, Cl\cdot \quad\quad\quad\quad\quad \text{Initiation} \quad k_1 \quad (1)$$
$$CCl_4 \xrightarrow{\gamma} Cl\cdot + C\cdot Cl_3 \quad\quad\quad\quad\quad\quad\quad k_2 \quad (2)$$

$$\cdot Cl + RH \rightarrow HCl + R\cdot \quad\quad \text{Propagation} \quad k_3 \quad (3)$$
$$R\cdot + Cl_2 \rightarrow RCl + Cl\cdot \quad\quad\quad\quad\quad\quad\quad k_4 \quad (4)$$

$$\cdot Cl + \cdot Cl \rightarrow Cl_2 \quad\quad\quad\quad\quad\quad\quad\quad\quad k_5 \quad (5)$$
$$\cdot Cl + R\cdot \rightarrow RCl \quad\quad\quad \text{Termination} \quad k_6 \quad (6)$$
$$\cdot Cl + X \rightarrow ClX\cdot \text{ (Inactive)} \quad\quad\quad\quad k_7 \quad (7)$$
$$R\cdot + X \rightarrow RX\cdot \text{ (Inactive)} \quad\quad\quad\quad k_8 \quad (8)$$

where RH is a molecule of PIB, and X is a molecule of an impurity such as oxygen.

As to the termination reaction, Equation 6 is very unlikely because of the competitive propagation reaction. $R\cdot$ is a coiled macroradical, and abstraction of a hydrogen atom by $Cl\cdot$ is assumed to be favored over a combination of $Cl\cdot$ with the radical site at $R\cdot$. In a system without any impurity, termination would therefore be expected mainly by combination of chlorine radicals (5).

Assuming steady-state conditions, the rate of chlorination is given by:

$$-\frac{d[RH]}{dt} = k_3 \sqrt{\frac{(k_1 + k_2)\, I}{k_5}}\, [RH] \quad\quad (9)$$

where I is the dose rate of radiation and $(k_1 + k_2)\, I$ is the initiation rate. According to this relation, the rate of chlorination should vary with the square root of the dose rate. When Equation 6 is not neglected in the kinetic treatment, proportionality of rate and I^2 is retained.

Figure 3 shows, on a logarithmic scale, dependence of the chlorination rate on the dose rate used in our experiments. The plot obtained is not linear, as would be the case if Equation 9 holds. The slope of the curve varies from about 1 at low dose rates to about 0.65 at the highest dose rate (1.3 rad/sec). We would like to interpret this

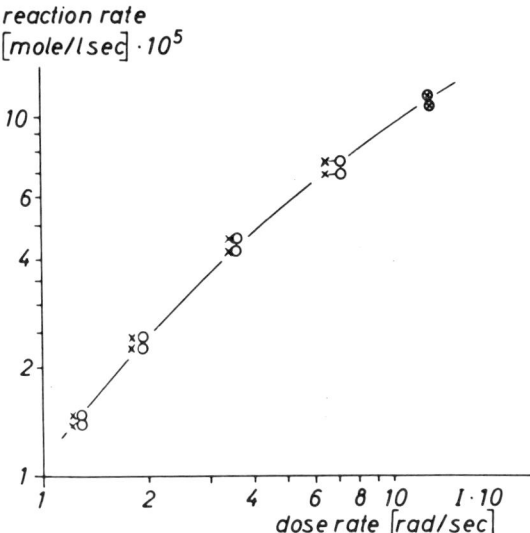

Figure 3. Logarithmic plot of reaction rate vs. dose rate; $[PIB]=3.45$ *g/l;* $[Cl_2]=34$ *g/l;* $t=5°C$
○ = *dose rate determined by Fricke dosimetry*
× = *dose rate determined by means of an ionization chamber*
All rates measured for constant conversion (~14 weight % Cl)

deviation from linearity by the fact that termination reactions other than Equation 5 occur. If we were to assume that, in spite of all our precautions, traces of oxygen in our samples still remained, termination would occur by combination of chlorine radicals (Equation 5) as well as by reaction of Cl· and R· with impurities (Equations 7 and 8), the ratio depending on the dose rate used. At very low dose rates, when the concentration of Cl· is still very small, Equations 7 and 8 will dominate. With increasing dose rates, the concentration of Cl· and, consequently, the bimolecular termination reaction will increase.

Assuming that the chain reaction is terminated by oxygen according to Equation 8, causing formation of $RO_2·$, the overall rate of chlorination is directly proportional to the dose rate according to:

$$-\frac{d[\text{RH}]}{dt} = \frac{k_4(k_1 + k_2) I}{k_8} \frac{[Cl_2]}{[O_2]} = \frac{K[Cl_2] I}{[O_2]} \qquad (10)$$

The same dependence is obtained when termination occurs by Equation 7. Our results show indeed that for low dose rates, a linear dependence of the reaction rate on the dose rate is obtained, but that with increasing dose rate, a square-root dependence is approached.

According to Equation 10, the chlorination rate should be directly proportional to chlorine concentration, as was in fact found in the experiments using low dose rates (Figure 4). When we assume that oxygen concentration in the sample is constant at least for a series of samples prepared in the same run, the $K/[O_2]$ is a constant, and its temperature dependence can be used to calculate the overall activation energy of the chlorination process. Using the chlorination rates for the three temperatures given in Figure 2, an overall activation energy of about 4 kcal/mole was calculated.

Degradation of the polyisobutene chain takes place parallel to the chlorination. This is shown in Figure 5, where the molecular weight $\overline{M}_n$ is plotted vs. the absorbed radiation dose. A similar effect was observed by McNeill and McGuchan (4) for the thermal chlorination of polyisobutene, which they explained by reaction of a polymer radical formed in the chlorination process:

$$\begin{array}{c} \cdot CH_2 \quad\quad CH_3 \\ | \quad\quad\quad | \\ -CH_2-C-CH_2-C- \\ | \quad\quad\quad | \\ CH_3 \quad\quad CH_3 \end{array} \xrightarrow{Cl_2} \begin{array}{c} CH_2 \\ \| \\ -CH_2-C \\ | \\ CH_3 \end{array} + \begin{array}{c} CH_3 \\ | \\ ClCH_2-C- \\ | \\ CH_3 \end{array} + Cl\cdot \quad (11)$$

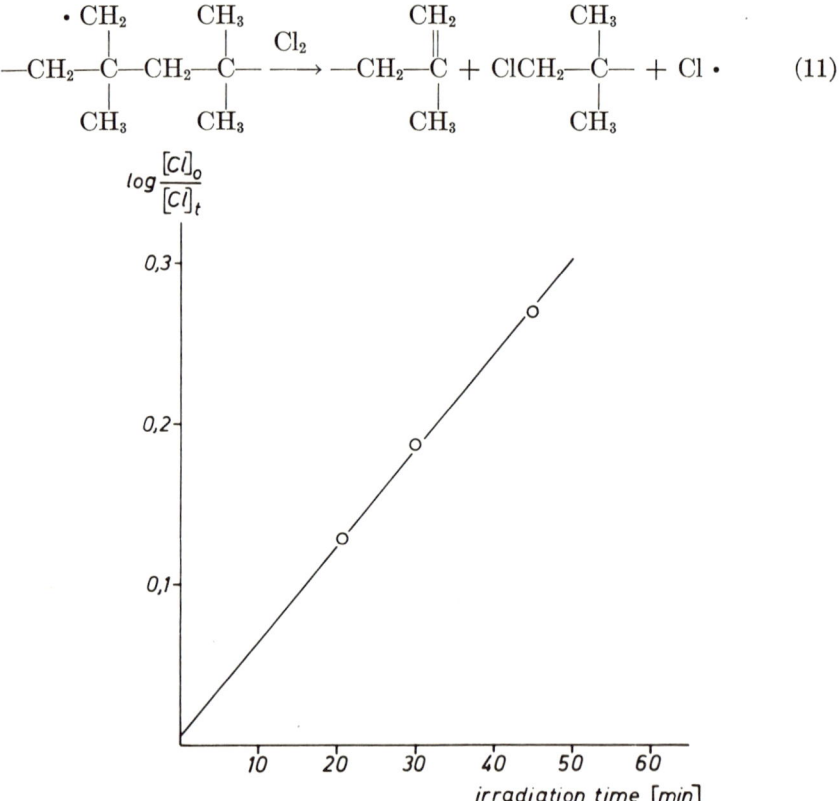

Figure 4. Plot of Cl_2 concentration vs. irradiation time; $[PIB]=15$ g/l; $[Cl_2]=15.5$ g/l; $t=5°C$; dose rate$=0.128$ rad/sec

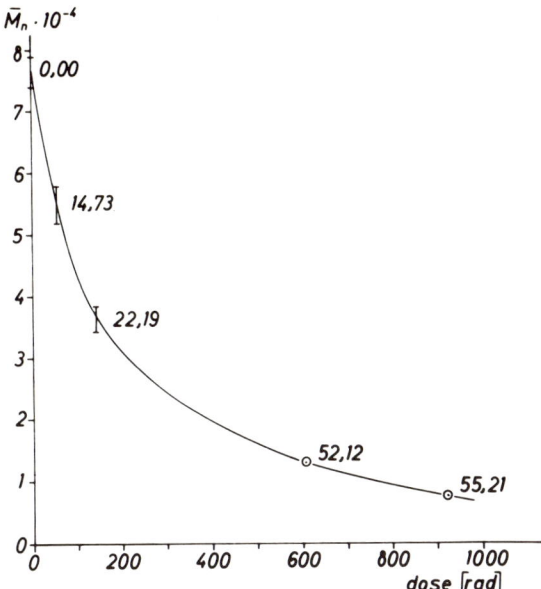

Figure 5. Plot of $\overline{M}_n$ of the chlorinated polyisobutene vs. absorbed radiation dose; [PIB]=3.5 g/l; [Cl₂]=33 g/l; t=25°C; dose rate=0.128 rad/sec
⊙ = vapor phase osmometry
I = membrane osmometry
The numbers at the experimental points refer to the respective chlorine content of the polymer (in weight %)

In radiation-induced chlorination, additional degradation of PIB by direct interaction with the C–C chain bond occurs (5). From a comparison with our results with literature data, we suggest that in our case, degradation takes place mainly by the more effective chlorination process.

The structure of the chlorinated products obtained is not yet clear. McNeill and McGuchan concluded from NMR data that, in the thermal chlorination process, both methyl and methylene groups were chlorinated but that the methylene groups were more readily substituted and that even some disubstitution of the methylene groups occurred. We studied our products by infrared-spectroscopy; Figure 6 shows the spectra of four chlorinated polyisobutenes with increasing chlorine content.

The spectra change characteristically and allow us to draw some conclusions with respect to the chlorination site. The absorption at 1282 cm⁻¹ belongs to the wagging CH_2Cl vibration and increases with increasing chlorine content. The absorption at 1366 cm⁻¹ is exclusively caused by the symmetrical in-phase vibrations δ_{si} of the geminal methyl groups of the isobutene units; it decreases with increasing chlorine con-

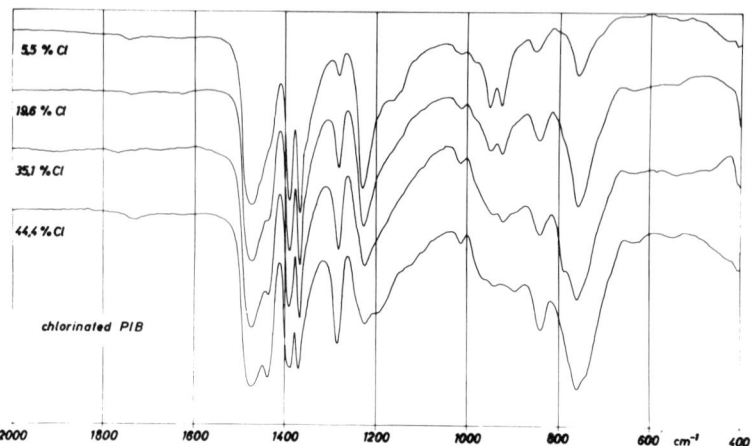

Figure 6. Infrared spectra of four chlorinated isobutenes with varying chlorine content

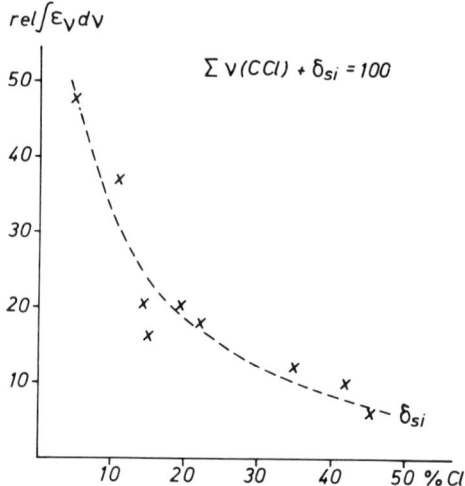

Figure 7. Plot of the relative integral extinction of the δ_{si} vibrations of geminal CH_3 groups vs. chlorine content

tent. The absorption at 1388 to 1399 cm^{-1} is assigned to the symmetrical out-of-phase vibrations of the geminal methyl groups and to deformation vibrations of isolated methyl groups. As the concentration of geminal methyl groups decreases with increasing chlorine content, the overlapping of the two absorptions becomes more distinct. The broad and complex absorption between 650 and 800 cm^{-1} belongs to all ν(CCl) vibrations.

When the relative integral extinction of the symmetrical in-phase deformation vibrations of the geminal methyl groups, $\delta_{si}(CH_3)$, is plotted against the chlorine content using the sum of $\delta_{si}(CH_3)$ and all $\nu(CCl)$ vibrations as an inner standard, the decrease in the concentration of those methyl groups is not linear (Figure 7). At the beginning of the chlorination process, there is a strong decrease, which slows down with increasing chlorine content. This may tentatively be explained by the fact that, at first, methyl groups are chlorinated preferably. Later on, chlorination of the methylene groups and some dichlorination become dominant. This result slightly contradicts previous work (4), and additional NMR studies are necessary.

Acknowledgment

We wish to thank Dr. Brück for the infrared analysis, and the Bundesministerium für Bildung und Wissenschaft for financial support.

Literature Cited

1. Wuckel, L., Savchenko, L., Seidel, A., *Z. Chem.* (1962) **2**, 371.
2. Dzhagatspanyan, R. V., Korolev, B. M., Zetkin, V. I., Fillipov, M. T., *Vysokomol. Soedin.* (1966) **8**, 1745.
3. Dzhagatspanyan, R. V., Korolev, B. M., Romanskii, I. A., Zetkin, V. I., *Vysokomol. Soedin., Ser. A* (1967) **9**, 1195.
4. McGuchan, R., McNeill, I. C., *J. Polym. Sci., A-1* (1967) **6**, 205.
5. Henglein, A., Schneider, C., *Phys. Chem. Neue Folge* (1959) **19**, 367.

RECEIVED May 9, 1972.

12

Some Aspects of Vinyl Ester Emulsion Polymerization

M. LITT

Division of Macromolecular Science, Case Western Reserve University, Cleveland, Ohio 44106

V. T. STANNETT

Department of Chemical Engineering, North Carolina State University, Raleigh 27607

E. VANZO

Xerox Corp., Rochester, N.Y. 14600

> *The emulsion polymerization of vinyl hexanoate has been studied to determine the effect of chain transfer on the polymerization kinetics of a water-insoluble monomer. Both unseeded and seeded runs were made. For unseeded polymerizations, the dependence of particle concentration on soap is much higher than Smith-Ewart predictions, indicating multiple particle formation per radical because of chain transfer. Once the particles have formed, the kinetics are much like those of styrene. The lower water solubility of vinyl hexanoate when compared with styrene apparently negates its increased chain transfer, since the monomer radicals cannot diffuse out of the particles.*

One of the basic questions in vinyl ester emulsion polymerization concerns the effect of chain transfer on the polymerization kinetics. Vinyl esters chain transfer easily; if the monomer radical can diffuse into the aqueous phase, the resulting kinetics should approximate Smith-Ewart Case I (*1*). However, vinyl acetate, the lowest member of the vinyl ester series, shows serious deviations from any Smith-Ewart Theory. This has been well documented in recent reviews and papers (*2, 3*). The deviation is probably because of its water solubility, which, in combination with its easy chain transfer, allows polymer growth in the aqueous phase. To isolate the effects of chain transfer, it is necessary to choose a vinyl ester of low water solubility. Vinyl hexanoate was therefore investigated, since it has a chain transfer constant to monomer higher than that of vinyl acetate (4), ~40 × 10^{-4} l/mole sec while the chain-transfer constant for styrene is ~0.6 × 10^{-4} l/mole sec. Its water solubility is even less than that of styrene: vinyl hexanoate, 7.3 × 10^{-4} mole/l at 60°; styrene, 5.3 × 10^{-3} mole/l at 60°C. These values are about twice those of Okamura (4), who measured solubilities at room temperature.

Experimental

Dilatometric Techniques for Kinetic Measurements. A dilatometric technique was used to follow the polymerization rate. This involved measuring the decrease in volume arising from conversion of the monomer to the denser polymer. This volume change was translated into the number of moles polymerized by considering the specific volumes of monomer and polymer. The dilatometer, adapted from that of Litt, Stannett, and Patsiga (3), permitted measuring the conversion without disturbing the polymerizing system. The volume was read by adjusting the level of the latex to a reference mark on the large neck, and reading the latex level in the precision-bore tubing. Determination of the per cent conversion was accurate to within 2%, depending upon the total volume contraction.

The emulsifier was sodium lauryl sulfate, an anionic soap of the micelle-forming type; the initiator was potassium persulfate. A suitably pure grade of vinyl hexanoate was not initially available. However, washing with water removed a substantial amount of impurity, and reproducible rates of polymerization were obtained. Even when prepolymerization techniques were used for purification, an apparent induction period of 5 to 10 minutes was observed in all polymerizations of vinyl hexanoate. By contrast, the vinyl acetate and styrene used had virtually no induction period. All reactions were carefully purged with nitrogen before and during the reaction. A "high purity" grade vinyl hexanoate obtained later in the investigation from Eastman Organic Chemicals gave identical rates without prepolymerization when carefully washed with water and distilled over a spinning-band fractionating column.

Particle-size measurements were accomplished with light-scattering techniques. Typical particle-size distributions were established for each polymer by electron microscopy (5). All polymerizations were conducted at 60.0°C in a thermostated bath having a temperature fluctuation of less than 0.03°C.

Results and Discussion

Kinetic Behavior of Styrene and Vinyl Acetate. To establish the applicability and accuracy of the techniques used, dependence of the rate of styrene polymerization on the number of particles in a seeded system was measured and compared with the results of Patsiga (6). The seed consisted of a polyvinyl acetate latex (No. 5-3+4) especially prepared for this purpose. Though the system was changed in total volume and organic volume fraction from the work of Patsiga, it agreed exactly with their determinations. A log-log plot of the rate vs. the number of particles gives a dependence of 0.95, compared with the expected value of 1.0. The slope of the curve leads to a value of k_p of 156 l/mole second, about 20% lower than the accepted literature value of 190 l/mole second. From the results of Kolthoff, a decrease is expected, since a drop occurs in the normally constant rate per particle in this range of particle concentration, 4 to 17 × 10^{14} particles/cc latex.

Shape of the Polymerization Curves. Since polymerization takes place within the polymer particle, the rate depends on monomer concentration in the particles; this concentration remains relatively constant as long as free monomer phase is present. At some point during the polymerization, the amount of polymer present is sufficient to absorb essentially all of the unreacted monomer. After this point in the polymerization, the monomer consumed by the radicals cannot be replaced and monomer concentration in the particle must decrease. It then follows that the rate must also decrease.

The conversion curve for styrene exhibits a drop in the rate in the calculated range according to the findings of Smith (1), Kolthoff (7), and others. Patsiga (3, 6), as well as many other investigators, have found the rates for vinyl acetate to be linear to around 80% conversion despite the fact that calculations indicate that all of the monomer is absorbed in the particles at about 40% conversion.

The solubility values for the various systems in question were determined in part I of this study (8). Table I lists these values along with the calculated values for the degree of conversion at which all of the monomer should be absorbed by the particles.

Table I. Calculated vs. Observed Point of Departure of Rate Curve from Linearity [a]

	Solubility of Monomer in Polymer Particles (moles/l)	Calculated Point of Disappearance (per cent conversion)	Observed Point of Departure from Linearity (per cent conversion)
Styrene	5.42	40	38
Vinyl hexanoate	4.28	36	38-43
Vinyl acetate	7.60	32	77

[a] Departures from linearity are measured on standard (nonseeded) polymerizations.

A range is given for vinyl hexanoate since the low percentage of monomer used makes determination of the point more difficult. From these observations and the shape of the polymerization curves, it seems that vinyl hexanoate behaves similarly to styrene in showing first-order dependence on monomer concentration during polymerization.

Kinetics of Vinyl Hexanoate. A series of standard polymerizations was conducted using 10 ml of vinyl hexanoate in a total volume of 140 ml with varying amounts of emulsifier and initiator. The effect of varying initiator, particle number, and organic volume fraction in seeded runs, as well as for unseeded polymerizations, were also studied. The results are summarized in Table II.

Table II. Data for Vinyl Hexanoate Polymerizations

Exp. No.	Initiator conc. $M/l \times 10^4$	Soap conc. %	Organic Vol / Aqueous Vol	$R_p{}^a$ M/l min $\times 10^2$	Particle conc. (N_p) No./ml $\times 10^{-14}$	Rate/ Particle M/min $\times 10^{20}$	Particle Volume cc $\times 10^{18}$
A-26 [b]	20	1.0	0.077	2.50	66.5	0.38	12
27	10	1.0	0.077	1.85	44.4	0.42	17
28	5	1.0	0.077	1.26	35.8	0.35	22
29	10	0.5	0.077	1.56	17.4	0.90	44
30	10	2.0	0.077	2.73	118.4	0.23	6.4
B-36 [c]	10		0.141	1.72	41.2	0.42	34
39	10		0.126	1.18	26.0	0.45	48
37	10		0.106	0.85	18.4	0.46	58
38	10		0.092	0.55	10.6	0.52	87
42	5		0.048	1.16	0.96	12.1	500
41	10		0.048	1.11	1.12	10.0	430
43	20		0.048	1.16	1.13	10.3	420

[a] Moles monomer polymerized per liter aqueous phase
[b] Unseeded polymerizations
[c] Seeded polymerizations

Inspection of Table II reveals some relationships important for understanding the kinetics. In the upper range of particle size, the rate is independent of initiator concentration once particles have been formed; compare the rate per particle of runs 41, 42, and 43. In this, it acts like styrene.

Log-log plots show these dependencies: $N_p \propto I^{0.47\pm.01}$ (runs 26, 27, and 28); $N_p \propto (\text{Soap})^{1.35\pm0.35}$ (runs 28, 29, and 30); $R_p \propto N_p{}^{0.75\pm.08}$ (all runs except 41, 42, and 43); and rate/particle $\propto$ (particle volume)$^{0.25}$ (all runs except 41, 42, and 43).

The dependency of N_p on soap is inaccurate because of the large scatter of the data. However, any possible dependency is much higher than the 0.6 power of the Smith-Ewart scheme. It therefore represents a deviation from the Smith-Ewart mechanism for particle formation. There is a reasonable rationale for this in the case of vinyl caproate. Chain transfer to monomer is large and would produce a species which, in the particle formation stage, could readily diffuse out of a micelle into another micelle and initiate polymerization there. Thus, each radical would initiate many chains and therefore many very small particles. Coalescence during further growth would then determine the final number of particles, rather than the "using up" of soap in the particle-formation stage.

Such a mechanism could also explain the apparent induction period found at the beginning of the polymerizations. With very small particles and large chain transfer, the polymerization could be acting as if it were in a homogeneous medium. This produces very low rates compared with a standard emulsion polymerization. As particle size increases, the rate rises because chain-transferred monomers would not diffuse into the aqueous phase.

Dependence of the polymerization rate on initiator concentration is confounded with variation in particle concentration. In the only place where a direct relationship can be found, runs 41, 42, and 43, there is zero dependence. For the other runs, there is a dependence on $N_p{}^{0.75}$; this gives $R_p \propto I^{0.12}$, using runs 26, 27, and 28 and correcting to constant N_p.

Runs 41, 42, and 43 have very much larger particles than do the other runs; these particles obviously change the polymerization kinetics. For all other runs, the rate per particle varies between 0.23 to 0.90 × 10^{-20} (average = 0.45 × 10^{-20}) moles/minute. For the above runs, the rate per particle was about 11 × 10^{-20} mole/minute, a factor of about 25 higher. One can calculate a rate constant of polymerization for any run, assuming one radical per particle. For run 42, "k_p" ≈ 280 l/mole sec. For run 36, the corresponding calculated "k_p" ≈ 9 l/mole sec. Since k_p has not been measured for any vinyl ester other than the acetate, we cannot reach strong conclusions. However, if we were to assume that radicals do not escape from the particle when it is large, runs 41, 42, and 43 should correspond to Smith-Ewart Case II, which is confirmed by the lack of dependence on initiator concentration. With 50% of the particles occupied by radicals at any one time, k_p ≈ 560 l/mole sec. This is a reasonable value; increasing length of a tail lowers k_p for alkyl acrylates from ~ 1500 l/mole sec for the methyl ester to ~ 14 l/mole sec for the butyl ester (9). Vinyl esters should behave similarly. Thus, for the smaller particles, only about 2% contain radicals.

Another factor that can influence polymerization rate is monomer concentration in the particles. This is a function of particle size; the larger the particle, the higher the monomer concentration (*10, 11*). At most, however, the change in monomer concentration caused by variation in particle size can double the rate per particle. However, the change in rate is a factor of 20 to 30. Therefore, particle size directly affects the rate per particle. The most likely hypothesis is that chain transfer with escape of the monomer radical from the particle is occurring part of the time with the smaller particles, but essentially never in the larger particles.

We can tentatively conclude, therefore, that the effect of chain transfer is still making itself felt in the polymerization of vinyl caproate in spite of its low water solubility. Except at the lowest particle concentrations, chain transfer is important. The polymerization in these regions is midway between Case I and Case II. When variables are considered separately, there is some dependence of polymerization rate on particle concentration, and also some dependence on initiator concentration. In addition, at constant organic volume, while the rate of polymerization increases as the particle concentration increases ($R_p \propto N_p^{0.75}$), the rate per particle decreases as the particles get smaller. This shows that transferred radicals are mainly trapped in the particles, but some diffuse out and can undergo termination with other growing radicals.

Vinyl caproate in emulsion polymerization behaves like styrene in most respects. The rate is first order in monomer. In the range of 10^{15} to 10^{-16} particles/cc, it depends on N_p to the 0.75 power. This is higher than that for styrene in this range, $R_p \propto N_p^{0.5}$, indicating that there is less diffusion into the aqueous phase for vinyl caproate. However, the mechanism of particle formation for vinyl caproate may not fit the Smith-Ewart mechanism because of the high chain transfer rate to monomer.

Literature Cited

1. Smith, W. V., Ewart, R. H., *J. Chem. Phys.* (1947) **16**, 592.
2. Vanderhoff, J. W., "The Mechanism of Emulsion Polymerization in Vinyl Polymerization, Part II," Marcel Dekker, New York (1969).
3. Litt, M., Stannett, V. T., Patsiga, R., *J. Polym. Sci.* (1970) **A-1, 8**, 3607.
4. Okamura, S., Motoyama, T., *J. Polym. Sci.* (1962) **58**, 221.
5. Vanzo, E., Marchessault, R. H., Stannett, V. T., *J. Colloid Sci.* (1964) **19**, 578.
6. Patsiga, R., Litt, M., Stannett, V. T., *J. Phys. Chem.* (1960) **64**, 801.
7. Bovey, F. A., Kolthoff, I. M., Medalia, A. I., Mehan, E. J., "Emulsion Polymerization," Interscience, New York (1955).
8. Vanzo, E., Marchessault, R. H., Stannett, V. T., *J. Colloid Sci.* (1965), **20**, 62.
9. Brandrup, J., Immergut, B. H., "Polymer Handbook," Interscience, New York (1966).
10. Gardon, J. L., *J. Polym. Sci.* (1968) **A-1, 6**, 2859.
11. Morton, M., Kaizerman, S., Altier, M. W., *J. Colloid Sci.* (1954) **9**, 300.

RECEIVED April 13, 1972. Work done at New York State College of Forestry, Syracuse, N. Y.

13

One-Step Synthesis of Cured Polyester

RUDOLPH D. DEANIN and VITTORINO G. DOSSI

Plastics Department, Lowell Technological Institute, Lowell, Mass. 01854

> *Unsaturated polyesters were polymerized and cured in a single simple step by mixing propylene oxide, maleic anhydride, phthalic anhydride, styrene, lithium bromide, and benzoyl peroxide in crown-cap bottles, heating and mixing to melt and dissolve, and casting to produce cured polyesters. Lithium bromide catalyzed an ionic ring-opening polyesterification, while benzoyl peroxide initiated styrene-maleic free-radical copolymerization to produce the cured polyester. Reaction times were comparable with commercial polyesterification and hand lay-up cycles. Cured properties were adequate though not equal to commercial polyesters. Problems of volatility, reaction rates, and higher cure would have to be solved before such a one-step process could be considered for commercialization.*

Glass-fiber-reinforced polyesters were first used commercially in 1944 and have now reached a market volume of 1.2 billion pounds per year (1). Prospects are good for continued growth, primarily in boats, building, and autos (2). At the same time, however, the base cost of these commodity resins has leveled off at about 18-22 cents per pound, keeping them out of the reach of still larger potential markets (3). For continued large-scale growth, it would be desirable to make their synthesis simpler and more economical to advance to a lower base cost and thus enter a greater variety of larger applications and markets.

These polyesters are primarily copolyesters of propylene glycol with maleic and phthalic acids, dissolved in styrene monomer, and cured by peroxides ± activators (1, 4, 5). In a typical general-purpose polyester, a 10% excess of propylene glycol is cooked with an equimolar mixture of maleic and phthalic anhydrides for eight or more hours at 150-200°C to produce a viscous liquid polyester of 1000-5000 molecular weight. This is then partially cooled, diluted about 60/40 with styrene monomer, cooled to room temperature, and stored. It is later mixed with peroxides ± activators, impregnated into glass mat or fabric, gelled at room temperature, and finally hot-molded or oven-cured.

These successive steps form a complex process involving considerable time, heat, power, and handling, all of which contribute to increasing the cost of the final product. If all of these separate steps could be combined into a single integrated process, the resulting saving in time, heat, power, and labor should significantly reduce the base cost of the polyester resin (Table I).

Table I. Base Cost of Polyester Resin

Ingredients	Lb. Monomer/ Lb. Polymer	Cents/Lb. Monomer (6)	Cents Monomer/ Lb. Polymer
Propylene Glycol	0.269	10.5	2.83
Maleic Anhydride	0.157	14	2.20
Phthalic Anhydride	0.238	8	1.91
Styrene	0.400	8	3.20
Raw Material Cost			10.14
Polyester Resin Price			18–22

The first difficulty is the long time, high temperature, and high viscosity involved in present condensation polymerization reactions to produce the initial polyester. A number of previous studies have indicated that alkylene oxides should react directly with dibasic acid anhydrides, by ionic ring-opening addition polymerization, to produce polyesters under milder conditions than current commercial practice (7–14).

Specifically, propylene oxide has reacted directly with maleic and phthalic anhydrides to produce unsaturated polyesters under these milder conditions (15, 16). This would certainly be a major first step toward simplifying the process and lowering the cost. Incidentally, use of propylene oxide in place of propylene glycol would also result in an additional saving of 1 cent per pound in total raw material cost as well (6). After polyesterification, the separate steps of cooling, dilution with styrene, catalysis, impregnation, gelation, and cure are a distinct operational and economic liability.

The idea put forth here is that these steps might be telescoped into one by running the ionic ring-opening polyesterification reaction in styrene as solvent and, simultaneously, running the free-radical copolymerization of styrene and maleic double bonds to produce crosslinking and cure at the same time. In commercial operation, all the ingredients (Table II) would be mixed to form a solution; this solution would be impregnated into the glass mat or fabric in the mold, and heat and pressure would be applied to produce polymerization and cure.

Table II. Ingredients of One-Step Preparation of Cured Polyester

Propylene Oxide
Maleic Anhydride
Phthalic Anhydride
Styrene
Catalyst for Ionic Ring-Opening Addition Polyesterification
Initiator for Free-Radical Styrene/Maleic Ester Copolymerization (Cross-linking Reaction)

Experimental

To test this proposal in the laboratory, unreinforced castings were made by charging all the ingredients (Table III) into crown-cap glass bottles, tumbling in a water bath at polymerization temperature to melt and dissolve the ingredients and initiate the reactions, and then standing in an oven at polymerization temperature to complete the polymerization, gelation, and cure. Generally, small exploratory runs used 18- or 36-gram charges in small soft drink bottles; larger runs for more thorough evaluation of properties were 54 or 90-gram runs in medium-sized soft-drink bottles. After polymerization and cure, each bottle was broken, and the cylinder of cured polyester was removed and used for testing.

In exploratory studies, the cure reaction was followed simply by measuring Barcol hardness. Several larger runs were then evaluated more thoroughly. When test pieces of specific size and shape were required, they were obtained by machining them out of the cyclindrical castings. Most properties were measured by ASTM standard test methods. Swelling was measured by immersing a 3-gram sample of cured polyester in 50 ml of acetone for 24 hours at room temperature. The sample was then removed from the acetone, wiped, and weighed as rapidly as possible.

Results

Series of runs were made to explore the effects of temperature, catalyst, and initiator on the cure rate (Figures 1 through 3). Taking a Barcol hardness of 63 as a reasonable level for comparison, standard formulations reached this level in 48 hours at 75°C, 24 hours at 85°C,

Table III. Standard Experimental Formulation

Propylene Oxide	20.5
Maleic Anhydride	15.7
Phthalic Anhydride	23.8
Styrene	40.0
LiBr	0.258
Benzoyl Peroxide	1.0
Reaction Temperature	85°C

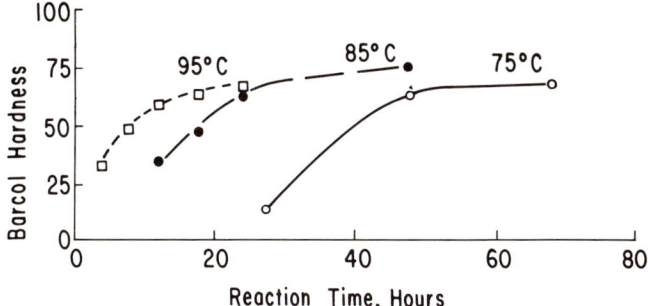

Figure 1. Effect of reaction temperature

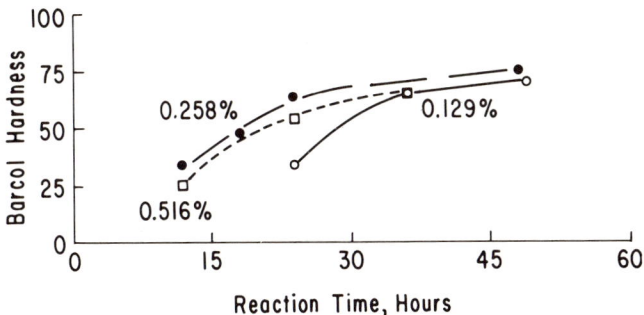

Figure 2. Effect of LiBr concentration

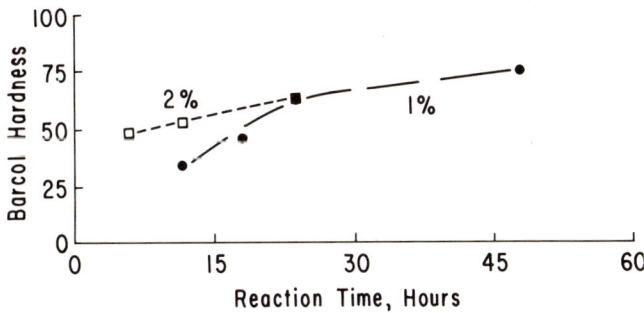

Figure 3. Effect of benzoyl peroxide concentration

and 18 hours at 95°C. These times are comparable with some polyesterification reactions and hand lay-up cycles, but do not approach the high speed of matched metal molding used in large-quantity production.

Lithium bromide was a satisfactory ionic catalyst for the ring-opening polymerization reaction (Figure 2). The standard concentration of 0.258% gave good reaction. A lower concentration gave a slower reaction, but higher concentration was no further help. This suggests either that a small amount was sufficient for the catalytic effect or that insolubility of the inorganic salt in the organic medium limited the amount that could be used effectively. Other catalysts reported in the literature (7–16) deserve evaluation to see if they could give faster reaction or higher cured properties.

Benzoyl peroxide initiator was more effective at a higher concentration (Figure 3). Other peroxides, and especially activators such as cobalt soaps or tertiary amines, could undoubtedly produce much faster cures. This would probably however, have to be balanced against the parallel rate of polyesterification.

Several formulations were prepared in larger quantity, to reasonable cure, for a greater variety of tests, and compared with a well-cured typical commercial general-purpose polyester, W. R. Grace Marco GR–941 (Table IV). At reasonably similar cure, reaction temperature had no significant effect on final properties. Low LiBr catalyst concentration gave a low Vicat softening point, suggesting incomplete polyesterification. High benzoyl peroxide initiator concentration gave a high Vicat softening point, suggesting more thorough styrene-maleic ester crosslinking.

Table IV. Properties of Cured Polyesters

Formulation	Barcol Hardness	Rockwell R Hardness	Compressive Strength, psi	Vicat Softening Point °C	Swelling in Acetone %
Standard	64	118	13,200	98	44
95°C	60			102	50
75°C	65			100	45
1/2 LiBr	65		15,100	67	56
2X LiBr	65		14,300	105	44
2X Bz_2O_2	63			140	40
Commercial	75		24,000	210	1

Cured properties of the one-step polyesters were generally inferior to conventional commercial polyester. The general belief is that, during conventional high-temperature commercial polyesterification, most of the maleic isomerizes into fumaric in the final polyester, and that the fumaric ester reacts much more readily with styrene by free-radical copolymerization, producing more thorough crosslinking. At the low temperatures used in our one-step study, such isomerization probably did not occur. The relative sluggishness of the styrene/maleic ester free-radical copolymerization would account for the lower final properties (*17*). Analysis for fumaric acid content would help in studying this possibility further (*18*).

Conclusions

Unsaturated polyesters are made and cured commercially by a series of difficult steps, resulting in fairly high cost and limited markets for the resulting glass-fiber-reinforced polyester products. The work reported here demonstrates that similar chemical compositions can be made in a single simpler step, potentially offering considerable savings both in raw material and processing costs. A number of major problems remain, however, before this process could be considered for commercialization:

(a) The volatility of propylene oxide would require processing in a closed pressurized system. Alternatively, a less volatile epoxide would have to be used.

(b) Reaction rates in this study were comparable with commercial polyesterification and hand lay-up cures of polyesters, but much slower than the high speed of matched metal molding used in large-quantity production. Reaction rates might be accelerated by screening of other polyesterification catalysts and could certainly be speeded greatly by using other peroxides and, especially, conventional activators such as cobalt soaps or tertiary amines.

(c) Low reactivity of maleic ester with styrene probably accounts for both slow reaction and low final properties. Higher temperatures and higher catalysis could certainly improve cure somewhat. If temperatures were high enough, they might even produce isomerization to fumarate ester, as in present commercial processes. Alternatively, other methods might have to be considered to improve cure and final properties.

Glass-fiber-reinforced polyesters have grown at a good rate to a large volume of commercial utilization, and this growth will undoubtedly continue even with modest improvements in present technology. If the above problems could be solved, however, the resulting one-step process could offer distinct savings in overall costs, producing faster and further growth into a greater variety of larger applications and markets.

Literature Cited

1. Lubin, G., "Handbook of Fiberglass and Advanced Plastics Composites," Van Nostrand Reinhold, New York (1969).
2. Rosato, D. V., Fallon, W. K., Rosato, D. V., "Markets for Plastics," Van Nostrand Reinhold, New York (1969).
3. Fullmer, G. E., *Soc. Plastics Eng. J.* (1968), **24**, 2, 22.
4. Billmeyer, Jr., F. W., "Textbook of Polymer Science," 2nd Ed., Wiley-Interscience, New York (1971).

5. Brydson, J. A., "Plastics Materials," Van Nostrand, New York (1966).
6. *Oil, Paint & Drug Reporter* (1971) **200**, 35–40.
7. Hayes, R. A., U.S. Patent 2,779,783 (1957).
8. Hayes, R. A., U.S. Patent 2,822,350 (1958).
9. Fischer, R. F., *Ind. Eng. Chem.* (1960) **52**, 321.
10. Fischer, R. F., *J. Polym. Sci.* (1960) **44**, 155.
11. Schwenk, E., Gulbins, K., Roth, M., Benzing, G., Maysenholder, R., Hamann, K., *Makromol. Chem.* (1962) **51**, 53.
12. Tsuruta, T., Matsuura, K., Inoue, S., *Makromol. Chem.* (1964) **75**, 211.
13. Hedrick, R. M., U.S. Patent 3,257,477 (1966).
14. Kern, R. J., Schaeffer, J., *J. Am. Chem. Soc.* (1967) **89**, 6.
15. Waddill, H. G., Milligan, J. G., Peppel, W. J., *Ind. Eng. Chem., Prod. Res. Develop.* (1964) **3**, 53.
16. Levine, L., ACS Div. of Org. Coatings & Plastics, Preprints (1971) **31**, 623.
17. Levine, L., private communication.
18. Funke, W., *Adv. Polym. Sci.* (1965) **4**, 157.

RECEIVED April 1, 1972.

14

Improved Process for Polycondensation of High-Molecular-Weight Poly(ethylene terephthalate) in the Presence of Acid Derivatives

TAKEO SHIMA, TAKANORI URASAKI and ISAO OKA

Products Development Institute, Teijin Ltd., Iwakuni, Japan

> *When a small amount of acid derivatives such as cyclic organoboronic anhydride, diphenyl dicarboxylate, diphenyl carbonate, or phenyl orthocarbonate was added to the polycondensation reaction system of poly(ethylene terephthalate) (PET) at the specified stage, the apparent rate of the polycondensation was markedly increased, and PET having a high molecular weight was readily obtained. The addition of certain diphenyl dicarboxylates and phenyl orthocarbonate further reduced the carboxyl group content of PET even further. Characteristics and properties of the PET obtained are similar to those of high-molecular-weight PET prepared conventionally.*

An increased degree of polymerization of poly(ethylene terephthalate) (PET) can lead to increased tenacity of the final oriented product (*1*); decreased carboxyl content improves the hydrolytic stability of the product (*2*). Both are important in tire-yarn properties.

In the manufacture of polyesters by melt polycondensation, the apparent polycondensation rate decreases with an increasing degree of polymerization, and the terminal carboxyls increase because of longer holdup time (*3*). Consequently, it is very difficult to obtain PET with an intrinsic viscosity above 0.9 and with carboxyls less than 15 eq/10^6 g polymer.

The melt polycondensation of PET is an equilibrium reaction—as shown by Reaction 1—and the apparent polycondensation rate depends on the rate of diffusion and elimination of liberated ethylene glycol (EG) through the reaction system.

$$2 \;-\!\!\overset{O}{\underset{I}{\overset{\|}{OC}}}\!\!-\!C_6H_4\!-\!\overset{O}{\overset{\|}{C}}OCH_2CH_2OH \;\underset{k_{-1}}{\overset{k_1}{\rightleftarrows}}$$

$$-\!\overset{O}{\overset{\|}{OC}}\!-\!C_6H_4\!-\!\overset{O}{\overset{\|}{C}}OCH_2CH_2O\overset{O}{\overset{\|}{C}}\!-\!C_6H_4\!-\!\overset{O}{\overset{\|}{C}}O\!-\; +\; HOCH_2CH_2OH \qquad (1)$$
$$\text{II}$$

As the viscosity of the reaction system increases, it becomes more difficult for ethylene glycol to diffuse through the reaction system and be eliminated from it. Thermal decomposition also occurs under melt polycondensation conditions, resulting in an increase of terminal carboxyl content and a decrease of hydroxyl ends that lower the apparent polycondensation rate even further.

Diffusion of ethylene glycol through the high-viscosity reaction system is the apparent rate-determining step in PET polycondensation. Preliminary examination of many univalent hydroxyl compounds shows that water and phenol diffuse much more readily through molten PET than does ethylene glycol. Therefore, a bifunctional reagent that re-

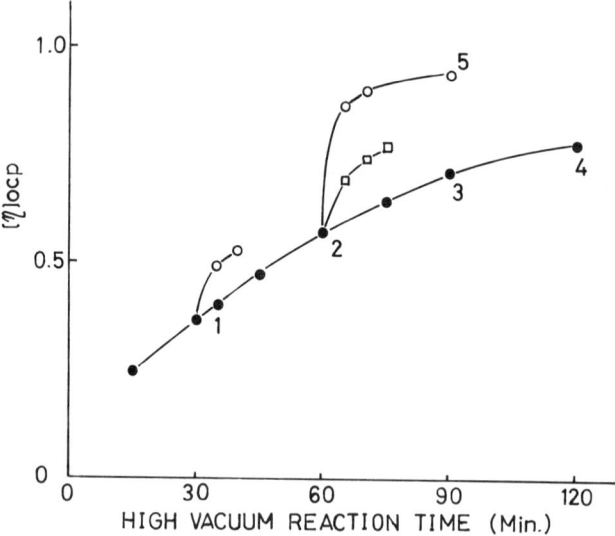

Figure 1. Reaction rates of PET promoted by BBA at 275°C

Symbol	Additive	Amount, mole %/DMT
●	—	0
□	BBA	0.25
○	BBA	0.50

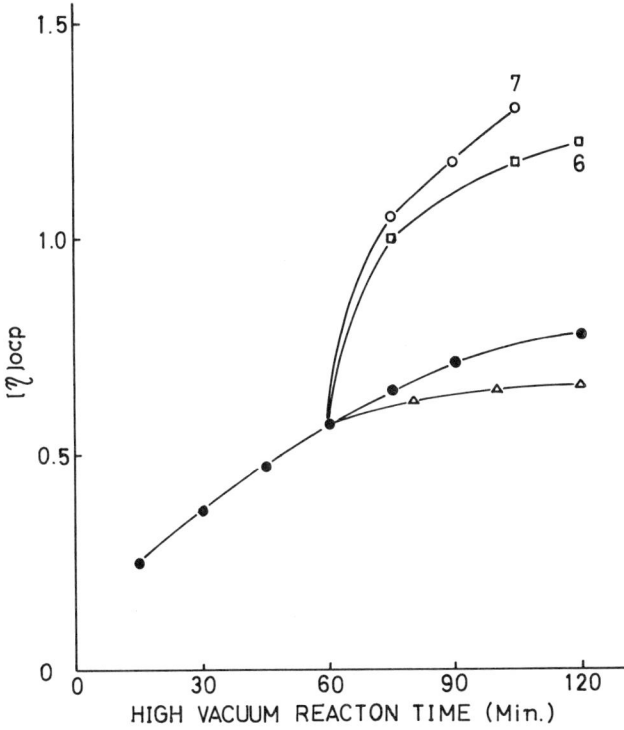

Figure 2. Reaction rates of PET promoted by DPT at 275°C

Symbol	Additive	Amount, mole %/DMT
●	—	0
□	DPT	0.60
○	DPT	1.00
△	DPT	2.00

acts rapidly with the terminal hydroxyl group in PET and eliminates water or phenol should form a high-molecular-weight PET in a shorter time:

$$-\underset{\|}{\overset{O}{C}}C_6H_4\underset{\|}{\overset{O}{C}}OCH_2CH_2O\!-\!H\ +\ H\!-\!OCH_2CH_2O\underset{\|}{\overset{O}{C}}C_6H_4\underset{\|}{\overset{O}{C}}O- \xrightarrow{\text{fast}}$$
$$+\ RO\!-\!A\!-\!OR$$

$$-\underset{\|}{\overset{O}{C}}C_6H_4\underset{\|}{\overset{O}{C}}OCH_2CH_2OAOCH_2CH_2O\underset{\|}{\overset{O}{C}}C_6H_4\underset{\|}{\overset{O}{C}}O-\ +\ 2ROH \qquad (2)$$

(R: H or C_6H_5, A: divalent radical)

Such a reagent may be considered a chain extender for PET. For high-molecular-weight PET manufactured with a chain extender to have the same composition and properties as that manufactured by the conventional method, the bivalent radical "A" in Reaction 2 must either be the terephthaloyl radical or one that will be converted to a volatile cyclic compound after chain extension.

We found that cyclic benzeneboronic anhydride (BBA), diphenyl carbonate (DPC) (4), diphenyl terephthalate (DPT) (5), diphenyl oxalate (DPO) (5), diphenyl malonate (DPM) (5), tetraphenyl orthocarbonate (POC) (6), hexaphenyl orthoterephthalate (POT) (7) are suitable compounds as chain extenders.

Experimental

Polycondensation Reaction. Transesterification was carried out in the conventional manner (4) with 97 grams (0.50 mole) of dimethyl terephthalate (DMT) and 69 grams (1.10 moles) of ethylene glycol in the presence of 0.088 gram of $Ca(OAc)_2 \cdot H_2O$ and 0.044 gram of Sb_2O_3 at 160°-230°C. After completing the transesterification, 0.080

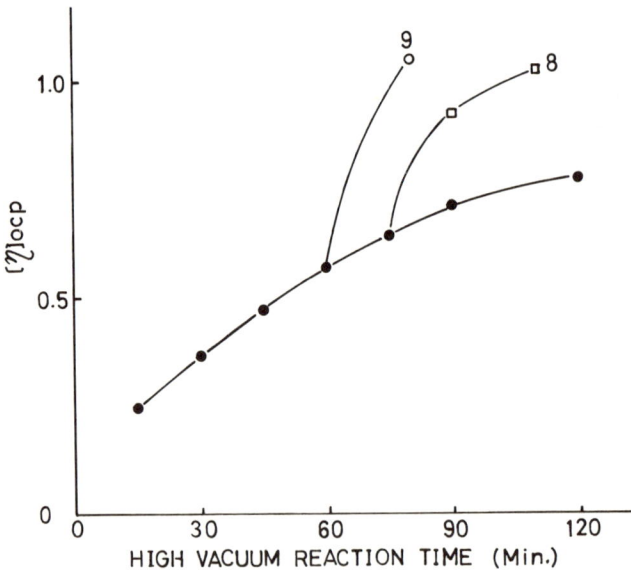

Figure 3. Reaction rates of PET promoted by DPM and DPO at 275°C

Symbol	Additive	Amount, mole %/DMT
●	—	0
□	DPM	0.80
○	DPO	1.00

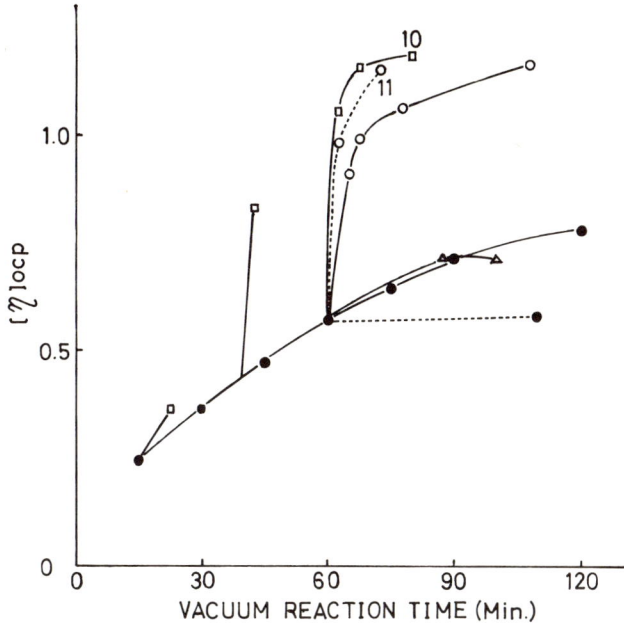

Figure 4. Reaction rates of PET promoted by DPC at 275°C

Symbol	Additive	Amount, mole %/DMT	Reaction pressure, mm Hg
——●	—	0	0.5
——□	DPC	2.00	0.5
——○	DPC	3.00	0.5
······●	—	0	20
······○	DPC	2.00	20
——△	$(C_6H_{11}O)_2CO$	2.00	0

gram of 50% aqueous solution of H_3PO_3 was added. The polycondensation was carried out by immersing the reaction vessel in a bath at the required temperature at atmospheric pressure with a nitrogen purge for the first 30 minutes. The pressure was gradually reduced to less than 1 torr during the following 30 minutes. The polycondensation was then continued until the desired degree of polymerization was obtained. The reaction system was then brought to atmospheric pressure and the chain extender added. The reaction system was then stirred for about two minutes at atmospheric pressure under nitrogen. The pressure was then gradually reduced to 20 torr within 2-3 minutes and to below 1 torr in the next 1-4 minutes. (The term "reaction time under high vacuum" used in this paper refers to the time at a pressure of less than 20 torr.)

Analytical Procedure. The intrinsic viscosity $[\eta]_{ocp}$ was calculated from the value measured in o-chlorophenol at 35°C. Reduced viscosity $\eta_{sp/c}$ was measured by a viscometer at 35°C after dissolving 120 mg of sample in 10 ml of o-chlorophenol at 100°C. The carboxyl content was obtained by the Conix method (8). Softening point (sp) was measured with a penetrometer after heat treatment at 140°C for one hour.

Product composition was determined from infrared and NMR spectra, gas chromatography, and elemental analysis. Distillate from the polycondensation system collected by cooling with ice water or Dry Ice–methanol was analyzed by gas chromatography and infrared and NMR spectra.

Results of Polycondensation

Results of the polycondensation of PET by the use of various additives are given in Figures 1-6 and in Tables I-III.

The chain extenders, when added to PET of a certain degree of polymerization accelerate the polycondensation but not when they react

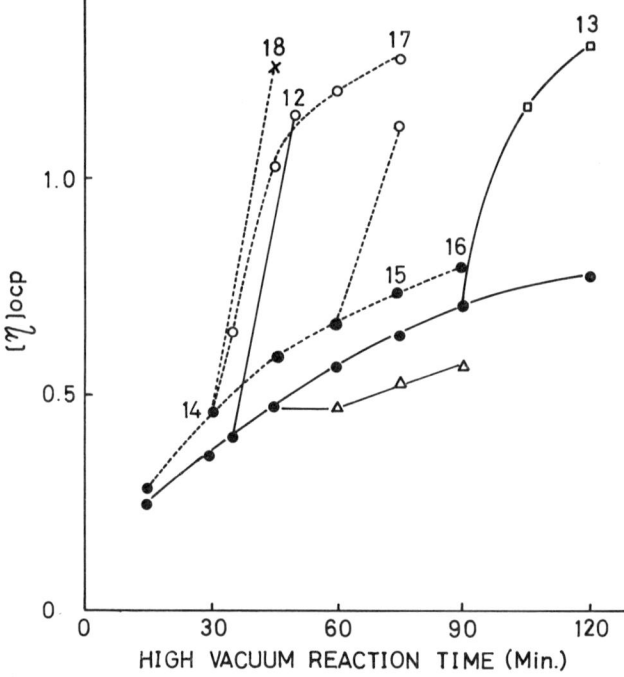

Figure 5. Reaction rates of PET promoted by POC at 275°C and 285°C

Symbol	Additive	Amount mole %/DMT	Reaction Temperature, °C
——●	—	0	275
——□	POC	0.50	275
——○	POC	1.00	275
——△	$(EtO)_4C$	1.00	275
······●	—	0	285
······○	POC	0.50	285
······×	POC	1.00	285

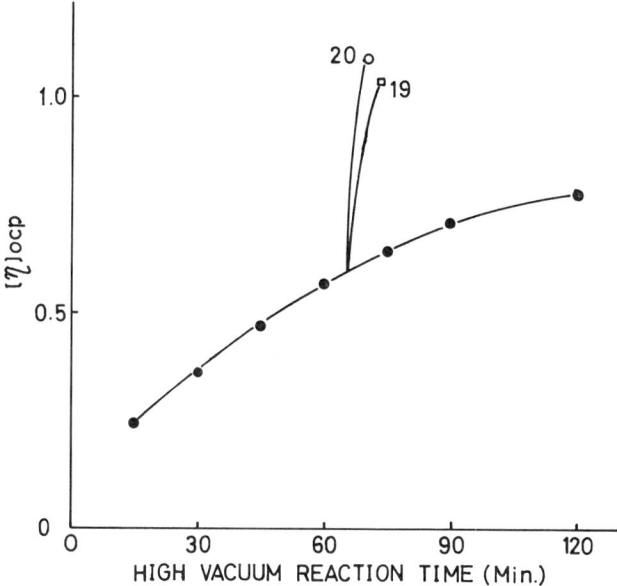

Figure 6. Reaction rates of PET promoted by POT and PPA at 275°C

Symbol	Additive	Amount mole %/DMT
●	—	0
□	POT	0.5
○	PPA	0.5

with monomer or PET with a low degree of polymerization. For example, diphenyl terephthalate did not accelerate the PET polycondensation when added to bis(β-hydroxyethyl) terephthalate in any proportion.

Experiments with Other Additives

Diphenyl sulfite decomposes PET immediately after its addition and lowered the degree of polymerization. Diphenylbenzene phosphite increased the degree of polymerization but 57% (calculated as P) of the added phosphite remained in PET, resulting in a lower softening point. Dicyclohexyl terephthalate lowered the degree of polymerization.

The physical properties of PET obtained by the use of various additives are the same as those obtained by the conventional method (Table I). This fact indicates that the additive is not copolymerized into the polymer chain as a third component. Diphenyl oxalate, diphenyl malonate, hexaphenyl orthoterephthalate, tetraphenyl orthocarbonate, and phenyl triphenoxyacetate not only have a remarkable effect in increasing the degree of polymerization in a short time but also decrease the terminal carboxyl content. Diphenyl carbonate decreases the terminal carboxyl content to some extent whereas diphenyl terephthalate only increases the degree of polymerization.

Table I. Characteristics of Resulting Polymers

Sample No.	Additive	$[\eta]_{ocp}$	SP (°C)	COOH Content (eq/10^6g)
1	—	0.40	264.0	12.4
2	—	0.57	262.3	11.3
3	—	0.71	261.3	12.6
4	—	0.78	260.9	15.1
5	BBA	0.94	260.0	15.4
6	DPT	1.22	257.3	15.2
7	DPT	1.30	257.1	15.7
8	DPM	1.03	259.0	8.5
9	DPO	1.05	258.9	4.0
10	DPC	1.18	257.6	10.1
11	DPC	1.15	258.0	13.5
12 [a]	POC	1.15	258.7	3.9
13	POC	1.31	257.6	4.0
14	—	0.46	262.3	12.3
15	—	0.74	261.1	15.6
16	—	0.80	260.9	18.1
17	POC	1.28	257.1	7.3
18	POC	1.26	256.7	3.8
19	POT	1.04	259.5	5.3
20	PPA	1.09	258.5	3.6
21 [b]	—	1.13	258.5	28.0

[a] Polymer was prepared as follows. Bis(β-hydroxyethyl) terephthalate (4/14 mole) reacted with terephthalic acid (TA) (3/14 mole) at 275°C using Sb_2O_3 as polycondensation catalyst. To the resulting polymer ($[\eta]$ 0.41, COOH content 51.3 eq/10^6g, and diethylene glycol content 0.60 mole %/TA) was added POC, followed by 15 minutes at high vacuum.

[b] This sample was prepared under the same conditions as sample No. 3, except that the amounts of DMT and EG charged was half that of the other experiments, and the high-vacuum reaction was for 320 minutes.

Table II shows the qualitative analytical data of the distillate obtained from the polymerization system. In all cases, the additives reacted with the terminal hydroxyl (or carboxyl) of PET, and phenol or water was distilled out. At the same time, the distillate contained the reaction product of the additives.

Table III shows analytical values of the polymer prepared. The presence of a small amount of terminal acetyl group was found with diphenyl malonate and small amounts of terminal formyl groups with phenyl triphenoxyacetate and diphenyl oxalate. A trace of phenol

Table II. By-Products Formed during the Polycondensation

Additive[b]	By-products[a]										
	EBB	H_2O	C_6H_5OH	HEF	HEA	EC	DPC	CO_2	PCL	EO	CEO
BBA	+	+									
DPT			+								
DPM		+		+			+				
DPO		+	+				+				+
DPC		+			+	(+)	+	+			
POC		+				+	+	+	+	+	
POT		+				+		+	+	+	

[a] Distillate other than by-products formed in conventional polycondensation without any additives. The components of negligible amount were excluded.
[b] EBB: Cyclic ethylene benzene-boronate
HEF: 2-Hydroxyethyl formate
HEA: 2-Hydroxyethyl acetate
EC: Ethylene carbonate
PCL: Phenyl Cellosolve
EO: Ethylene oxide
CEO: Cyclic ethylene oxalate

and phenyl Cellosolve was found in some of the products. There was no significant difference in the content of diethylene glycol from PET obtained by the conventional method.

Scale-Up Data of the Chain-Extender Method

The polycondensation reactions using some of the chain extenders described were carried out in a reaction system having a polymer depth of 750 mm. The results are given in Figure 7 and Table IV.

The number 27 in Figure 7 was the result of the polycondensation with successive additions of 0.47 mole % of diphenyl terephthalate and 0.23 mole % of tetraphenyl orthocarbonate in the same reaction system. These chain extenders were added directly to the polycondensation reaction system under high vacuum.

Discussion

Molecular Weight Distribution. The ratio of weight-average molecular weight to number-average molecular weight of PET obtained by the addition of the chain extenders measured by gel permeation chromatography is shown in Table V.

The relationship between number-average molecular weight (M_n) and weight-average molecular weight (M_w) is the same as in PET obtained by the conventional method, regardless of the kind of chain extender used or the intrinsic viscosity of PET. This indicates that the

Table III. Composition of PET

Chain Extender

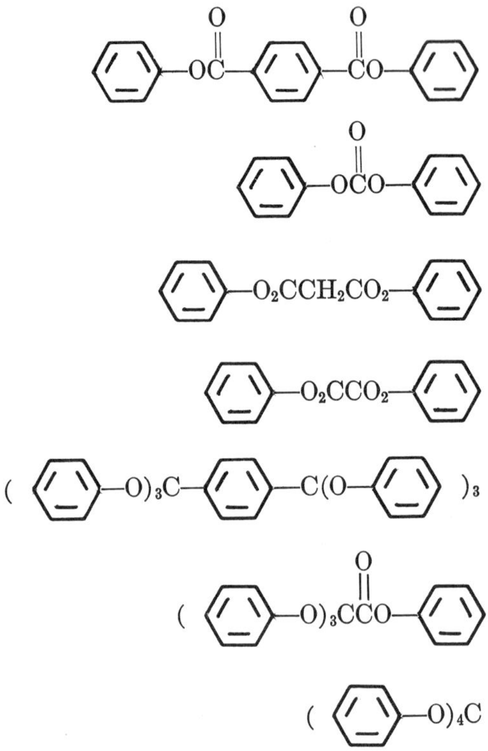

⟨C₆H₅⟩—OC(O)—⟨C₆H₄⟩—C(O)O—⟨C₆H₅⟩

⟨C₆H₅⟩—OC(O)O—⟨C₆H₅⟩

⟨C₆H₅⟩—$O_2CCH_2CO_2$—⟨C₆H₅⟩

⟨C₆H₅⟩—O_2CCO_2—⟨C₆H₅⟩

(⟨C₆H₅⟩—O)₃C—⟨C₆H₄⟩—C(O—⟨C₆H₅⟩)₃

(⟨C₆H₅⟩—O)₃CC(O)O—⟨C₆H₅⟩

(⟨C₆H₅⟩—O)₄C

polymer is not substantially branched. Molecular-weight distribution of PET obtained by this method was measured by the fractionation method (9) and is compared with that of PET obtained by the conventional method in Figure 8.

A solution of 10 grams of the sample dissolved with heating in 990 grams of a mixed solvent of phenol and tetrachloroethane (6:4 by weight) was placed in a thermostat at 30°C and fractionally precipitated with petroleum benzine to obtain a cumulative weight distribution. The sample used for this measurement was a PET of $\eta_{sp/c}$ 0.83, obtained by the reaction of PET of $\eta_{sp/c}$ 0.50 *via* the addition of 1.0 mole % of diphenyl carbonate to the terephthalic acid component and with the reaction carried out *in vacuo* for 3 minutes, then cooled immediately.

Made with Chain Extenders [a]

Weight %/Polymer

Phenol	Phenyl Cellosolve	Formic Acid	Acetic Acid	Diethylene Glycol
—	—	—	—	0.67
0.01	—	—	—	0.65
0.01	0.12	—	—	0.71
0.01	—	—	0.08	0.67
0.01	—	0.07	—	0.66
0.02	—	—	—	0.67
0.02	—	0.05	—	0.69
0.01	0.04	—	—	0.73

[a] Excluding terephthalic acid and ethylene glycol

There is no difference in molecular-weight distribution from that of PET of $\eta_{sp/c}$ 0.83 obtained by the conventional method.

Melt Viscosity. Figure 9 compares the relationship between melt viscosity and intrinsic viscosity of PET obtained by polycondensations using various additives and that of PET manufactured by the conventional method. The melt viscosity was measured with a rotary viscometer after the sample was melted at 280°C in a nitrogen stream. There is no difference in this relationship between the two kinds of PET.

Thermal Stability Polycondensation of PET was carried out at 280°C by the method shown above on a 0.5 molar scale. The reaction system was held under a nitrogen purge for one hour without stirring after the intrinsic viscosity of the polymer reached $[\eta]_{ocp}$ 0.90. The

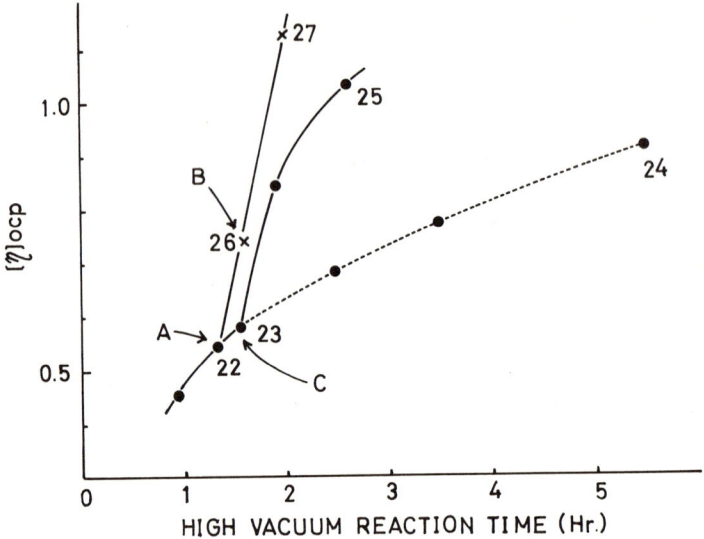

Figure 7. Manufacture of PET by chain extender method on a large scale. Depth of molten polymer: 750 mm. Reaction temperature: 275°–280°C.

		Mole %/DMT
A:	⟨C₆H₄⟩–OC(=O)–⟨C₆H₄⟩–CO(=O)–⟨C₆H₄⟩	0.47
B:	(⟨C₆H₄⟩O)₄C	0.23
C:	⟨C₆H₄⟩–OC(=O)–⟨C₆H₄⟩–CO(=O)–⟨C₆H₄⟩	0.70

Table IV. Characteristics of Resulting Polymers

Sample No.	Additive	$[\eta]_{ocp}$	SP (°C)	COOH Content (eq/10^6g)
22	—	0.55	262.4	11
23	—	0.59	262.3	12
24	—	0.93	260.1	35
25	DPT	1.03	259.3	21
26	DPT	(0.73) [a]		
27	POC	1.12	258.8	4

[a] Value assumed from the melt viscosity

Table V. Relationship between Weight-Average and Number-Average Molecular Weight

Additives (mole %/TA)	$[\eta]_{ocp}$	M_w/M_n
None	0.64	2.18
None	0.78	2.26
None	1.07	2.42
Ph–OC(O)–C₆H₄–C(O)CO–Ph (0.5)	0.98	2.29
Ph–OC(O)O–Ph (0.5)	0.95	2.30
(Ph–O)₄C (0.5)	0.98	2.21
Branched PET [a]	0.66	3.16

[a] C(CH₂OH)₄ 0.4 mole %/TA
Instrument: Waters Assoc., Inc. (Model 200)
Solvent: *m*-cresol

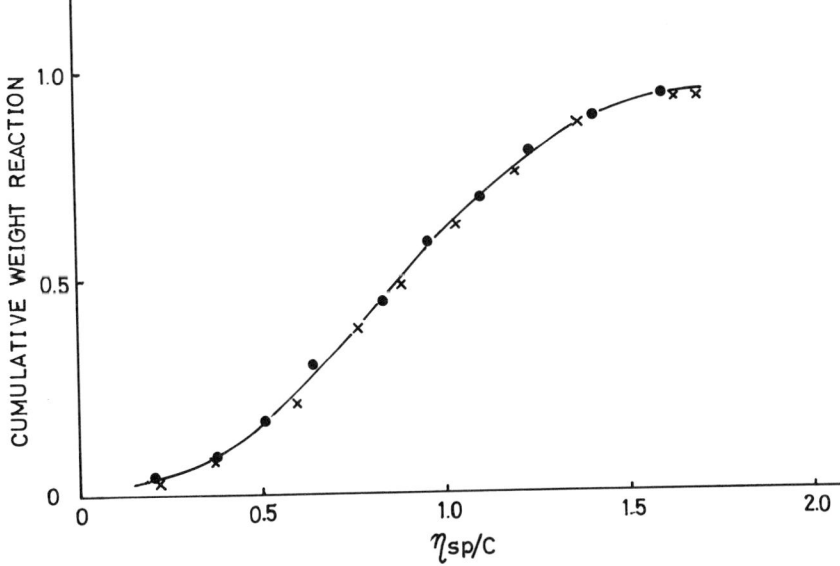

Figure 8. *Molecular weight distribution of PET obtained by the chain extender method*

×: PET manufactured by the conventional method
●: PET polymerized rapidly in vacuum using diphenyl carbonate
$\eta_{sp/c}$: Reduced viscosity of PET measured in o-chlorophenol

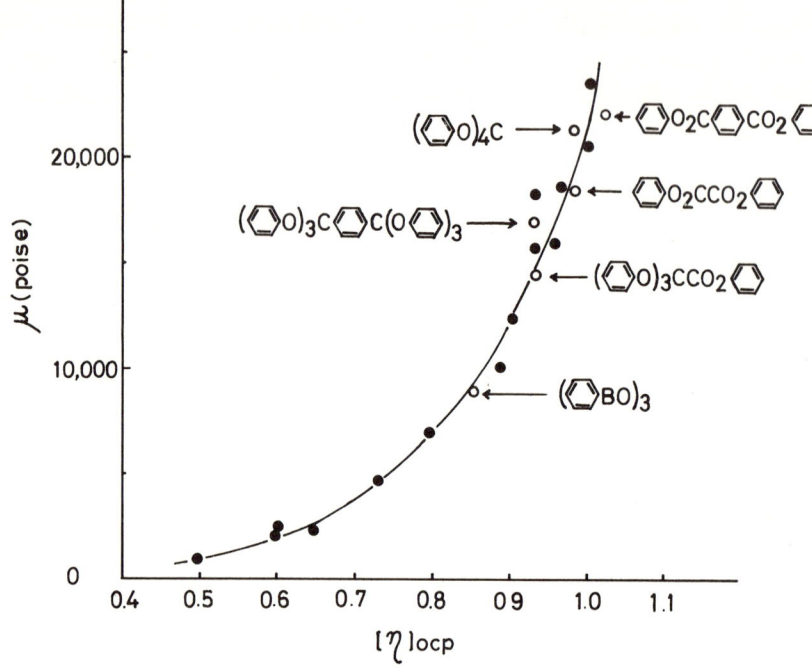

Figure 9. Melt viscosity vs. solution viscosity for PET obtained by the chain extender method

●conventional method

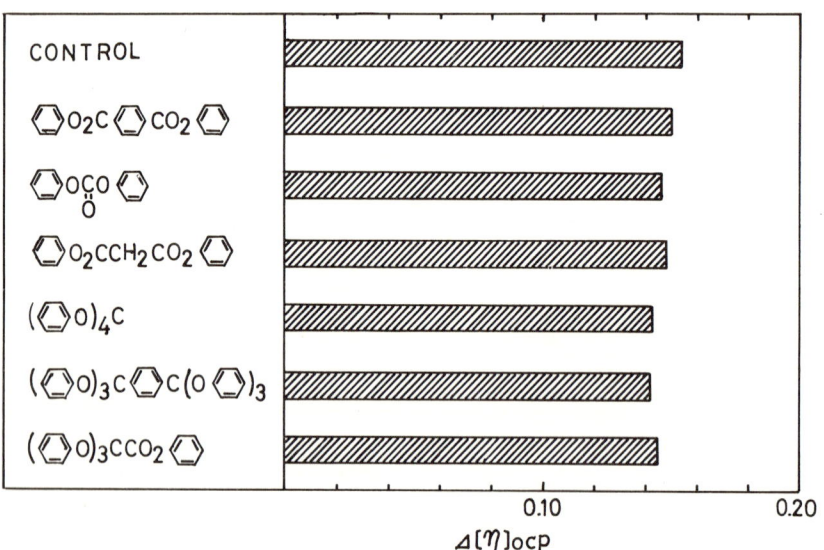

Figure 10. Thermal stability of PET. $\Delta[\eta]_{ocp}$ means the difference of $[\eta]_{ocp}$ between the initial and final value

results shown in Figure 10 indicate that heat stability in the molten state of PET produced by the chain extender method is somewhat better than that of PET produced conventionally.

Hydrolytic Stability. Hydrolytic stability of PET filaments obtained by the chain extender method was examined by holding the filaments at 150°C for six hours at 100% relative humidity. The results shown in Figure 11 indicate that the hydrolytic stability of PET decreases with increasing carboxyl content.

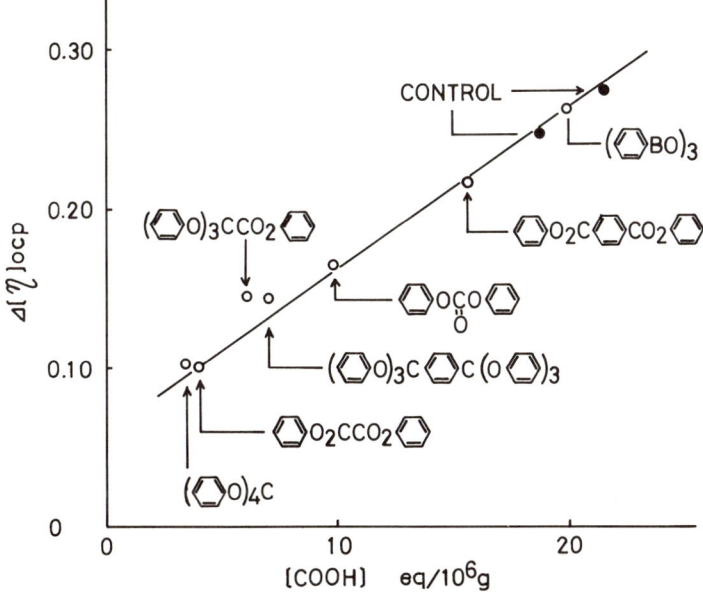

Figure 11. Hydrolytic stability of PET obtained by the chain extender method

$\Delta[\eta]_{ocp}$: lowering of intrinsic viscosity after treatment, measured in o-chlorophenol
[COOH]: carboxyl content before treatment

Thermal Resistance. Thermal resistance of high-tenacity PET filaments under moist conditions was determined by first conditioning a sample, placing in a sealed tube at 20°C and 65% relative humidity, and heating for 48 hours at 150°C. The results are shown in Figure 12. The breaking strength retained (%) was calculated from:

Breaking strength retained (%)

$$= \frac{\text{breaking strength of filament after test}}{\text{breaking strength of filament before test}} \times 100$$

Figure 12 also indicates that thermal stability of high-tenacity PET

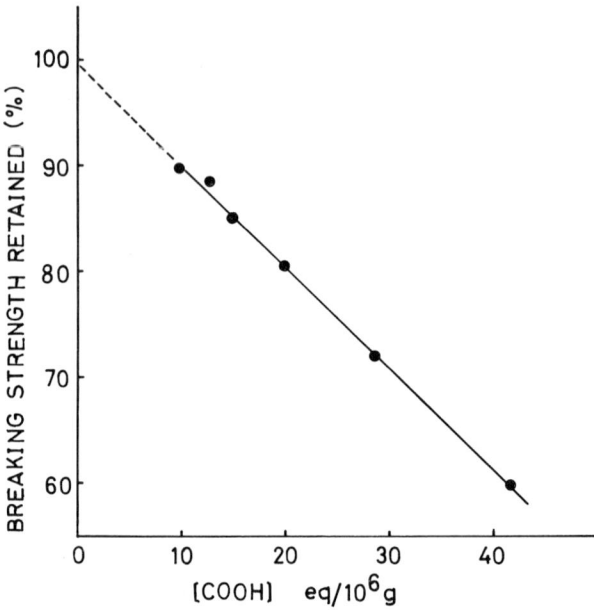

Figure 12. *Thermal stability of poly(ethylene terephthalate) filament*

Samples: high-tenacity PET filament
Treatment: 48 hours at 150°C

filament under moist conditions decreases with an increase in carboxyl content.

Reaction Mechanisms

The effect of the chain extenders in increasing the polycondensation rate and in decreasing terminal carboxyl content is explained by several reaction mechanisms.

Elimination of Water. Effect of benzeneboronic anhydride is shown by Reactions 3, 4, and 5. The reaction products, cyclic ethylenebenzeneboronate and water, diffused more readily through the PET polycondensation system than did ethylene glycol.

$$2 \text{—OOC—C}_6\text{H}_4\text{—COOCH}_2\text{CH}_2\text{OH} + \text{H}_5\text{C}_6\text{—B} \begin{array}{c} \text{O} \\ \diagup \diagdown \\ \text{O} \end{array} \text{B—C}_6\text{H}_5 \quad \text{(BBA)}$$

I

$$\rightleftarrows \quad -\text{O}\overset{\overset{\text{O}}{\|}}{\text{C}}-\text{C}_6\text{H}_4-\overset{\overset{\text{O}}{\|}}{\underset{\text{C}_6\text{H}_5}{\text{C}}}\text{OCH}_2\text{CH}_2\text{OBOCH}_2\text{CH}_2\text{O}\overset{\overset{\text{O}}{\|}}{\text{C}}-\text{C}_6\text{H}_4-\overset{\overset{\text{O}}{\|}}{\text{C}}\text{O}-$$

$$+ \left[\text{HO}-\underset{\text{C}_6\text{H}_5}{\text{B}}-\text{O}-\underset{\text{C}_6\text{H}_5}{\text{B}}-\text{OH} \right] \quad (3)$$

$$\rightleftarrows \quad -\text{O}\overset{\overset{\text{O}}{\|}}{\text{C}}-\text{C}_6\text{H}_4-\overset{\overset{\text{O}}{\|}}{\text{C}}\text{OCH}_2\text{CH}_2\text{O}\overset{\overset{\text{O}}{\|}}{\text{C}}-\text{C}_6\text{H}_4-\overset{\overset{\text{O}}{\|}}{\text{C}}\text{O}-$$

II

$$+ \quad \underset{\underset{\underset{\text{C}_6\text{H}_5}{|}}{\text{B}}}{\overset{\text{CH}_2-\text{CH}_2}{\underset{\text{O}\quad\text{O}}{|\quad|}}} + 2/3 \text{ BBA} + \text{H}_2\text{O} \uparrow \quad (4)$$

$1/3 \text{ BBA} + \text{HOCH}_2\text{CH}_2\text{OH}$

$$\rightleftarrows \quad \underset{\underset{\underset{\text{C}_6\text{H}_5}{|}}{\text{B}}}{\overset{\text{CH}_2-\text{CH}_2}{\underset{\text{O}\quad\text{O}}{|\quad|}}} + \text{H}_2\text{O} \uparrow \quad (5)$$

Elimination of Phenol. The equilibrium constants of the reactions indicated by Reactions 1 and 6 were measured at 275°C to explain the remarkable effect of diphenyl terephthalate in accelerating the PET polycondensation by elimination of phenol (10).

$$-\text{OC}(=\text{O})-\text{C}_6\text{H}_4-\text{C}(=\text{O})\text{OCH}_2\text{CH}_2\text{OH} + \text{C}_6\text{H}_5\text{OC}(=\text{O})-\text{C}_6\text{H}_4-\text{C}(=\text{O})\text{OC}_6\text{H}_5$$

I III

$$\underset{k_{-2}}{\overset{k_2}{\rightleftarrows}} -\text{OC}(=\text{O})-\text{C}_6\text{H}_4-\text{C}(=\text{O})\text{OCH}_2\text{CH}_2\text{OC}(=\text{O})-\text{C}_6\text{H}_4-\text{C}(=\text{O})\text{OC}_6\text{H}_5 + \text{C}_6\text{H}_5\text{OH} \qquad (6)$$

IV

In Reactions 1 and 6, the values obtained were $k_2/k_{-2} = 156 \pm 27$, $k_1/k_{-1} = 1.92 \pm 0.27$, and $k_1/k_2 = 0.21 - 0.15$.

DIPHENYL TEREPHTHALATE. In Reaction 7, the effect of diphenyl terephthalate in accelerating the polycondensation reaction can be explained by assuming $k_3/k_{-3} \simeq k_2/k_{-2}$.

$$-\text{OC}(=\text{O})-\text{C}_6\text{H}_4-\text{C}(=\text{O})\text{OCH}_2\text{CH}_2\text{OC}(=\text{O})-\text{C}_6\text{H}_4-\text{C}(=\text{O})\text{OC}_6\text{H}_5 + \text{OC}(=\text{O})\text{C}_6\text{H}_4\text{C}(=\text{O})\text{OCH}_2\text{CH}_2\text{OH}$$

IV I

$$\underset{k_{-3}}{\overset{k_3}{\rightleftarrows}} -\text{OC}(=\text{O})-\text{C}_6\text{H}_4-\text{C}(=\text{O})\text{OCH}_2\text{CH}_2\text{OC}(=\text{O})-\text{C}_6\text{H}_4-\text{C}(=\text{O})\text{OCH}_2\text{CH}_2\text{OC}(=\text{O})-\text{C}_6\text{H}_4-\text{CO}-$$

II

$$+ \text{C}_6\text{H}_5\text{OH} \uparrow \qquad (7)$$

DIPHENYL MALONATE AND DIPHENYL OXALATE. Above 200°C, the half ester of malonic acid decomposes to carbon dioxide and an ester of acetic acid, and the half ester of oxalic acid decomposes to carbon dioxide and an ester of formic acid. Diphenyl malonate and diphenyl oxalate are chain extenders that decrease the terminal PET COOH content by these decomposition reactions.

The reaction mechanism with phenyl oxalate is shown in Reaction 8. The reaction products, phenol, β-hydroxyethyl formate and cyclic ethylene oxalate, are more diffusible through the PET polycondensation system than is ethylene glycol.

$$2 -\text{OC}(=\text{O})-\text{C}_6\text{H}_4-\text{C}(=\text{O})\text{OCH}_2\text{CH}_2\text{OH} + \text{C}_6\text{H}_5-\text{OC}(=\text{O})\text{C}(=\text{O})-\text{C}_6\text{H}_5 \rightleftarrows$$

I V

$$-\overset{O}{\underset{\|}{O C}}-C_6H_4-\overset{O}{\underset{\|}{C}}OCH_2CH_2O\overset{OO}{\underset{\|\|}{CC}}OCH_2CH_2O\overset{O}{\underset{\|}{C}}-C_6H_4-\overset{O}{\underset{\|}{C}}O-$$

VI

$+ 2C_6H_5OH$ (8)

$$VI \rightarrow -\overset{O}{\underset{\|}{OC}}-C_6H_4-\overset{O}{\underset{\|}{C}}OCH_2CH_2O\overset{O}{\underset{\|}{C}}-C_6H_4-\overset{O}{\underset{\|}{C}}O-$$

II

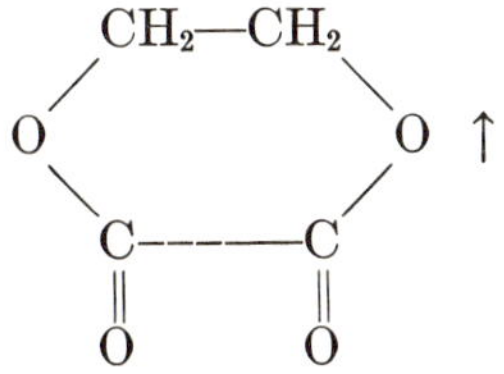

$$+ \;\; O\underset{\underset{\|}{\overset{}{C}}\text{----}\underset{\|}{\overset{}{C}}}{\overset{CH_2-CH_2}{\diagup\diagdown}} O \uparrow \qquad (9)$$

$$VI + -\overset{O}{\underset{\|}{OC}}-C_6H_4-\overset{O}{\underset{\|}{C}}OH \rightarrow -\overset{O}{\underset{\|}{OC}}-C_6H_4-\overset{O}{\underset{\|}{C}}OCH_2CH_2O\overset{OO}{\underset{\|\|}{CC}}OH$$

VII VIII

$$+ -\overset{O}{\underset{\|}{OC}}-C_6H_4-\overset{O}{\underset{\|}{C}}-OCH_2CH_2O\overset{O}{\underset{\|}{C}}\overset{O}{\underset{\|}{C}}_6H_4\overset{O}{\underset{\|}{C}}O- \qquad (10)$$

II

$$VIII \rightarrow -\overset{O}{\underset{\|}{OC}}-C_6H_4 \;\; \overset{O}{\underset{\|}{C}}OCH_2CH_2O\overset{O}{\underset{\|}{C}}H + CO_2 \uparrow \qquad (11)$$

IX

$$IX + I \rightarrow -\overset{O}{\underset{\|}{OC}}-C_6H_4-\overset{O}{\underset{\|}{C}}OCH_2CH_2O\overset{O}{\underset{\|}{C}}-C_6H_4-\overset{O}{\underset{\|}{C}}O-$$

(II)

$$+ \; HOCH_2CH_2O\overset{O}{\underset{\|}{C}}H \uparrow \qquad (12)$$

The mechanism with diphenyl malonate is indicated by Reactions

13 through 16. In these reactions, the phenol formed is more diffusible through the PET polycondensation system than is ethylene glycol.

$$2\ \mathrm{-OC\!-\!C_6H_4\!-\!COCH_2CH_2OH} + \mathrm{C_6H_5OCCH_2COC_6H_5} \rightleftarrows$$
$$\text{I} \qquad\qquad\qquad\qquad \text{X}$$

$$\mathrm{-OC\!-\!C_6H_4\!-\!COCH_2CH_2OCCH_2COCH_2CH_2OC\!-\!C_6H_4\!-\!CO-}$$
$$\text{XI}$$

$$+\ 2\mathrm{C_6H_5OH} \qquad\qquad (13)$$

$$\mathrm{XI} + \mathrm{-OC\!-\!C_6H_4\!-\!COH} \rightarrow$$
$$\text{VII}$$

$$\mathrm{-OC\!-\!C_6H_4\!-\!COCH_2CH_2OC\!-\!C_6H_4\!-\!CO-} +$$
$$\text{II}$$

$$\mathrm{HO\!-\!CCH_2COCH_2CH_2OC\!-\!C_6H_4\!-\!CO-} \qquad (14)$$
$$\text{XII}$$

$$\mathrm{XII} \rightarrow \mathrm{CH_3COCH_2CH_2OC\!-\!C_6H_4\!-\!CO-} + \mathrm{CO_2} \uparrow \qquad (15)$$
$$\text{XIII}$$

$$\mathrm{XIII} + \mathrm{I} \rightarrow \mathrm{II} + \mathrm{CH_3COOCH_2CH_2OH} \uparrow \qquad (16)$$

DIPHENYL CARBONATE. The reaction mechanism with diphenyl carbonate as the chain extender is shown in Reactions 17, 18, and 19.

$$\mathrm{-OC\!-\!C_6H_4\!-\!COCH_2CH_2OH} + \mathrm{C_6H_5OCOC_6H_5} \rightleftarrows$$
$$\text{I} \qquad\qquad\qquad\qquad \text{XIV}$$

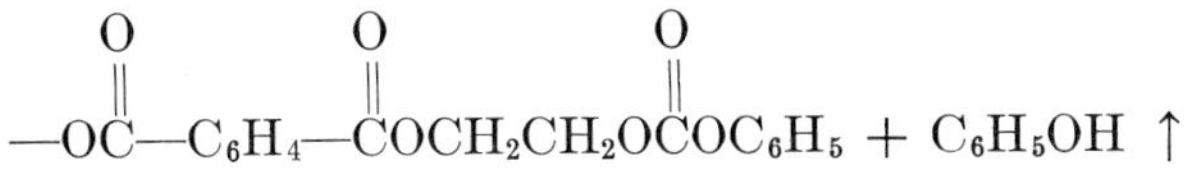

$$-\overset{O}{\overset{\|}{C}}-C_6H_4-\overset{O}{\overset{\|}{C}}OCH_2CH_2O\overset{O}{\overset{\|}{C}}OC_6H_5 + C_6H_5OH \uparrow \qquad (17)$$

XV

XV + I ⇌

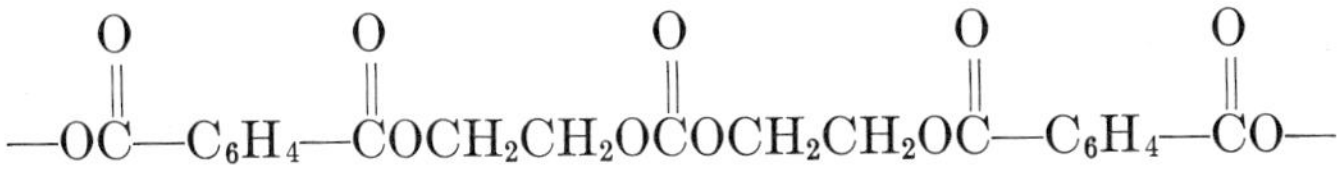

$$-\overset{O}{\overset{\|}{O C}}-C_6H_4-\overset{O}{\overset{\|}{C}}OCH_2CH_2O\overset{O}{\overset{\|}{C}}OCH_2CH_2O\overset{O}{\overset{\|}{C}}-C_6H_4-\overset{O}{\overset{\|}{C}}O-$$

XVI

$$+ C_6H_5OH \uparrow \qquad (18)$$

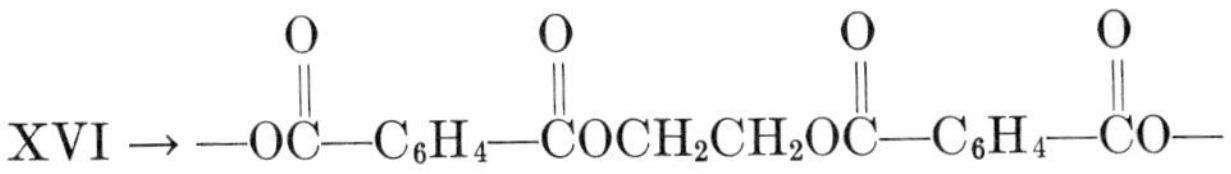

$$\text{XVI} \rightarrow -\overset{O}{\overset{\|}{O C}}-C_6H_4-\overset{O}{\overset{\|}{C}}OCH_2CH_2O\overset{O}{\overset{\|}{C}}-C_6H_4-\overset{O}{\overset{\|}{C}}O-$$

II

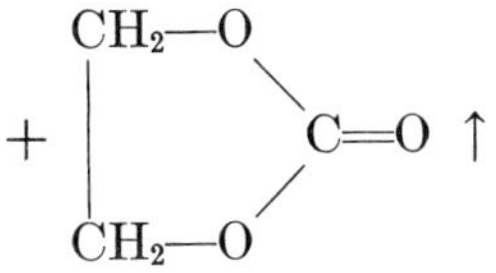

$$+ \quad \begin{matrix} CH_2-O \\ | \\ CH_2-O \end{matrix} \!\!\!\! \diagdown \!\!\! C=O \uparrow \qquad (19)$$

Reaction 19 was presumed from the following model experiment.

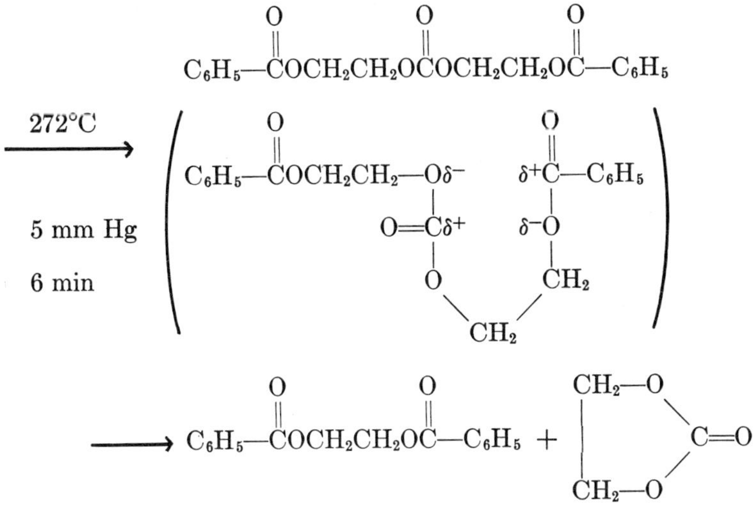

$$C_6H_5-\overset{O}{\overset{\|}{C}}OCH_2CH_2O\overset{O}{\overset{\|}{C}}OCH_2CH_2O\overset{O}{\overset{\|}{C}}-C_6H_5$$

$$\xrightarrow[\substack{5 \text{ mm Hg} \\ 6 \text{ min}}]{272°C} \left(\begin{matrix} C_6H_5-\overset{O}{\overset{\|}{C}}OCH_2CH_2-O\delta^- & \delta^+\overset{O}{\overset{\|}{C}}-C_6H_5 \\ O=C\delta^+ & \delta^-O \\ | & | \\ O & CH_2 \\ \diagdown & \diagup \\ & CH_2 \end{matrix} \right)$$

$$\longrightarrow C_6H_5-\overset{O}{\overset{\|}{C}}OCH_2CH_2O\overset{O}{\overset{\|}{C}}-C_6H_5 + \begin{matrix} CH_2-O \\ | \\ CH_2-O \end{matrix} \!\!\!\! \diagdown \!\!\! C=O \qquad (20)$$

When a large quantity of diphenyl carbonate is used, or when the reaction is carried out at a low vacuum, (for example, at 100 torr), side Reaction 21 occurs and produces a phenyl cellosolve ester.

$$XV \rightarrow -O\overset{O}{\overset{\|}{C}}-C_6H_4-\overset{O}{\overset{\|}{C}}OCH_2CH_2OC_6H_5 + CO_2 \uparrow \quad (21)$$

Reaction 21 would interfere in increasing the degree of polymerization. The reaction, however, is minimized by selecting the suitable Reaction 21 occurs and produces a phenyl Cellosolve ester.

Reaction 22 will decrease terminal carboxyl content and intermediate IV will accelerate the polycondensation Reaction 7.

$$-O\overset{O}{\overset{\|}{C}}-C_6H_4-\overset{O}{\overset{\|}{C}}OH + XIV \rightarrow$$
$$\text{VII}$$

$$-O\overset{O}{\overset{\|}{C}}-C_6H_4-\overset{O}{\overset{\|}{C}}OC_6H_5 + C_6H_5OH + CO_2 \uparrow \quad (22)$$
$$\text{IV}$$

ORTHOESTERS. As model experiments (1) phenyl Cellosolve and tetraphenyl orthocarbonate, and (2) hydroxyethyl benzoate and tetraphenyl orthocarbonate reacted at 275°C. Diphenyl carbonate was obtained from the latter reaction but not from the former. The facts that the distillate from the latter reaction contained ethylene oxide and that diphenyl carbonate was formed quantitatively from the reaction of carboxylic acid and tetraphenyl orthocarbonate suggest the following reaction mechanism:

$$-O\overset{O}{\overset{\|}{C}}-C_6H_4-\overset{O}{\overset{\|}{C}}OCH_2CH_2OH + R-C(OC_6H_5)_3 \rightarrow$$
$$\text{I} \qquad\qquad\qquad\qquad \text{XVII}$$

$$(R: -OC_6H_5, -\overset{O}{\overset{\|}{C}}OC_6H_5, -C_6H_4-C(OC_6H_5)_3)$$

$$-O\overset{O}{\overset{\|}{C}}-C_6H_4-\overset{O}{\overset{\|}{C}}OCH_2CH_2O-\overset{R}{\underset{\underset{C_6H_5}{O}}{C}}-OC_6H_5 + C_6H_5OH \uparrow \quad (23)$$
$$\text{XVIII}$$

$$\text{XVIII} \rightarrow \begin{pmatrix} \begin{matrix} & \text{O} & & \text{O} & & \\ & \| & & \| & & \\ -\text{OC}-\text{C}_6\text{H}_4-\text{C}\delta+ & & \delta-\text{O}-\text{C}_6\text{H}_5 \\ & | & & | & \\ & \text{O}\delta- & & \delta+\text{C}-\text{R} \\ & | & & / \quad \backslash \\ & \text{CH}_2 & & \text{O} \quad \text{OC}_6\text{H}_5 \\ & \backslash & & / \\ & & \text{CH}_2 & \end{matrix} \end{pmatrix}$$

$$\rightarrow \begin{matrix} \text{O} & & \text{O} \\ \| & & \| \\ -\text{OC}-\text{C}_6\text{H}_4-\text{COC}_6\text{H}_5 \end{matrix} + \begin{matrix} \text{CH}_2-\text{O} & & \text{R} \\ & \diagdown \quad \diagup & \\ & \text{C} & \\ & \diagup \quad \diagdown & \\ \text{CH}_2-\text{O} & & \text{O}-\text{C}_6\text{H}_5 \end{matrix} \quad (24)$$

$$\text{XIX}$$

$$\text{XIX} \rightarrow \text{C}_6\text{H}_5\text{O}\overset{\text{O}}{\underset{\|}{\text{C}}}-\text{R} + \underset{\text{O}}{\underset{\diagdown \diagup}{\text{CH}_2-\text{CH}_2}} \uparrow \quad (25)$$

$$\text{XX}$$

$$\begin{matrix} \text{O} & & \text{O} \\ \| & & \| \\ -\text{OC}-\text{C}_6\text{H}_4-\text{COH} \end{matrix} + \text{R}-\text{C}(\text{OC}_6\text{H}_5)_3$$

VII XVII

$$(\text{R}: -\text{OC}_6\text{H}_5, -\overset{\text{O}}{\underset{\|}{\text{C}}}\text{OC}_6\text{H}_5, -\text{C}_6\text{H}_4-\text{C}(\text{OC}_6\text{H}_5)_3)$$

$$\rightarrow \begin{matrix} \text{O} & & \text{O} \\ \| & & \| \\ -\text{OC}-\text{C}_6\text{H}_4-\text{COC}_6\text{H}_5 \end{matrix} + \text{R}-\overset{\text{O}}{\underset{\|}{\text{C}}}\text{OC}_6\text{H}_5 + \text{C}_6\text{H}_5\text{OH} \uparrow \quad (26)$$

IV XX

For the reason given earlier, compounds IV and XX accelerate the polycondensation reaction of PET.

When tetraphenyl orthocarbonate reacts with low terminal carboxyl PET, a small amount of diethylene glycol is formed by Side Reactions 27 and 28.

$$2\text{I} + \text{C}(\text{OC}_6\text{H}_5)_4 \rightleftarrows$$

XXI

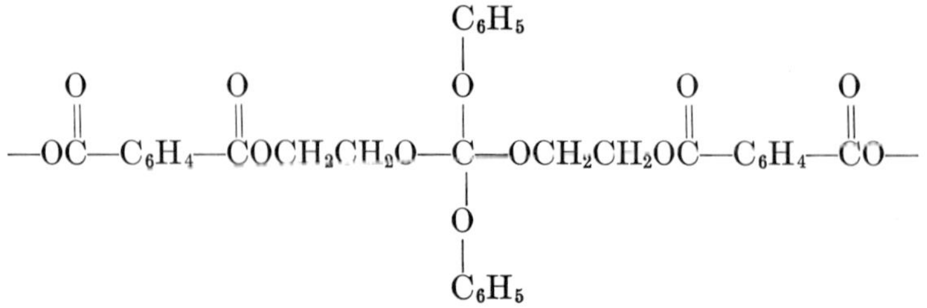

$$\text{—OC}-\text{C}_6\text{H}_4-\overset{\overset{O}{\|}}{\text{C}}\text{OCH}_2\text{CH}_2\text{O}-\underset{\underset{\underset{C_6H_5}{|}}{O}}{\overset{\overset{C_6H_5}{|}}{\underset{|}{\text{C}}}}-\text{OCH}_2\text{CH}_2\text{O}\overset{\overset{O}{\|}}{\text{C}}-\text{C}_6\text{H}_4-\overset{\overset{O}{\|}}{\text{C}}\text{O—}$$

XXII

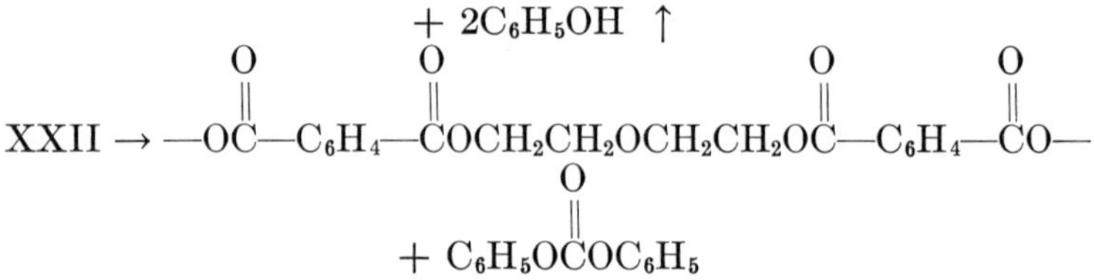

$$+ \ 2\text{C}_6\text{H}_5\text{OH} \uparrow \quad (27)$$

$$\text{XXII} \rightarrow \text{—OC}-\text{C}_6\text{H}_4-\overset{\overset{O}{\|}}{\text{C}}\text{OCH}_2\text{CH}_2\text{OCH}_2\text{CH}_2\text{O}\overset{\overset{O}{\|}}{\text{C}}-\text{C}_6\text{H}_4-\overset{\overset{O}{\|}}{\text{C}}\text{O—}$$

$$+ \ \text{C}_6\text{H}_5\text{O}\overset{\overset{O}{\|}}{\text{C}}\text{OC}_6\text{H}_5 \quad (28)$$

XXIII

Side Reaction 28 can be explained by assuming reaction intermediates XXII' or XXII".

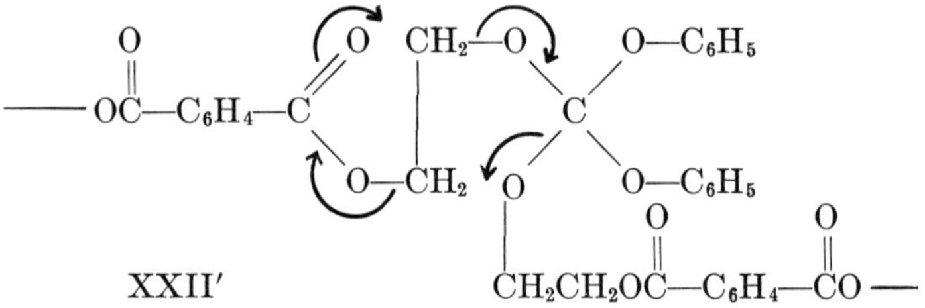

XXII'

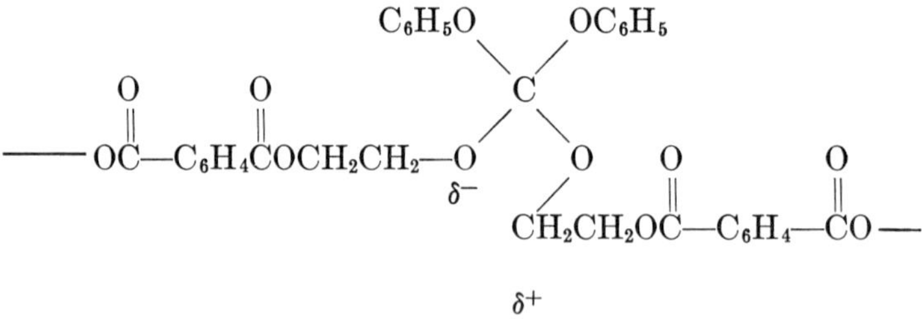

XXII"

As a model experiment to see the effect of orthoesters in reducing the content of terminal carboxyl, aliphatic long-chain dicarboxylic acid and a large excess of various additives reacted in biphenyl at 250°C, and the rate constants of the pseudo-first-order reaction were compared. The reaction rate constant decreases in the order of tetraphenyl orthocarbonate, hexaphenyl orthoterephthalate> diphenyl oxalate, diphenyl malonate ≫ diphenyl carbonate.

Summary

Several kinds of chain extenders that permit rapid production of poly(ethylene terephthalate) of high molecular weight and low carboxyl content have been found. The composition, molecular-weight distribution, and chemical properties of poly(ethylene terephthalate) obtained by this chain-extender method are quite similar to those of the polymer obtained by the conventional method, except for a lower carboxyl content.

Literature Cited

1. Daniels, W. W., U.S. Patent **3,051,212** (1962).
2. Ravens, D. A. S., Ward, I. M., *Trans Faraday Soc.* (1961) **57**, 150.
3a. Buxbaum, H., *Angew. Chem. Intl. Ed.* (1968) **7**, 182.
3b. Zimmerman, H., *Faserforsch. Textiltech.* (1962) **15**, 481.
4. Shima, T., Netherlands Patent Application **6,407,013** (1964); U.S. Patent **3,444,141** (1969).
5. Shima, T., Yamada H., Netherlands Patent Application **6,505,915** (1965); U.S. Patent **3,433,770** (1969).
6. Shima, T., Urasaki, T., Oka, I., Netherlands Patent Application **7,103,731** (1971).
7. Shima, T., Urasaki, T., Kobayashi, T., Oka, I., Suzuki, K., German Offen. **2,259,531** (1973).
8. Conix, A., *Makromol. Chem.* (1958) **26**, 226.
9. Gordijenko, A., *Faserforsch. Textiltech.* (1953) **4**, 499.
10. Uno, F., unpublished data.
11. Shima, T., Urasaki, T., Oka, I., German Offen. **1,945,594** (1970); U.S. Patent **3,637,910** (1972).

RECEIVED April 1, 1972.

15

Hexacyanometalate Salt Complexes as Catalysts for Epoxide Polymerizations

R. J. HEROLD and R. A. LIVIGNI

Research and Development Division, The General Tire & Rubber Co., Akron, Ohio 44329

> *The discovery, compositions, and physical and catalytic properties of a new class of complexes of certain hexacyanometalate salts are reported here. A typical example consists of the nonstoichiometric complex $Zn_3[Co(CN)_6]_2 \cdot x[glyme] \cdot yZnCl_2 \cdot zH_2O$. These complexes are catalysts for preparing high-molecular-weight, amorphous, poly(propylene oxide); poly(propylene ether) polyol having a narrow molecular-weight distribution; and polyesters from the copolymerization of epoxides and cyclic anhydrides. The data presented indicate that these catalysts operate according to a coordinate mechanism.*

This paper describes the discovery, means of preparation, and catalytic behavior of certain hexacyanometalate salt complexes (*1*). These materials are useful in polymerizing propylene oxide to high-molecular-weight (*2*), preparing poly(propylene ether) polyols from propylene oxide and compounds containing active hydrogens (*3*), and preparing polyesters from propylene oxide and cyclic anhydrides (*4*). Some of the known catalysts are useful for two of these reactions; however, no other catalyst system having activity for all three reactions has been described. As examples, alkali metal hydroxides are useful for preparing polyols (*5*) and polyesters (*6*), and zinc alkyls are useful for preparing polyesters (*7*) and high-molecular-weight polymers (*8*).

Subsequent to the report (*9*) by Pruitt and co-workers that certain iron catalysts could be used to prepare high-molecular-weight poly(propylene oxide), a number of other catalysts were found useful for this purpose (*10, 11, 12*). These catalysts were necessarily used in amounts larger than would be required, assuming that each metal atom were active. Furukawa (*13*) and Vandenberg (*11*) concluded that polymeric species containing metal-oxygen-metal groupings were present in the metal–alkyl–water type of catalysts. Light-scattering studies (14, 15) have shown the presence of relatively large particles in clear "solutions" of metal–alkyl–water catalysts, thus suggesting that the catalyst consists of large particles with active sites on their surfaces. The high-molecular-weight poly(propylene oxide) prepared with the catalysts previously reported was at least partially crystalline and isotactic (16). The fact that so many varied structures gave crystalline polymers suggested to Osgan and Price (10) that no special geometric feature of a crystalline surface was responsible, and that steric hindrance in the neighborhood of some of the active sites was a more likely requirement.

The hexacyanometalate salt complexes are characterized by an insensitivity to prolonged exposure to atmospheric moisture and by their very high catalytic activity. Although sufficient data have not been obtained regarding the detailed nature of the catalysis, the amount of catalyst required for the preparation of high-molecular-weight polymer is in the range expected if each metal atom were associated with one polymer molecule. Another unique characteristic of this system is the fact that the high-molecular-weight poly(propylene oxide) prepared from these catalysts is completely amorphous.

Experimental

Preparation of Catalysts. PREPARATION OF ZINC–HEXACYANOFERRATE (II, III)-BASED CATALYST. A solution of 17.7 grams (0.0538 mole) of $K_3Fe(CN)_6$ in 125 ml of water was added rapidly with stirring to a solution of 12.1 grams (0.0888 mole) of $ZnCl_2$ in 47 ml of water at room temperature, producing an orange precipitate; 187 ml of distilled diglyme (dimethylether of diethylene glycol) was added and the system stirred for 20 minutes. It was then alternately centrifuged and stirred with additional 200 ml portions of diglyme containing 10% water until the supernatant was free of chloride ion, determined by adding $Cu(NO_3)_2$ to precipitate $Fe(CN)_6^{2-}$, centrifuging, and adding $AgNO_3$ to the resulting clear liquid. To the chloride-free solids, 1000 ml of hexane and 400 ml of diglyme were added. This mixture was refluxed using a Dean-Stark trap to remove water. The slurry was centrifuged and the excess diglyme removed from the solids by extraction with n-pentane, followed by vacuum drying at 30°C. A fluffy yellow powder resulted, having an apparent density of 0.15 g/ml. The yield was 17.8 grams. Anal.: Calcd. for $Zn_3[Fe(CN)_6]_2 \cdot Zn_2Fe(CN)_6 \cdot ZnCl_2 \cdot 1.56KCl \cdot 3.31$diglyme$\cdot 0.1H_2O$; Zn, 21.5; Fe, 9.2; Cl, 8.7; C, 28.2; N, 13.8; H, 3.2; O, 11.2; K, 4.2. Found: Zn, 22.33; Fe, 9.43; Cl, 9.11; C, 27.51; N, 14.11; H, 3.33; O, 11.4; K (by difference), 2.6. At a level of 0.023 weight % of this catalyst (based on monomer), 34% conversion of propylene oxide was obtained in 20 hours at 30°C; $[\eta]^{25}_{benzene} = 9.0$ dl/g.

PREPARATION OF ZINC–HEXACYANOCOBALTATE(III)-BASED CATALYST. A solution of 88 grams of calcium hexacyanocobaltate(III) (0.265 mole) in 640 ml of water and 32 ml of glyme (1,2-dimethoxyethane) was added to a solution of 70 grams of $ZnCl_2$ (0.51 mole) in 80 ml of water and 160 ml of glyme. This was diluted to 1850 ml with distilled water and treated in a Webcell continuous dialyzer (Brosites Machine Co.) with a 25% solution of glyme in water for 15 hours. The liquid was recycled through cationic and anionic ion exchange resin (Rohm and Haas A-15 and A-21, respectively). Centrifugation of the deionized slurry and vacuum drying of the solid residue gave 92 grams of a fluffy white powder having an apparent density of 0.33 g/ml. Anal.: Calcd. for $Zn_3[Co(CN)_6]_2 \cdot 2.4$glyme$\cdot 0.85$ $ZnCl_2 \cdot 4.4$ H_2O: Zn, 24.3; Co, 11.4; C, 25.0; N, 16.2; O, 14.2; Cl, 5.8; H, 3.2. Found: Zn, 25.0; Co, 11.7; C, 24.7; N, 16.1; O, 14.2; Cl, 5.9; H, 3.0; Ca, 0.2. At a level of 0.005% of this catalyst, based on monomer, 36% conversion of propylene oxide was obtained in 24 hours at 25°C; $[\eta]^{25}_{benzene}$ = 6.50 dl/g.

An alternative method of preparing highly active zinc hexacyanocobaltate catalyst involved the use of hexacyanocobaltic acid prepared by passing the potassium salt through cationic exchange resin (Rohm and Haas A-15).

POLYMERIZATION TECHNIQUES. Small-scale experiments were carried out in citrate bottles with perforated caps lined with rubber and Teflon. The catalyst and other solid ingredients, when used—for example, maleic anhydride—were weighed in air and transferred to the bottle. The bottle was then evacuated and pressured with nitrogen to 1 atm. Propylene oxide and other liquid ingredients were then admitted through the perforated caps. Thereafter, the caps were usually changed with nitrogen flushing to perforated caps having unbroken Teflon seals. During polymerization, the bottles were rotated in a constant-temperature bath. As a result of very fast polymerization rates immediately after the induction period, a number of heat shock explosions occurred in soft-glass bottles. This was particularly true in preparations of high-molecular-weight polymers and polyols. When the catalyst was adjusted to the amount required to give conversions of 40% or less in 24 hours, no explosions occurred in soft-glass bottles. However, it was found advisable to use borosilicate glass bottles when testing new catalysts or monomer combinations, even with as little as 15 grams of propylene oxide in a 12-oz bottle.

Larger-scale experiments were done in 1-gallon and 5-gallon jacketed stainless-steel stirred autoclaves. The catalyst is these cases was usually charged as a slurry with solvent or, in some cases, with solutions of the monomer and initiator (such as diol or triols). A violent explosion occurred in attempting to charge a mixture of propylene oxide and maleic acid from a bottle to the reactor. Unlike fumaric or phthalic acids, which have to be heated to react with propylene oxide, maleic acid reacts rapidly with propylene oxide within 10 minutes at room temperature.

STEPWISE SOLUTION POLYMERIZATION OF PROPYLENE OXIDE TO HIGH-MOLECULAR-WEIGHT POLYMER (see Figure 6). Propylene oxide (150 grams), tetrahydrofuran (1800 grams), phenyl-β-naphthylamine (0.1 gram), and zinc hexacyanocobaltate–glyme–zinc chloride catalyst (0.30 gram) were mixed in a 1-gallon autoclave, and they reacted at 50°C for 19 hours. At this point, conversion to polymer was 99%. A series of samples was withdrawn then, and monomer additions were made as indicated here:

Product removed, g	Propylene oxide added, g	Additional time, hr at 50 °C	Conversion [a], per cent
650	100	4	82
330	75	17	100
241	58	4	84
194	46	16.5	100

[a] Of added monomer plus the monomer remaining from the previous period

A similar run was made using 600 grams of benzene initially instead of tetrahydrofuran. The reaction began in a very short time at 50°C and within the first hour, 2400 grams more benzene were added. After a total of two hours, 90% conversion had occurred. Again a series of withdrawals and monomer additions was made:

Product removed, g	Propylene oxide added, g	Additional time, hr at 50°C	Conversion, per cent
1050	100	4	54
556	75	6.5	47
383	58	10	85
300	50	5	69

PREPARATION OF POLY(PROPYLENE ETHER) TRIOL. A charge consisting of 1500 grams of propylene oxide, 4.0 grams of zinc hexacyanocobaltate, and 1022 grams of 1,2,3-tri(hydroxypropoxy)propane was placed in a 5-gallon autoclave. Heating this mixture for three hours at 77°C resulted in a slow reduction of the initial pressure from 3.0 to 1.0 atm. An additional 9900 grams of propylene oxide were then charged continuously over four hours at the same temperature; at the end of that time, the pressure was 1.2 atm. After heating for an additional three hours, the pressure had dropped to 0.93 atm. The resulting triol had a viscosity of 613 cP at 27°C, and a hydroxyl content of 0.09 mmole/g. The expected hydroxyl content from the charge was 0.93 mmole/g.

PREPARATION OF POLY(PROPYLENE MALEATE). A charge of 4900 grams of maleic anhydride, 510 grams of fumaric acid, 6.0 grams of zinc hexacyanocobaltate–zinc chloride–glyme complex catalyst, and 1800 grams of propylene oxide was placed in a 5-gallon autoclave. This mixture was heated at 77°C for two hours and an additional 1860 grams of propylene oxide then added over the next two hours. After an additional hour at 77°C, the temperature was raised to 104°C for two hours. The product of this reaction had a hydroxyl content of 1.07 mmoles/g, and an acid content of 0.004 mmole/g. The expected end-group content from the charge was 0.98 mmole/g.

DETERMINATION OF DIISOCYANATE EQUIVALENTS OF DIOLS. Fifty grams of the diol, 0.017 gram of stannous octoate, and variable amounts of toluene-2,4-diisocyanate were added by syringe to 4-oz screw-capped cans equipped with butyl-rubber gaskets and Teflon liners. The cans were rotated at room temperature for 20 minutes to obtain uniform mixing and then heated for 40 hours at 120°C. Williams plasticity measurements were made on samples cut from the center of the resulting rubber cakes. (The Williams plasticity is the height in mils of a 1-gram sample of material after a three-minute period under a 5-kg load.) The maximum value in this test could be obtained by closer and closer approach to the equivalence point, but in practice is accomplished by extrapolation. Before extension, the experimental and newly obtained commercial diols were dried under vacuum for four hours at 100°C with vigorous stirring and then stored in stainless-steel containers under nitrogen. The values shown in Figure 10 for the commercial diols were the highest of those tested. The commercial polyols used were from Union Carbide (PPG-2025, -3025), Dow Chemical (P-2000, -3000), Wyandotte Chemical (P-2010), Jefferson Chemical (PPG-2000, -3000), and Olin Mathieson (G-2020-P, -3030-PG, -4030-PG).

Results and Discussion

Discovery. These catalysts were discovered during a study of the use of transition metal cyanides in combination with metal alkyl and hydride reducing agents in polymerizations. The combination of nickel cyanide and lithium aluminum hydride complexed very strongly with tetrahydrofuran. A similar complexing action occurred with propylene oxide and nickel hexacyanoferrate(II)–lithium aluminum hydride. This led to speculation as to the role of the double-metal cyanide itself.

Reasoning from the success of Price (*16*) and others with zinc compounds as catalysts for epoxide polymerizations, the zinc hexacyano-

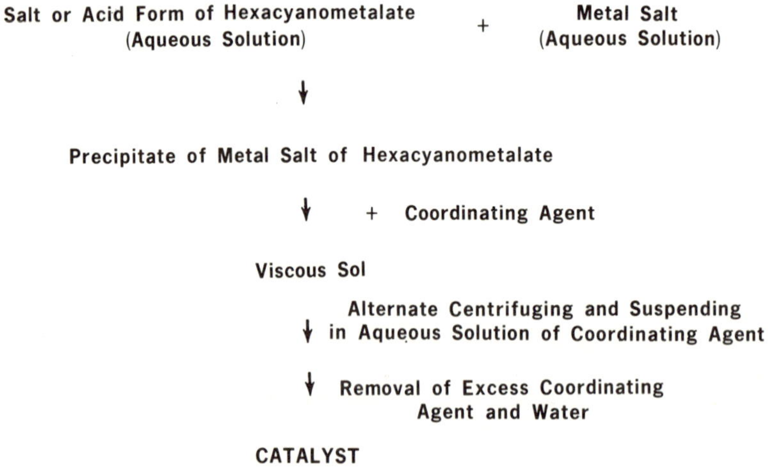

Salt or Acid Form of Hexacyanometalate   +   **Metal Salt**
(Aqueous Solution)   (Aqueous Solution)

↓

Precipitate of Metal Salt of Hexacyanometalate

↓   +   Coordinating Agent

Viscous Sol

↓   Alternate Centrifuging and Suspending
    in Aqueous Solution of Coordinating Agent

↓   Removal of Excess Coordinating
    Agent and Water

CATALYST

Figure 1. Generalized scheme for preparation of hexacyanometalate salt complex catalysts

ferrates (II, III) were among the first to be prepared and tried. Since these salts were precipitated from aqueous solutions, removal of water was necessary. This was first done by washing with acetone and alcohol and then heating in flowing nitrogen. Early work showed that excessive heating (above 60°C) degraded the catalytic action of some of these materials. However, washing with acetone and certain other electron donors has a great activating effect. This discovery was thus the result of a fortuitous choice of laboratory technique.

Preparation and Properties of Hexacyanometalate Salt Complexes. A generalized scheme for preparing hexacyanometalate salt complexes is given in Figure 1. Catalysts prepared this way varied from translucent gels to fluffy powders. They generally had low apparent densities, down to a minimum of 0.08 g/ml. X-ray analysis showed the catalysts to be largely amorphous with low-intensity crystalline peaks shifted only slightly from those of the pure salts.

Many compositional variations give complexes with catalytic activity, some of which are shown in Table I.

Table I. Examples of Catalyst Component Variations

Components	Orders of Decreasing Catalyst Activity
Cations	$Zn^{2+} > Fe^{2+} \sim Co^{2+} \sim Ni^{2+}$
Anions	$Co(CN)_6^{3-} > Fe(CN)_6^{3-,2-} > Cr(CN)_6^{3-}$
Coordination Agents	Glyme $\sim$ 1,3-Dimethoxypropane > Dioxane $\sim$ Acetone
Activating Salts	$ZnCl_2 > ZnBr_2 > Zn(NO_3)_2 \sim ZnSO_4$

The zinc cation gives by far the most active catalyst. Iron, cobalt, and nickel cations also gave salts with considerable catalytic activity. Cadmium, because of its chemical similarity to zinc, and aluminum, because of its use in other epoxide polymerization catalysts, were considered as likely candidates to give active catalysts. However, complexes of the salts of these cations were only slightly catalytic. The salts used as cation sources in catalyst preparations also affected catalytic activity. Zinc salts, especially zinc chloride and zinc bromide, were retained in considerable amounts in the finished complexes, and the use of these salts gave the most active catalysts.

Hexacyanometalate anions containing cobalt(III) and a combination of iron(II) and iron(III) gave the most active catalysts. The particular hexacyanometalate compound used as a precipitant often had an effect on the catalyst preparation. Potassium coprecipitated with the zinc salts and use of potassium hexacyanometalate resulted in less-active catalysts. This effect was reduced by using relatively dilute solutions of the reagents. Calcium hexacyanometalate salts did not cause such problems. Use of the hexacyanometalic acids was complicated by differences in their stabilities.

The preferred methods of preparation of catalysts based on the hexacyanocobaltate and hexacyanochromate ions were with the respective acids. However, because of the instability of the acid of the latter ion, successful use depended upon using it immediately after preparation. No satisfactory preparation of a catalyst using the hexacyanoferric acid was developed.

Shown here are typical compositions of two highly active catalysts, one based on zinc hexacyanoferrate(II, III) and the other on zinc hexacyanocobaltate(III):

$$0.75 \ Zn_3[Fe(CN)_6]_2 \cdot 0.25 Zn_2Fe(CN)_6 \cdot 0.82 \ ZnCl_2$$
$$\cdot 1.1 KCl \cdot 2.3 \ \text{diglyme} \ 0.6 H_2O$$

$$Zn_3[Co(CN)_6]_2 \cdot 0.91 ZnCl_2 \cdot 2.21 \ \text{glyme} \cdot 4.29 H_2O$$

The first of these has particular value in making high-molecular-weight poly(propylene oxide), while the other is best for preparing polyether polyols and polyesters. The iron-containing catalyst was prepared using the potassium salt of hexacyanoferrate(II) anion. Partial reduction of the anion to hexacyanoferrate(II) was shown by chemical and spectroscopic evidence. Retention of zinc chloride in such large amounts in these materials appears to be associated with use of the coordinating agents. Repeated washings of such a catalyst with 10% glyme in water, a solution in which zinc chloride is readily soluble, did not reduce the zinc chloride content appreciably, whereas washing with water alone removed the zinc chloride.

No particular significance is attached to the fact that the molar amounts of glyme and diglyme in these two catalysts are almost the same. Extensive vacuum drying of the cobalt catalyst, which reduced the water level from 4.29 to less than 1.0 mole per mole of the zinc salt (and, presumably, reduced the glyme to a similar or larger extent), gave no appreciable effect on catalytic activity. This result emphasizes the nonstoichiometric nature of the catalytic forms of the hexacyanometalate salt complexes.

The infrared spectra of the zinc hexacyanometalate–glyme complexes were examined with the aim of observing differences associated with catalytic activity. The absorptions of most interest were those of the cyanide group at 2100-2200 cm^{-1}, glyme at 1300-850 cm^{-1}, the metal-carbon bond of the hexacyanometalate group at 500-600 cm^{-1}, and the zinc chloride bond at 350-300 cm^{-1}. Figure 2 shows the spectra of typical hexacyanoferrate and hexacyanocobaltate catalysts, together with the spectra of zinc hexacyanocobaltate made in the absence of glyme.

Some changes occur in all of the absorptions mentioned with catalysts of different activities. A study of the spectra of zinc hexacy-

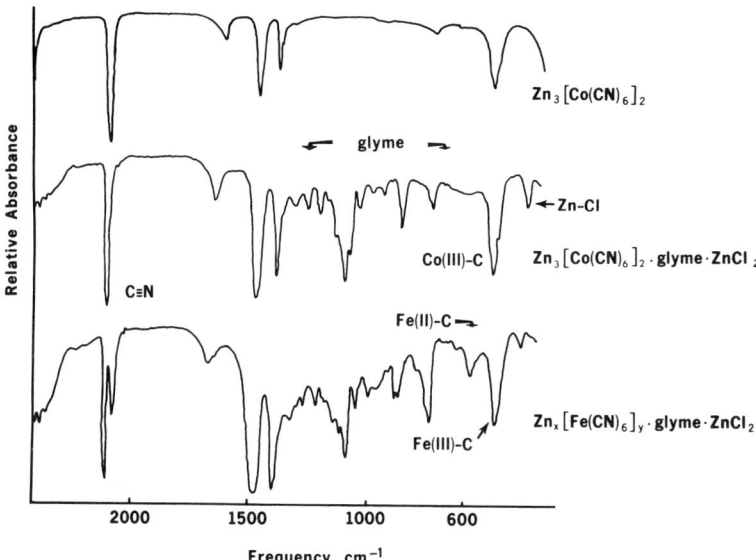

Figure 2. Infrared spectra of Nujol mulls of hexacyanometalate salt complexes

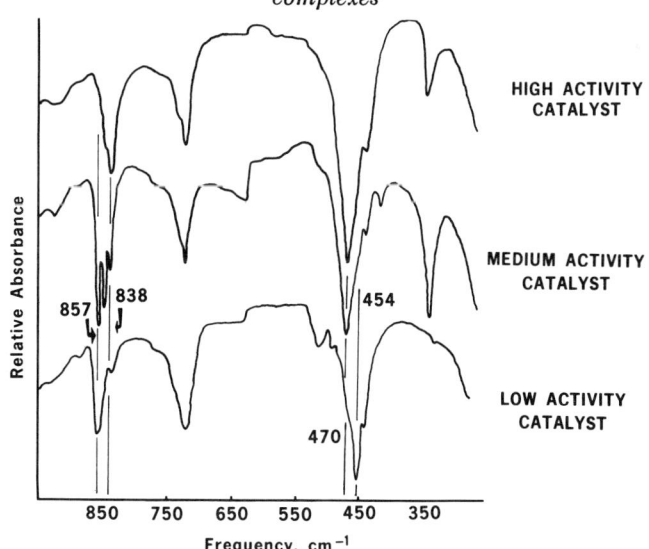

Figure 3. Infrared spectra of Nujol mulls of zinc hexacyanocobaltate · glyme complexes of different activities

anocobaltate–glyme complexes showed that changes in one of the absorptions in the region associated with glyme is the most clearly associated with activity. In liquid glyme, the strong absorption at 851 cm^{-1} has been assigned to the methylene rocking mode (*17*). Figure 3 shows that in highly active catalysts, this absorption was shifted to 838 cm^{-1}, while in the least active catalyst, it was at 857 cm^{-1}. A catalyst of intermediate activity showed both of these absorptions plus another at about 844 cm^{-1}. The 838 cm^{-1} absorption was observed only in highly active catalysts containing zinc chloride. However, the absorption at 857 cm^{-1} was present in inactive catalysts regardless of whether they contained zinc chloride. Figure 3 also shows a shift in the absorption caused by the cobalt-carbon bond from 450 to 470 cm^{-1}. This feature appears to be associated with the presence or absence of zinc chloride, and not specifically associated with activity.

It may be assumed that the glyme in these catalysts is coordinated to the zinc cations. The fact that the glyme absorptions vary suggests that changes occur in the environment of the zinc atoms, causing the variation. Coupling this with zinc chloride's being required for highest activity suggests that the glyme may be coordinated to zinc ions associated with both hexacyanocobaltate ions and chloride ions. The absorptions at 857, 844, and 838 cm^{-1} in the middle curve of Figure 3 show that there are at least three structures possible responsible for this spectral feature.

Qualitative differences in catalytic activities are referred to above. It is beyond the scope of this paper to present a detailed description of the interrelation of the differences in composition and preparative methods with catalytic activity. However, to give some perspective, results on the activation of one such catalyst by the combination of a coordinating agent and a salt is given in Table II.

Table II. Effects of Glyme and Zinc Chloride on Activity of Zinc Hexacyanocobaltate for Bulk Polymerization of Propylene Oxide at 30°C

Zinc Hexacyanocobaltate Complexed with:		Activity
$ZnCl_2$	Glyme	g Polymer/g Catalyst/20 hr
+	+	7000
+	−	800
−	+	500
−	−	300

Activities of the catalysts for polymerizing propylene oxide in bulk at 30°C were compared on the basis of the weight of polymer formed per weight of catalyst per unit time. Zinc hexacyanocobaltate itself was a fairly active catalyst, giving 300 grams of polymer per gram of catalyst in 20 hours. Under similar conditions, metal–alkyl–water catalysts gave yields of no more than 50 grams of polymer per gram of catalyst (R. J. Herold, unpublished data). Complexing with glyme and with zinc chloride raised the activity of zinc hexacyanocobaltate by factors of 1.7 and 2.7, respectively. However, with a combination of these two agents, the activity was raised 23-fold.

Preparation and Properties of High-Molecular-Weight Poly(propylene oxide). Figure 4 shows a typical conversion-time plot for polymerization of propylene oxide by a hexacyanometalate salt complex catalyst. This reaction is characterized by an initial period during which almost no conversion occurs, followed by a period of rapid polymerization. The initial period, termed the induction period, is highly

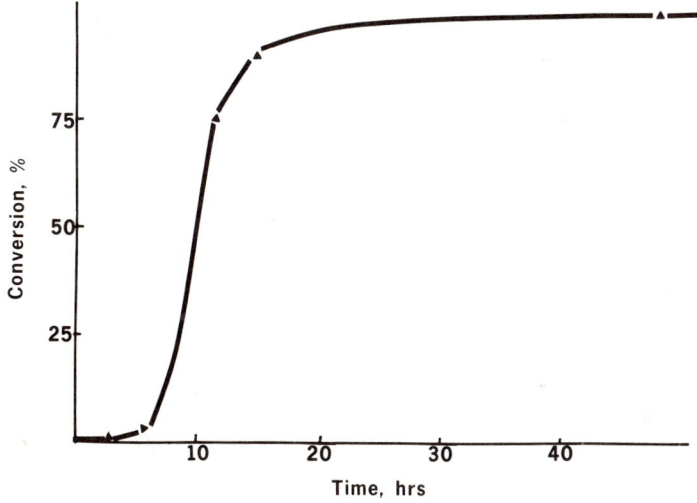

Figure 4. Typical time-conversion plot. Bulk polymerization of propylene oxide with $Zn_x[Fe(CN)_6]_y \cdot$ acetone $\cdot ZnCl_2$ (0.20 wt %) at 25°C

temperature-dependent. With the particular catalyst shown in Figure 4, the induction period varied from 18 hours at 10°C to six minutes at 60°C. The induction period is associated with a change in dispersion of the catalyst. These catalysts are insoluble in monomer and in all common organic solvents. When the catalysts were used in a bulk polymerization of propylene oxide, the particles initially settled out within a few seconds after agitation ceased. At the end of the induction period, the monomer-catalyst mixture appeared homogeneously cloudy, and no settling was observed for up to a minute. Soon after polymerization started, the reaction solution became clear.

When the amount of catalyst was less than that necessary to give complete conversion, the general shape of the reaction curve remained the same—that is, an induction period followed by rapid propagation up to a plateau of limited conversion. Table III shows the results of experiments with various concentrations of catalyst; in each of those experiments, a less than quantitative conversion was reached. Increases in the amount of catalyst produced proportionate increases in the yields. The fact that limiting conversions were reached is evidence for the presence of a termination process. That the intrinsic viscosities and, presumably, the molecular weights were relatively similar over a considerable variation of catalyst concentration suggests that the termination process was not caused by an impurity originally present in the monomer.

Table III. Bulk Polymerization of Propylene Oxide with Zinc Hexacyanocobaltate–Glyme-Zinc Chloride Catalyst at 30°C

Catalyst Concn, wt %	*% Conversion, 24 hr*	*Yield, g/g Catalyst*	$[\eta]^{25}_{Benzene}$ *dl/g*
0.0037	12	3060	9.1
0.0072	23	3060	10.3
0.012	37	3040	11.6
0.020	57	2840	11.4
0.055	94	1750	9.3

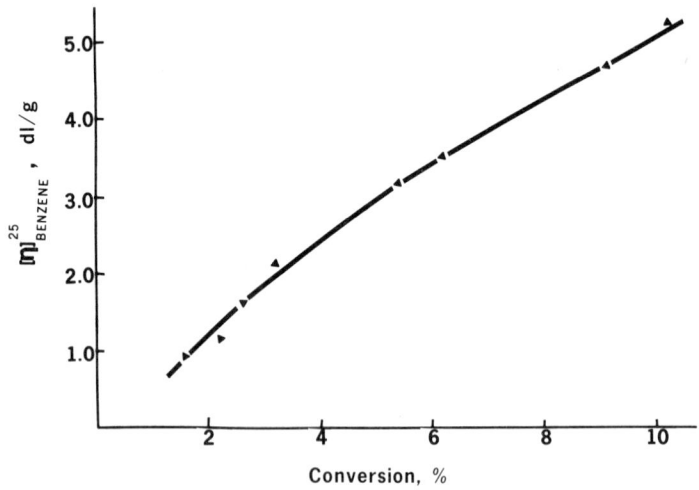

Figure 5. Variation of intrinsic viscosity with conversion during bulk polymerization of propylene oxide with $Zn_3[Co(CN)_6]_2 \cdot$ glyme $\cdot ZnCl_2$ (0.01 wt %) at 30°C

Figure 5 shows that the intrinsic viscosity increased regularly during the initial stages of a bulk polymerization. However, all of these polymers had broad distributions of molecular weight. To explain the increasing intrinsic viscosity, it is necessary to assume that some chains continue to grow. Additionally, either the chain-termination process mentioned or further initiation must be invoked to explain the continued high proportion of low-molecular-weight polymer.

Figure 6 shows the results of stepwise polymerizations carried out in tetrahydrofuran and benzene. In these cases, excesses of catalyst were used. Catalyst, solvent, and the first part of the propylene oxide were charged initially, and the polymerizations allowed to go to completion. Samples were taken, and a further amount of propylene oxide then added and allowed to polymerize. The latter procedure was repeated three times in each case. The experiment in tetrahydrofuran covered three days, while the experiment in benzene took about 30 hours.

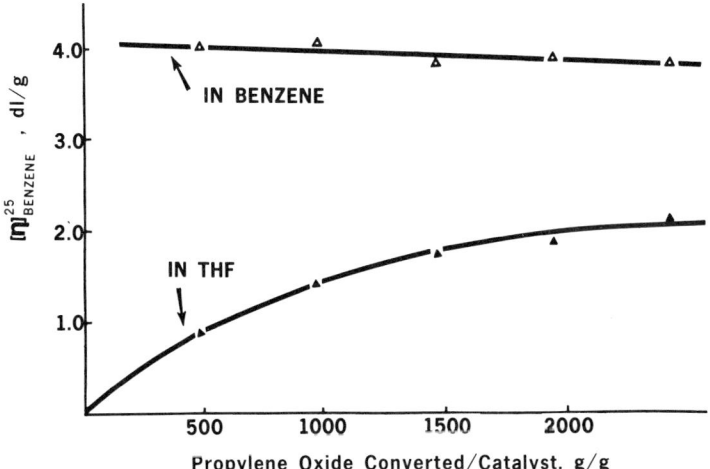

Figure 6. Variation of intrinsic viscosity with conversion in stepwise polymerization of propylene oxide in benzene and tetrahydrofuran with $Zn_3[Co(CN)_6]_2 \cdot glyme \cdot ZnCl_2$ at 50°C

The intrinsic viscosities of the polymers prepared in tetrahydrofuran increased throughout the experiment. This system thus exhibits some of the aspects of living polymerization—that is, catalyst activity over an extended period, and increasing viscosity average molecular weights with added amounts of monomer. The rather broad molecular-weight distributions of these polymers, however, differentiates this system from that of the classical case in which polymerization proceeds in the complete absence of a termination process.

In benzene, the intrinsic viscosity was high initially and did not change appreciably with further monomer additions. We speculate that in this nonpolar solvent, the catalyst remained aggregated to a greater extent than when used in bulk or in tetrahydrofuran. This would have led to the higher initial molecular weight observed as the result of the smaller ratio of catalyst sites to monomer molecules. To explain the constant intrinsic viscosity, it is assumed that a limiting molecular weight was reached, and that new chains were continually initiated.

Table IV. Properties of Cured [a] Copolymers of Propylene Oxide and Allyl Glycidyl Ether

Copolymer Prepared With:	$Zn_x[Fe(CN)_6]_y$ · acetone · $ZnCl_2$		$ZnEt_2$–H_2O	
	Room-Temperature Properties			
	Gum	Black	Gum	Black
Tensile, psi	450	2400	2490	2820
Elongation, %	595	745	885	670
Modulus, 300%, psi	135	925	135	1075
	Properties at 205°F			
Tensile, psi		1430		1450

[a] Curing recipe: Copolymer containing 3 mole % allyl glycidyl ether: 100 parts; HAF black: 0 or 40 parts per hundred of rubber (phr); ZnO: 5.0 phr; stearic acid: 2.0 phr; sulfur: 2.0 phr; mercaptobenzothiazole: 1.0; tetramethylthiuram disulfide: 1.0 phr

The high-molecular-weight poly(propylene oxide) produced with hexacyanometalate salt complexes shows no crystallinity. Moreover, it was shown by Price et al. (18) and confirmed in our laboratory that these polymers have more than 95% head-to-tail enchainment. The amorphous fractions of partially crystalline polymers made with metal-alkyl and ferric–chloride-based catalysts were shown by those authors to have considerable head-to-head enchainment. They postulated that this was the cause of the amorphous nature of these fractions. It seems clear, however, that the amorphous nature of the polymers prepared with hexacyanometalate salt complexes must be the result of their low degrees of tacticity.

Practical interest in high-molecular-weight poly(propylene oxide) centers in its potential use as an elastomer (19). Copolymerization of propylene oxide with allyl glycidyl ether gives a copolymer with double bonds suitable for sulfur vulcanization. Table IV shows the properties of elastomers made with a copolymer prepared with a zinc hexacyanoferrate–acetone–zinc chloride complex. Also shown are the properties of elastomers made from partially crystalline copolymers prepared with zinc diethyl–water catalyst. Of particular interest are the lower room-temperature tensile strengths of the rubber made with the hexacyanometalate salt complex, and the fact that the hot tensile strengths of the black stocks [above the 70°C melting point of crystalline poly(propylene oxide)] are the same. The amorphous nature of the copolymer made with the hexacyanoferrate complex is thus evident.

Preparation of Poly(propylene ether) Polyols. The polymerization of propylene oxide with zinc hexacyanocobaltate complexes in the presence of proton donors results in the production of low-molecular-weight polymers. Table V shows the variety of types of compounds that have been found to act this way. Since these compounds end up in the polymer chains, it seems reasonable to call them chain initiators. Thus, in essence, each of these compounds is activated by the catalyst to react with propylene oxide to form a hydroxylpropyl derivative. Thereafter, the reaction continues on the same basis, with the proton of the hydroxyl group reacting with further propylene oxide. This sequence is shown here with 1,5-pentanediol as the initiator. The hydroxyl

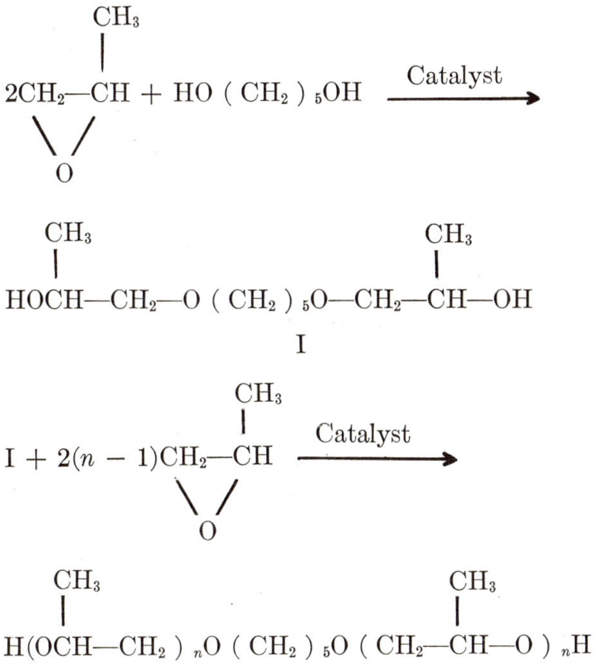

$$2CH_2\overset{\overset{\displaystyle CH_3}{|}}{-}CH + HO(CH_2)_5OH \xrightarrow{\text{Catalyst}}$$
$$\underset{O}{\diagdown\diagup}$$

$$HO\overset{\overset{\displaystyle CH_3}{|}}{C}H-CH_2-O(CH_2)_5O-CH_2-\overset{\overset{\displaystyle CH_3}{|}}{C}H-OH$$
$$I$$

$$I + 2(n-1)CH_2\overset{\overset{\displaystyle CH_3}{|}}{-}CH \xrightarrow{\text{Catalyst}}$$
$$\underset{O}{\diagdown\diagup}$$

$$H(O\overset{\overset{\displaystyle CH_3}{|}}{C}H-CH_2)_nO(CH_2)_5O(CH_2-\overset{\overset{\displaystyle CH_3}{|}}{C}H-O)_nH$$

groups were determined by Siggia's acetylation method (20) to be predominantly (>95%) secondary as indicated.

It should be emphasized at the outset of the description of these reactions that the molar amount of initiator is greatly in excess of the

Table V. Initiators for Polymerization with Hexacyanometalate Salt Complex Catalysts

Water	Alcohols
Mercaptans	Carboxylic Acids
Ketones	Aromatic Amines
Malonic Esters	Acetylenic Compounds
Caprolactam	

molar amount of zinc in the catalyst. For example, in the preparation of 2000-molecular-weight diol using 1,5-pentanediol as the initiator, there are about 1000 equivalents of hydroxyl end groups for each gram atom of zinc in the catalyst. Thus, whatever the structure or the composition of the active site, it seems certain that each site must become involved with many different molecules during the reaction if the moles of initiator used are equivalent to the moles of resulting polymer. In this sense, the reaction involves chain transfer.

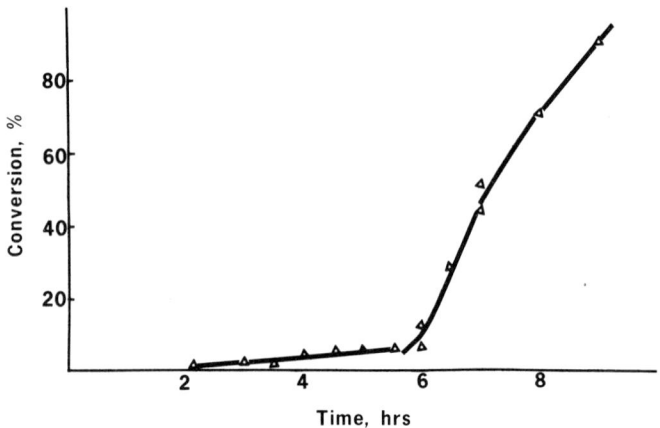

Figure 7. Time-conversion plot for 1,5-pentanediol initiated preparation of poly(propylene ether) diol in pentane (propylene oxide/pentane wt ratio = 3) with $Zn_3[Co(CN)_6]_2 \cdot glyme \cdot ZnCl_2$ at 50°C

The rate of formation of the low-molecular-weight polymer using 1,5-pentanediol as the initiator is shown in Figure 7. As in the preparation of the high-molecular-weight polymer, there was an initial slow reaction followed by a rapid one. Completion of the first part of the reaction was, in this case, characterized by disappearance of the initiator (21), as well as by dispersion of the catalyst. When this reaction was carried to complete consumption of monomer, addition of further monomer resulted in immediate resumption of the rapid rate. Once again, we see a similarity to a living polymerization. Both the rapid rate obtained in the first case and that caused by addition of further monomer showed a first-order dependence on monomer concentration up to relatively high conversions. The apparent activation energy of polymerization carried out in this manner was 13.7 kcal/mole.

Having demonstrated that the polymerization could be continued by incremental monomer addition, studies of the reaction products of such experiments were made. The results of one such study are shown in Table VI. It shows a comparison of the observed and expected number-average molecular weights for an experiment involving two incre-

Table VI. Comparison of Calculated and Observed Molecular Weights of Poly(propylene ether) Diols

Sample	$\overline{M}_n(VPO)$	$\overline{M}_n(Calcd.)$ [a]
9410–IX06–5B	1690	1783
9410–IX–6–5C	2350	2209
9410–IX–6–5A	3130	3049

[a] $\overline{M}_n(Calcd.) = \dfrac{\text{Wt of polyetherdiol obtained}}{\text{Moles 1,5-pentanediol charged}}$

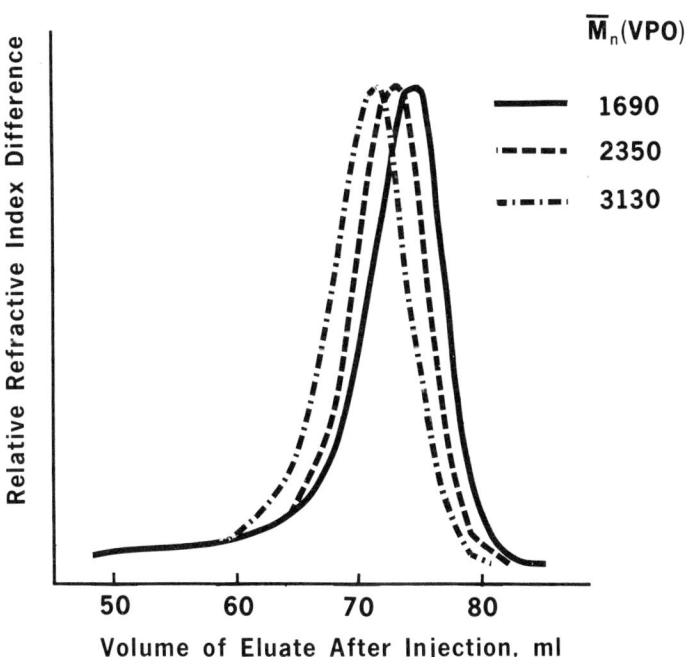

Figure 8. Gel-permeation chromatograms of poly(propylene ether) diols prepared stepwise with $Zn_3[Co(CN)_6]_2 \cdot glyme \cdot ZnCl_2$

mental additions of monomer after the initial polymerization. The molecular weights at each stage show that the number of moles of polymer chains is equivalent to the moles of 1,5-pentanediol charged. From these data, we conclude no significant number of chains were formed directly from the catalyst alone or by side reactions. Strong-base catalysis of propylene oxide polymerization has been shown to cause side reactions with the monomer to give products capable of forming additional chains (22).

Figure 8 shows the characterization of these poly(propylene ether) diols by gel-permeation chromatography. There is a shift in the peak position to lower elution volumes, in accord with an increase in molecular weight with each monomer increment polymerized. The molecular-weight distributions of the three diols are similar and remain narrow after the addition of monomer increments. Since all of the molecules apparently continued to grow, this polymerization must proceed with very little chain termination under these conditions.

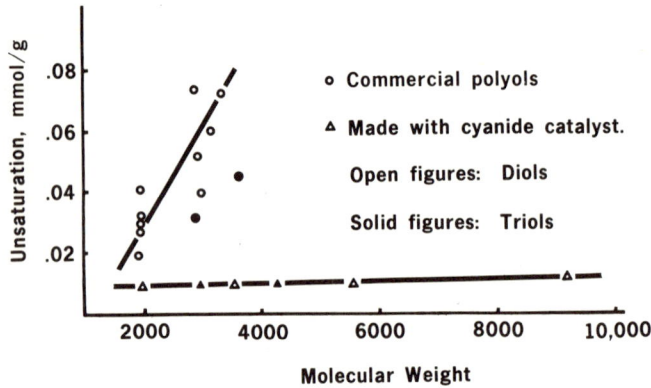

Figure 9. Unsaturation vs. molecular weight for poly(propylene ether) polyols made with $Zn_3[Co(CN)_6]_2 \cdot glyme \cdot ZnCl_2$ and made commercially

The point made earlier concerning the activation of many polymer chains by a single catalyst site can now be carried one step further. Not only must each catalyst site "service" a large number of polymer chains, but these data make it apparent that only a few monomer units at most can be added during each of these polymer-catalyst associations. In the preparation of a 2000-molecular-weight diol, 17 monomer units must be added per hydroxyl. Assuming the addition of three monomer units per association, about 10 chain transfers would occur for the completion of each molecule.

The major uses of poly(propylene ether) polyols involve reaction of their hydroxyl groups with diisocyanates to form foams, adhesives, coatings, and elastomers. Polyols prepared with the zinc hexacyanocobaltate complex using polyhydroxyl initiators had functionalities (average number of hydroxyl groups per polymer molecule) very close to those of the initiators. This was determined by the correspondence of the molecular weights as calculated from hydroxyl end-group analysis and as measured by vapor pressure osmometry, measuring the amounts of unsaturated end groups (the major alternative end group), and following the extension reaction of these polyols with diisocyanates.

Figure 9 shows that the unsaturation of poly(propylene ether) diols and triols remained constant as the molecular weight was varied. By comparison, the unsaturation of commercial polyols (presumably made with potassium hydroxide) increased appreciably as the molecular weight increased. Figure 10 shows that upon varying the amount of diiso-

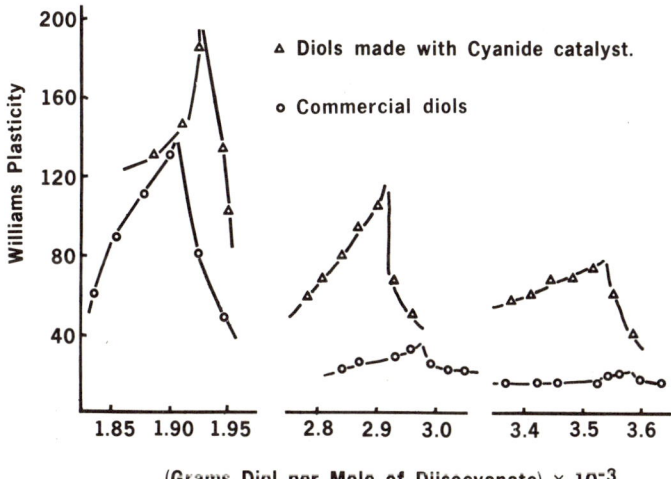

Figure 10. Variation of Williams plasticity of polyurethanes made with diols of various molecular weights; cyanide catalyst used: $Zn_3[Co(CN)_6]_2 \cdot glyme \cdot ZnCl_2$

cyanate, the Williams plasticity of the extended products made with these diols goes through a maximum. The height of this maximum is a relative measure of the functionality of the diol when compared with other diols of the same molecular weight and chemical structure. Extensions of commercial diols with the same hydroxyl end-group content as the experimental diols have lower maxima. These results demonstrate the relative lack of side reactions in the preparation of poly(propylene ether) polyols using the zinc hexacyanocobaltate complex catalyst.

Copolymerization of Epoxide and Cyclic Anhydrides. The zinc hexacyanocobaltate complex is also an efficient catalyst for copolymerizing epoxides and cyclic anhydrides. Here again, relatively high-molecular-weight copolymers were obtained without intentionally added initiators. The highest molecular weights obtained were about 50,000. Whether this limitation stemmed from the nature of the reaction or from the difficulty in preparation of anhydrides in the purity required for making still-higher-molecular-weight copolymers is not known. Copolymerization in the presence of added initiators gave copolymers of the molecular weights expected from the amounts of initiators and monomers used.

During the preparation of these polyesters, considerable concentrations of carboxyl groups are probably present. This emphasizes the fact that this type of catalyst is compatible with an acid environment. As discussed above, carboxylic acids as well as the weakly basic aromatic amines were suitable initiators. However, more basic aliphatic amines acted as inhibitors of these catalysts.

Mode of Action of the Catalysts. No detailed mechanism of the action of these catalysts has been developed. There are, however, two groups of related observations that may be of great significance in arriving at a mechanism.

The first of these relates to the differences between reactions catalyzed by these materials in the presence and absence of hydroxyl compound initiators. Polymerization with added amounts of such initiators proceeds slower than without them at the same catalyst concentration. This is in contrast to polymerization of epoxides and to copolymerization of epoxides and anhydrides using potassium hydroxide as the catalyst. In the latter cases, little or no reaction occurs in the absence of an initiator. Hydroxyl compounds also retard the activation of the catalyst; that is, the induction periods are longer in their presence. Finally, the molecular-weight distribution of the polymer prepared without the initiators is much broader than that of the polymers made with the initiators.

It can be assumed that without added initiator, adventitious amounts of water acts as the initiator. Thus the reaction can be regarded generally as a process involving activation of catalyst, addition to the initiator of a single monomer unit to form a hydroxypropoxy derivative, and a propagation reaction of the latter product and monomer (Equations 1-3). It must be assumed further that an equilibrium, Equation 4, exists between excess initiator and the activated catalyst-polymer species. This is required because the number of growing chains in the preparation of polyols far exceeds the potential number of catalyst sites, and because as the molecular weight of the polyol increases, its molecular

$$\text{Cat} + \text{XH or } \underset{O}{\overset{CH_3}{\underset{|}{CH_2-CH}}} \longrightarrow \text{XH} \cdot \text{Cat}^* \text{ or } (\underset{O}{\overset{CH_3}{\underset{|}{CH_2-CH}}}) \cdot \text{Cat}^* \quad (1)$$

$$\text{XH} \cdot \text{Cat}^* + \underset{O}{\overset{CH_3}{\underset{|}{CH_2-CH}}}$$

$$\searrow$$

$$X(CH_2-\overset{CH_3}{\underset{|}{CH}}-O)H \cdot \text{Cat}^* \quad (2)$$

$$\nearrow$$

$$(\underset{O}{\overset{CH_3}{\underset{|}{CH_2-CH}}}) \cdot \text{Cat}^* + \text{XH}$$

$$X(CH_2-\overset{CH_3}{\underset{|}{CH}}-O)H \cdot \text{Cat}^* + (n-1)\underset{O}{\overset{CH_3}{\underset{|}{CH_2-CH}}} \longrightarrow$$

$$X(CH_2-\overset{CH_3}{\underset{|}{CH}}-O)_n H \cdot \text{Cat}^* \quad (3)$$

$$X(CH_2-\overset{CH_3}{\underset{|}{CH}}-O)_n H \cdot \text{Cat}^* + X'H \rightleftarrows$$

$$X'H \cdot \text{Cat}^* + X(CH_2-\overset{CH_3}{\underset{|}{CH}}-O)_n H \quad (4)$$

weight distribution remains narrow. In this case, X′H is usually another hydroxy-terminated polymer molecule. In the event the equilibrium is rapid with respect to the rate of Equation 3, a narrow-molecular-weight distribution would result.

However, the preceding sequence is inadequate to explain the higher rates in the absence of initiator. A plausible mechanism to explain this effect would be that excess initiator forms a complex with the activated chains so that the epoxide must displace it before reaction (Equations 5 and 6).

$$X(CH_2-\underset{\underset{CH_3}{|}}{CH}-O)_nH \cdot Cat^* + X'H \longrightarrow$$

$$[X(CH_2-\underset{\underset{CH_3}{|}}{CH}-O)_nH \cdot Cat^* \cdot X'H] \quad (5)$$

$$[X(CH_2-\underset{\underset{CH_3}{|}}{CH}-O)_nH \cdot Cat^* \cdot X'H] + CH_2-\underset{\underset{CH_3}{|}}{CH} \longrightarrow$$
$$\underset{O}{\diagdown\diagup}$$

$$[X(CH_2-\underset{\underset{CH_3}{|}}{CH}-O)_nH \cdot Cat^* \cdot CH_2-\underset{\underset{CH_3}{|}}{CH}] + X'H \quad (6)$$
$$\underset{O}{\diagdown\diagup}$$

We have no direct evidence for this type of behavior. However, it appears reasonable to expect that hydroxyl oxygens would compete with ether oxygens for coordination. Thus, displacement of the hydroxyl complexes, Equation 6, might be rate controlling.

The second group of observations concerns data presented that provide insight into the nature of the catalyst sites. The catalysts are active in acidic media, and inhibited by basic compounds. This suggests that the active sites are electron-accepting. However, the chains are exclusively head-to-tail and the hydroxyl groups are predominantly secondary. Cationic catalysis tends to give random enchainment, as evidenced by the work of Price and co-workers (18), which showed that such catalysts give some head-to-head enchainment. Thus the sites, although electron-accepting, do not act strongly cationic and may be coordinate cationic. As suggested by infrared analysis, the zinc ion is the most likely locus for this coordination.

Acknowledgment

The authors are indebted to many past and present colleagues of General Tire's Research and Development Division. In addition to those mentioned in the references, we wish to acknowledge the help and encouragement of R. A. Briggs in all phases of this effort; I. G. Hargis and R. E. Bingham for many helpful discussions; E. F. Kalafus and R. J. Emerson for compounding and evaluating the unsaturated copolymer; P. T. Suman and J. R. Adams for interpreting infrared spectra; O. C. Elmer and J. S. Duncan for evaluating the functionality of the polyether polyols; and J. L. Cowell, T. C. Neubert, and J. A. Wilson for assisting in the experimental work.

Literature Cited

1. Belner, R. J., Herold, R. J., Milgrom, J., U.S. Patents **3,427,256, 3,427,334** and **3,427,335** (1969).
2. Belner, R. J., Herold, R. J., Milgrom, J., U.S. Patents **3,278,457, 3,278,458** and **3,278,459** (1966).
3. Netherlands Application **6,614,355**; German Application **1,803,383**.
4. Herold, R. J., U.S. Patent **3,538,053** (1970).
5. Fife, H. R., Robert, F. H., British Patent **601,604** (1948).
6. Hayes, R. A., U.S. Patent **2,822,350** (1958).
7. Tsuruta, T., Matsuura, K., Inoue, S., *Makromol. Chem.* (1965) **83**, 289.
8. Furukawa, J., Tsuruta, T., Saegusa, T., Sakata, R., Kakogawa, G., Kawasaki, R., Harada, I., *J. Chem. Soc. Japan, Ind. Chem. Soc.* (1959) **62**, 1269.
9. Pruitt, M. E., Baggett, J. M., U.S. Patents **2,706,181** and **2,706,189** (1955).
10. Osgan, M., Price, C. C., *J. Polym. Sci.* (1959) **34**, 153.
11. Vandenberg, E. J., *J. Polym. Sci.* (1960) **47**, 486.
12. Lal, J., U.S. Patent **3,345,308** (1967).
13. Furukawa, J., Tsuruta, T., Sakata, R., Saegusa, T., Kawasaki, A., *Makromol. Chem.* (1959) **32**, 90.
14. Herold, R. J., Aggarwal, S. L., Neff, V., *Can. J. Chem.* (1963) **41**, 1368.
15. Booth, C., Higginson, W. C. E., Powell, E., *Polymer* (1964) **5**, 479.
16. Price, C. C., Osgan, M., *J. Amer. Chem. Soc.* (1956) **78**, 4787.
17. Iwamoto, R., *Spectrochim. Acta* (1971) **27A**, 2385.
18. Price, C. C., Spector, R., Tumolo, A. L., *J. Polym. Sci.* (1967) **A1 5**, 407.
19. Gruber, E. E., Meyer, D. A., Swart, G. H., Weinstock, K. V., *Ind. Eng. Chem., Prod. Res. Devel.* (1964) **3**, 194.
20. Siggia, S., Hanna, J. G., *Anal. Chem.* (1961) **33**, 896.
21. Pavelich, W. A., Livigni, R. A., *J. Polym. Sci.* (1968) **C-21**, 215.
22. Simons, D. M., Verbanc, J. J., *J. Polym. Sci.* (1960) **44**, 303.

RECEIVED April 13, 1972.

16

Preparation and Properties of Poly(arylene oxide) Copolymers

GLENN D. COOPER and JAMES G. BENNETT, JR.

Plastics Department, General Electric Co., Selkirk, N.Y. 12158

ARNOLD FACTOR

Research and Development Center, General Electric Co., Schenectady, N.Y. 12301

> *Poly(arylene oxide) copolymers were prepared by simultaneous and sequential oxidation of 1:1 mixtures of 2,6-dimethylphenol (DMP), 2-methyl-6-phenylphenol (MPP), and 2,6-diphenylphenol (DPP), and methods were developed for determination of their structure. DMP and DPP yielded either random copolymers or block copolymers with crystallizable DMP and DPP blocks, depending on the order of oxidation and reaction conditions. Four types of copolymers were produced from MPP and DPP: random copolymers, block copolymers with crystallizable DPP blocks, "short block" copolymers with DPP segments too short to permit crystallization, and "mixed block" copolymers containing DPP blocks and randomized MPP-DPP segments. Redistribution is so facile in the DMP-MPP system that only random copolymers were obtained, even on oxidation of a mixture of the two homopolymers.*

The reaction of 2,6-disubstituted phenols with oxygen in the presence of a suitable catalyst yields poly(1,4-arylene oxides). For example:

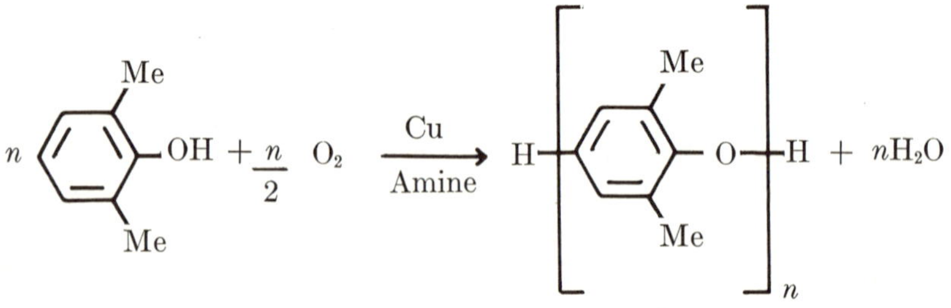

$$n \; \text{[2,6-Me}_2\text{C}_6\text{H}_3\text{OH]} + \frac{n}{2} \text{O}_2 \xrightarrow{\text{Cu} \atop \text{Amine}} \text{H}\!-\!\!\left[\text{[2,6-Me}_2\text{C}_6\text{H}_2\text{O]}\right]_n\!\!-\!\text{H} + n\text{H}_2\text{O} \quad (1)$$

I

This reaction has been actively studied since it was first reported by Hay in 1959 (*1*), but most of the extensive literature, which includes several recent reviews (*2-8*), deals primarily with the complex polymerization mechanism. Few copolymers have been prepared by oxidative coupling of phenols, and only one copolymer system has been examined in any detail. Copolymers of 2,6-dimethylphenol (DMP) and 2,6-diphenylphenol (DPP) have been prepared and the effect of variations in polymerization procedure on the structure and properties of the copolymers examined (*4, 9*); this work has now been extended to copolymers of each of these monomers with a third phenol, 2-methyl-6-phenylphenol (MPP). This paper presents a study of the DMP-MPP and MPP-DPP copolymers and a comparison with the DMP-DPP system previously reported.

Reactivity of Phenols

The reactivity of each of the phenols in homopolymerization was determined by following the rate of oxygen absorption in a closed system. In each case, a plot of oxygen absorption against time was linear over at least 80% of the total reaction. Measurements were made at 25°C with a cuprous chloride–pyridine catalyst and at 60°C with a more active catalyst, cuprous bromide–tetramethylethylenediamine (TMEDA). Relative rates, from the slope of the linear portion of the oxygen absorption curves, are summarized in Table I. DMP is about 30 times more reactive than DDP at 25°C and five times more reactive at 60°C. MPP is intermediate in reactivity (as expected from its structure) at both temperatures but is comparable at the lower temperature with DMP and at 60°C with DPP (about a third slower than DMP at 25°C and 50% faster than DPP at 60°C).

Table I. Relative Reactivity of Phenols in Homopolymerization

Phenol	Reactivity	
	25°C, Py–CuCl	*60°C, CuBr–TMEDA*
DMP	1.0	1.0 [a]
MPP	0.67	0.34
DPP	0.033	0.22

[a] Product largely diphenoquinone rather than polymer

The disappearance rate of each monomer on oxidation of an equimolar mixture of phenols is shown for the three monomer pairs in Figure 1. In each pair, the phenol more reactive in homopolymerization is more rapidly consumed. Even for the DMP–DPP and MPP–DPP pairs, however, where the phenols differ in reactivity by a factor of 20 or 30, the less reactive phenol is consumed much more rapidly than would be expected on the basis of its reactivity because of redistribution between phenols and low oligomers. This helps remove the less reactive phenol and return the more reactive phenol to the system. For example:

$$\text{(2)}$$

The redistribution is a very facile process and largely determines the structure of poly(arylene oxides) prepared by the oxidative coupling reaction. A mixture of a high-molecular-weight polymer with a phenol may be converted, in presence of an initiator capable of generating aryloxy radicals, to a mixture of low oligomers, each having a terminal unit corresponding to the added monomer (10):

$$\text{ArOH} + \text{H}-[]_n-\text{H} \longrightarrow$$

$$\text{ArO}-[]_m-\text{H} \qquad m = 1, 2 .. \qquad (3)$$

Homopolymer Properties

Structural assignment of the copolymers was based largely on their NMR spectra, but thermal and solubility characteristics were sometimes useful, particularly in distinguishing block copolymers from homopoly-

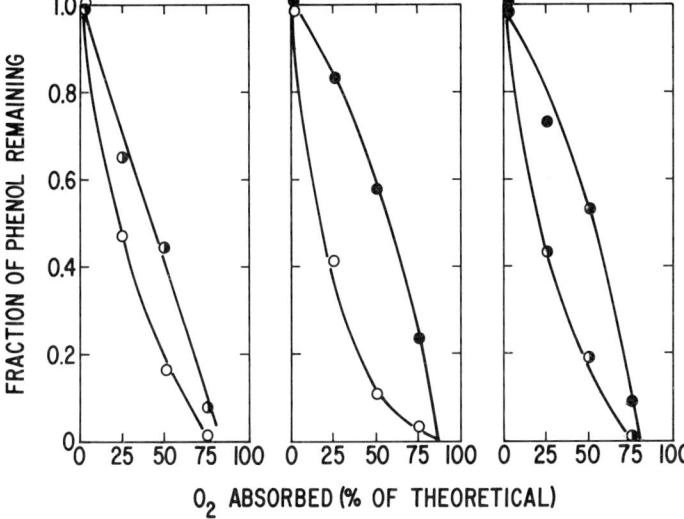

○ DMP
◐ MPP
● DPP

Figure 1. Disappearance of phenols on oxidation of equimolar mixtures at 25°C; pyridine–CuCl catalyst

Table II. Properties of Homopolymers

Polymer	T_g	T_m	T_c	CH_2Cl_2	m-Xylene
DMP	220	258	—	Ppt	Stable
MPP	171	—	—	Stable	Stable
DPP	228	490	300	Stable	Ppt

mer blends. The thermal properties of the three homopolymers—and their solubility behavior in methylene chloride and m-xylene—are summarized in Table II. These solvents are extremely useful. DMP homopolymer dissolves readily in methylene chloride, but on standing at room temperature, precipitates almost quantitatively in the form of a crystalline polymer–CH_2Cl_2 complex (11). DPP homopolymer dissolves readily in m-xylene but precipitates on heating for a few hours on a steam bath; this seems to be simply because of solvent-induced crystallization rather than formation of a polymer–solvent complex. Each solvent affects only one of the polymers, so that a mixture of any of the three homopolymers can be separated into its components with the aid of one or both of these solvents.

Experimental

The homopolymers of DMP and DPP and DMP–DPP copolymers were prepared by methods previously reported (9, 12); details of the preparation of DMP–DPP copolymers and MPP-DPP copolymers are described in the sections of this paper devoted to these copolymers. A typical procedure for homopolymerization of MPP is given here.

Poly(2-methyl-6-phenyl-1,4-phenylene)oxide. A mixture of 1.08 grams of cuprous bromide, 7.73 ml of diethylamine, and 5 grams of anhydrous magnesium sulfate in 140 ml of benzene was vigorously stirred at 25°C, with a stream of oxygen introduced near the bottom of the flask. After five minutes, 22.8 grams of MPP were added, and oxidation continued for two hours. The mixture was stirred with 20 ml of 50% aqueous acetic acid, filtered to remove the magnesium sulfate, and the polymer isolated by precipitation in methanol; the yield was 20.8 grams (93%), and the intrinsic viscosity was 0.43 dl/g in chloroform at 30°C.

NMR spectra were taken in deuteriochloroform solution, using a Varian HA100 spectrometer. Thermal measurements were made with a Perkin-Elmer DSC 1B differential scanning calorimeter at 40°C/min. Near-infrared spectra were measured in carbon disulfide solution with a Beckman DK 2A spectrophotometer. Gas-chromatographic analyses of reaction mixtures were carried out after conversion of the phenols to trimethylsilyl ethers by reaction with bis(trimethylsilyl)acetamide.

Solubility behavior of the copolymers in methylene chloride was examined by preparing a 10% solution in warm methylene chloride and allowing the solution to stand for two days at room temperature. The precipitate, if any, was filtered off under nitrogen pressure using a 0.45μ Millipore filter, washed with cold methylene chloride, and dried under vacuum. The soluble fraction was recovered by precipitation with methanol. A similar procedure was followed with m-xylene: a 10% solution of the copolymer was heated overnight on a steam bath, the precipitate filtered off, dissolved in chloroform, and reprecipitated with methanol.

DMP–DPP Copolymers

The DMP-DPP system has been previously described but will be reviewed here to permit comparison with the DMP–MPP and MPP–DPP copolymers (which are the principal subject of this report), and because the procedures developed for preparing and characterizing the DMP–DPP copolymers were followed, as far as possible, with the new copolymers.

Polymerization Procedure. Copolymers were prepared by simultaneous and sequential oxidation of the two phenols. A number of different catalyst systems and reaction conditions were used, but with only four basic procedures:

(a) Both phenols were added at the start of reaction and oxidized simultaneously.

(b) The more reactive phenol (DMP) was oxidized alone until the increase in solution viscosity showed that high-molecular-weight polymer had been produced, the DPP was added, and oxidation continued. In a variation of this procedure, DPP and the homopolymer of DMP, separately prepared and isolated, were oxidized together.

(c) The DPP was oxidized alone, DMP was added, and oxidation continued, or, as in B, DMP was oxidized with the separately prepared homopolymer of DPP.

(d) Low-molecular-weight ($DP = 50$ to 100) homopolymers of each of the phenols were prepared and isolated separately. The homopolymers were mixed and further oxidized to high molecular weight copolymer.

Characterization of Copolymers. The NMR spectrum, particularly in the region of the methyl protons, is probably the most useful tool for characterizing the copolymers. In a random copolymer of DMP and DPP, a methyl-substituted ring may be situated between two methyl-substituted rings (MMM), between two phenyl-substituted rings (PMP), or between one of each type; in this case, the phenyl-substituted ring may be in the direction of the head of the chain (MMP) or in the opposite direction (PMM). The NMR spectra of polymers prepared by methods a or b have four methyl proton signals, with the chemical shifts predicted from a study of model compounds (3) for the four possible arrangements. These copolymers must accordingly have a random structure. Copolymers prepared by methods c or d have a single methyl signal, corresponding to the MMM sequence; the spectrum is almost identical with that of a blend of the two homopolymers. These must be either block copolymers or homopolymer blends. The solubility behavior, discussed elsewhere in this paper, shows that they are largely block copolymers. The spectra of a random copolymer, a block copolymer, and a blend of homopolymers are compared in Figure 2.

The homopolymer of DMP dissolves readily in methylene chloride but precipitates on standing as a crystalline polymer–CH_2Cl_2 complex, providing a method for distinguishing between block copolymers and mixtures of homopolymers. Random copolymers prepared by methods a and b form stable solutions in methylene chloride. Copolymers with a 1:1 ratio of DMP and DPP prepared by methods c and d also yield stable methylene chloride solutions. Since the NMR spectrum shows that the DMP portion of these materials is present as a block and the solubility in methylene chloride shows that DMP homopolymer is absent, these copolymers have the block structure. They can be separated by crystallization from m-xylene into an insoluble DPP-rich fraction and a soluble DMP-rich fraction, both fractions having the NMR spectra characteristic of block copolymers. A typical 1:1 copolymer prepared by adding DMP to growing DPP polymer yielded 35% of insoluble material

236   POLYMERIZATION KINETICS AND TECHNOLOGY

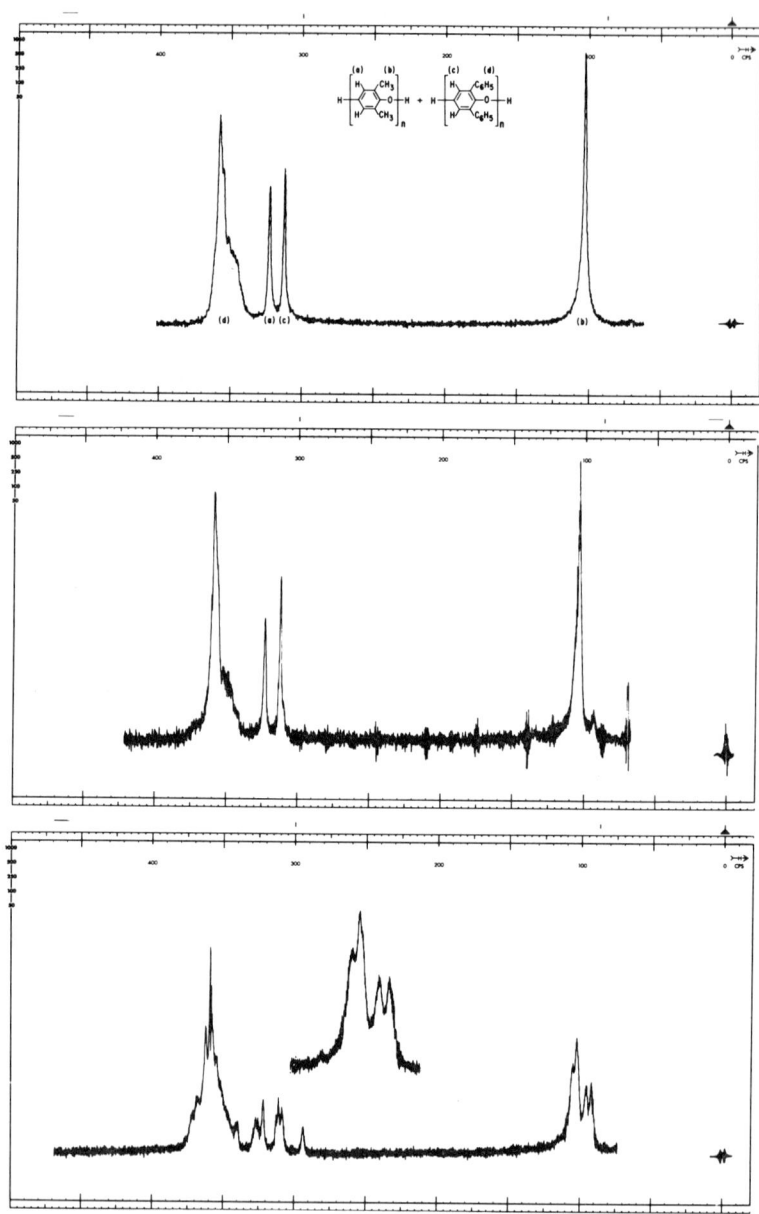

Figure 2. NMR spectra of 1:1 DMP-DPP copolymers; top: homopolymer blend; bottom: random copolymer

with a composition of 77 mole % DPP and 23 mole % DMP; the soluble portion contained 65 mole % DMP with 35 mole % DPP.

Thermal Properties. The DPP portion of block copolymers crystallizes on heating at ~290°C and then melts at ~480°C. The DMP portion of block copolymers does not crystallize thermally but can be caused to crystallize by treatment with a suitable solvent, such as a mixture of toluene and methanol; the crystallized DMP then melts at ~258°C. The glass-transition temperatures of the homopolymers are too close (221°C for DMP, 228° for DPP) to permit observation of separate transitions, either in block copolymers or blends of the homopolymers.

Redistribution and Polymer Structure. The structure of DMP–DPP copolymers is probably determined by the relative rates of the polymerization reaction and the monomer-polymer redistribution reaction. In the DMP–DPP system, structure may be predicted simply by observing the effect on solution viscosity of the addition of one of the monomers to the growing polymer derived from the other monomer. When DPP is added to a DMP reaction mixture, the solution viscosity drops immediately almost to the level of the solvent, as redistribution converts the polymer already formed to a mixture of low oligomers:

$$n \; \text{Ph-C}_6\text{H}_3(\text{Ph})\text{-OH} + \text{H-[C}_6\text{H}_2(\text{Me})_2\text{-O]}_n\text{-H} \longrightarrow$$

$$n \; \text{Ph-C}_6\text{H}_2(\text{Ph})_2\text{-O-C}_6\text{H}_2(\text{Me})_2\text{-OH} \quad \text{etc} \tag{4}$$

Continued oxidation of this mixture should yield a random copolymer, which was the result when method b was used. When DMP was added to growing DPP polymer, the solution viscosity did not drop but continued to increase at a faster rate than before the reactive DMP monomer was added. If redistribution between DMP and DPP homopolymer occurred, the rate was small compared with the rate of polymer growth. This at least allows the possibility of producing block copolymers, the result obtained when procedure c was followed. The effect of variations, in the polymerization procedure on the structure of 1:1 DMP–DPP copolymers is summarized in Table III.

Table III. Structure of 1:1 DMP–DPP Copolymers

Procedure	Catalyst	Product
DMP and DPP	TMBDA–CuBr	Random Copolymer
DMP and DPP	py–CuCl	Random Copolymer
DMP and DPP	DEA–CuBr	Random Copolymer
DPP, then DMP	TMBDA–CuBr	Block Copolymer
DPP, then DMP	DEA–CuBr	Block Copolymer
DMP, then DPP	TMBDA–CuBr	DMP Homopolymer and Randomized Copolymer
DMP, then DPP	DEA–CuBr	Random Copolymer
DPP block and DMP	TMBDA–CuBr	Block Copolymer
DPP block and DMP	DEA–CuBr	Block Copolymer
DMP block and DPP	DEA–CuBr	Random Copolymer
DMP block and DPP	TMBDA–CuBr, 60°C	Random, Low MW
DMP block and DPP block	TMBDA–CuBr	Block Copolymer

DMP–MPP Copolymers. Because DMP and MPP are quite close in reactivity at 25°C, as shown in Table I and Figure 1, copolymers produced by simultaneous oxidation of a mixture of the phenols should be random, and formation of block copolymers by sequential oxidation should be less likely than in the DMP–DPP system. This prediction is supported by observing the effect on solution viscosity of adding monomer to growing polymer. When either monomer is added to the growing polymer from the other, redistribution occurs and the solution viscosity decreases, then increases as oxidation continues. The decrease in viscosity is smallest and the subsequent rate of increase greatest when DMP is added to growing MPP polymer, with an active catalyst such as tetramethylethylenediamine–cuprous bromide; if block copolymers can be prepared by sequential oxidation of the two monomers, this would appear to be the most promising method.

Polymerization

The examples here are typical of the procedures used for copolymerization of DMP and MPP.

Procédure 1: Polymerization of a Mixture of DMP and MPP. A mixture of 0.432 gram of cuprous bromide, 3.09 ml of diethylamine, and 5.0 grams of anhydrous magnesium sulfate was stirred at 25°C in 140 ml of benzene, with a vigorous stream of oxygen introduced near the bottom of the flask. After five minutes, 4.9 grams (0.04 mole) of 2,6-

dimethylphenol and 7.3 grams (0.04 mole) of 2-methyl-6-phenylphenol were added, and oxidation continued for five hours. Dilute acetic acid was added to terminate the reaction, the mixture filtered, and the copolymer (10.6 grams, 88%) was isolated by precipitation in methanol. The polymer had an intrinsic viscosity of 0.61 dl/g.

Procedure 2: Addition of DMP to Polymerizing MPP. This was the same as procedure 1, but with only the MPP added initially. After four hours, the DMP was added. A gradual decrease in solution viscosity was observed over 10 minutes. Polymerization was continued for 90 minutes after DMP addition. The polymer was isolated in 92% yield and had an intrinsic viscosity of 0.72 dl/g.

Procedure 3: Addition of MPP to Polymerizing DMP. Procedure 1 was followed with only the DMP initially present. After two hours, the MPP was added, causing a prolonged decrease in solution viscosity; oxidation was continued for three hours after MPP addition. The copolymer was obtained in 92% yield, with an intrinsic viscosity of 0.58 dl/g.

Procedure 4: Oxidation of DMP with Redissolved MPP Homopolymer. Catalyst was prepared from 0.144 gram of cuprous bromide and 0.18 ml of N,N,N',N'-tetramethyl-1,3-butanediamine in 50 ml of benzene containing 5 grams of magnesium sulfate. Oxygen was introduced with vigorous stirring, and a solution of 3.7 grams of MPP homopolymer ($DP = 80$) in 50 ml of benzene was added, followed immediately by 2.5 grams of DMP. After 45 minutes, the reaction was stopped and the copolymer isolated in 93% yield; intrinsic viscosity was 0.82 dl/g.

Procedure 5: Oxidation of MPP with Redissolved DMP Homopolymer. To the catalyst solution described in procedure 1 was added 4.9 grams of a DMP homopolymer ($DP = 40$) and 7.3 grams of MPP. After three hours, the copolymer was isolated in 97% yield; it had an intrinsic viscosity of 0.61 dl/g.

Procedure 6: Polymerization of a Mixture of Homopolymers. Following procedure 1, a mixture of 7.3 grams of a homopolymer of MPP ($DP = 145$) and 4.9 grams of a homopolymer of DMP ($DP = 40$) was oxidized for 90 minutes. The copolymer was isolated in 98% yield and had an intrinsic viscosity of 1.08 dl/g.

Procedure 7: Polymerization of a Mixture of Homopolymers at $-25°$. A homopolymer of DMP ($DP = 40$) and a homopolymer of MPP ($DP = 90$) were cooxidized at $-25°$ in trichloroethylene, using 0.099 gram of cuprous chloride and 0.127 gram of N,N,N',N'-tetraethylethylenediamine. The polymer was isolated in 95% yield and had an intrinsic viscosity of 0.67 dl/g.

Characterization of Copolymers

NMR Spectra. Most of the information concerning the structure of DMP-MPP copolymers has been obtained from NMR spectra although the analysis is not as simple as with the DMP–DPP system. The methyl proton region, which clearly distinguished random from block copolymers of DMP and DPP, is almost useless in the DMP–MPP system. Both types of rings have methyl substituents, so that a methyl proton may have—considering only the effect of the neighboring rings—a total of eight different magnetic environments, four for each type of ring. The chemical shifts of the methyl protons in the two homopolymers are almost identical (δ 2.08 ppm), and the effect of the neighboring rings appears to be small, so that the methyl proton signal of random copolymers produced by simultaneous oxidation of the two monomers does not show separate peaks corresponding to the various possible sequences of rings. The major difference between a random copolymer and a blend of the two homopolymers in this region is that the methyl peak is broadened in the copolymer.

The region of the aromatic backbone protons is more useful. A blend of the two homopolymers shows three NMR signals in this region, at δ 6.46 ppm, coresponding to the protons of the DMP homopolymer, and at δ 6.31 and 6.36 ppm for the two protons in the MPP homopolymer. The random copolymer produced by simultaneous oxidation of the monomers has a much more complex spectrum, with five peaks in this region. Twelve could be observed if each type of proton could be distinguished in its four possible sequences of dimethyl- and methylphenyl-substituted rings, but the chemical shifts of many of these apparently do not differ enough to allow resolution.

The NMR spectra of copolymers prepared by simultaneous oxidation of the two phenols and those prepared by sequential oxidation, in either order, are almost identical. The methyl peak is broadened, as is the peak caused by the protons of the pendant phenyl rings centered at δ 7.20 ppm, and all show the same peaks for aromatic backbone protons in about the same intensity ratios. The polymer obtained by oxidizing a mixture of DMP and the separately prepared homopolymer of MPP with a cuprous bromide–tetramethylbutanediamine catalyst, the procedure considered to have the best chance of producing a block copolymer, was completely random.

Redistribution in Polymer Coupling. Monomer-polymer redistribution occurs most easily when the monomeric phenol and the phenol of the polymer are identical or, at least, very similar in reactivity (2). The homopolymers of DMP and MPP obviously redistribute very rapidly with either of the two monomers, so that sequential oxidation of DMP and MPP can produce only random copolymer. The redistribution reaction and its relation to the overall polymerization mechanism have been the subject of many previous investigations (2, 10, 13, 14), but the extraordinary facility of redistribution in the DMP–MPP system leads to results that could not be observed in other systems examined.

The reactive intermediates in oxidative coupling of phenols are aryloxy radicals. Growth may occur by the successive addition of aryloxy units as shown in Reaction 5.

[Scheme/Reaction (5)]

The growth characteristics of the reaction, however, are those of a polycondensation process, in which polymer molecules couple with other polymer molecules (15).

$$2 \; H\!-\!\!\left[\!\!\begin{array}{c}\text{Ar}\!-\!O\end{array}\!\!\right]_{n}\!\!\!-\!H + 1/2 \, O_2 \longrightarrow$$

$$H\!-\!\!\left[\!\!\begin{array}{c}\text{Ar}\!-\!O\end{array}\!\!\right]_{2n}\!\!\!-\!H + H_2O \tag{6}$$

In one of the mechanisms advanced to explain the coupling of polymeric aryloxy radicals to a polymeric phenol polymer, polymer redistribution plays an essential part. Polymeric aryloxy radicals probably couple to form an unstable quinone ketal, which dissociates either to regenerate the radicals from which it was formed or to produce two new radicals, one having one more, and the other one less, aryloxy units than the original. For the simplest case, the reaction of two dimers, the reaction is:

(7)

Aryloxy radicals react with phenolic species, either by direct hydrogen transfer or by oxidation–reduction reactions with the catalyst as carrier, to form new aryloxy radicals, which continue the redistribution.

$$\text{ArO} \cdot + \text{Ar'OH} \rightleftarrows \text{Ar'O} \cdot + \text{ArOH} \quad (8)$$

Redistribution is a free-radical chain reaction that does not consume oxygen or change the overall degree of polymerization. However, the net result of redistribution between polymeric phenols to form a monomeric phenol or phenoxy radical, followed by coupling of the monomer as in reaction (5) is the same as if two polymer molecules combined in a single step.

$$\begin{aligned} X_m + X_n &\rightleftarrows X_{m+1} + X_{n-1} \\ X_{m+1} + X_{n-1} &\rightleftarrows X_{m+2} + X_{n-2} \text{ etc.} \\ X_m + X_n &\rightleftarrows X_{m+n-1} + X_1 \\ X_{m+n-1} X_1 + 1/2\, O_2 &\rightarrow X_{m+n} + H_2O \end{aligned} \quad (9)$$

A second proposed coupling mechanism also involves the intermediate formation of a quinone ketal but postulates rearrangement rather than dissociation. The carbonyl oxygen of the ketal is close to the para position of one of the succeeding rings; formation of a new carbon-oxygen bond simultaneously with the breaking of one of the ketal bonds forms a new quinone ketal, with the second ring becoming the dienone ring. When, after a succession of such rearrangement reactions, one of the terminal rings becomes the dienone ring, tautomerization yields the coupled phenol.

$$(10)$$

Both growth mechanisms have been demonstrated by studying the initial products of the oxidation of simple model compounds, particularly dimers and trimers of DMP; see structure I in Reaction 1, $n=1, 2$ (2, 13, 14, 16). At normal polymerization temperatures, coupling occurs primarily via the redistribution sequence, but at $-25°C$, the coupling products observed are those predicted by the rearrangement mechanism. A major objective of this work was to study the coupling reaction of high-molecular-weight polymers by examining the product obtained on reoxidation of a mixture of two different homopolymers. The DMP–MPP system is particularly useful as mixtures of the two homopolymers can be reoxidized to high-molecular-weight copolymers under a variety of conditions. The first steps in the formation of a copolymer from the homopolymers by the redistribution sequence are:

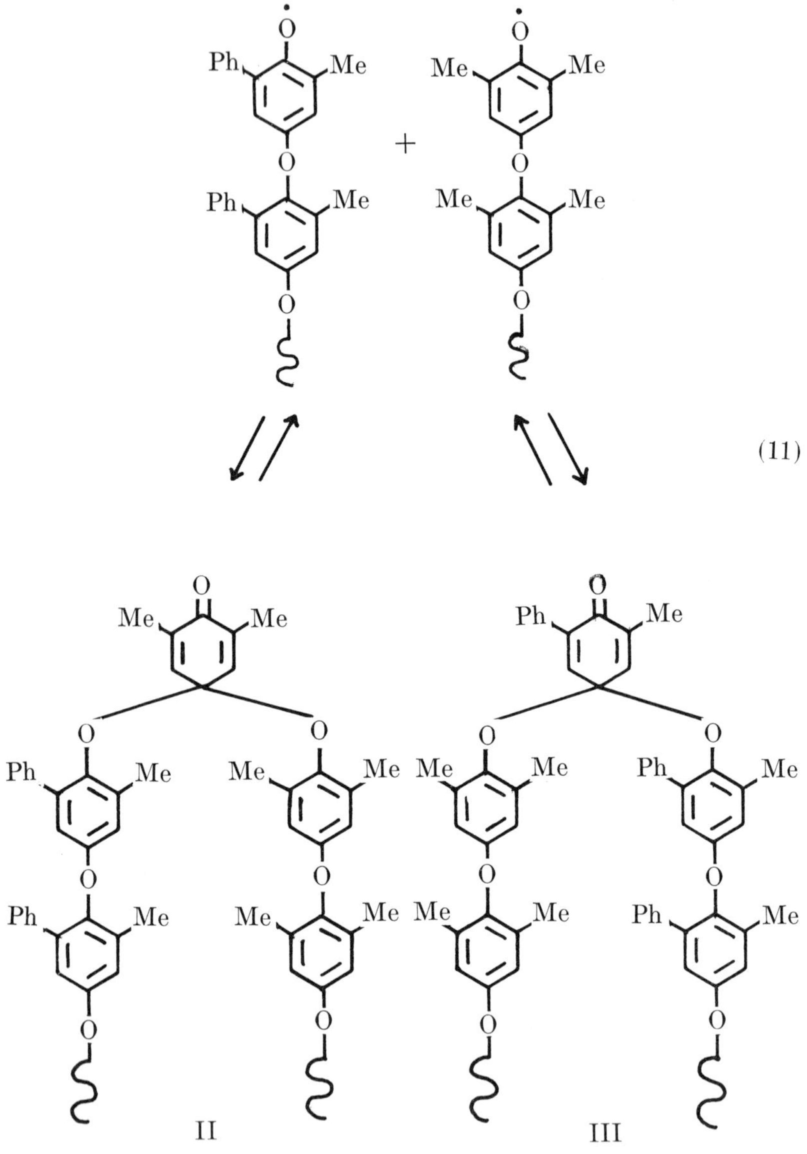

(11)

II   III

II ⟶ [structure: –O–C₆H₂(Ph)(Me)–O–C₆H₂(Me)₂–O·]   (12)

III ⟶ [structure: –O–C₆H₂(Me)₂–O–C₆H₂(Ph)(Me)–O·]   (13)

Continuation of this process, with monomer produced by redistribution and then removed by coupling, would lead to a random copolymer. Alternatively, if polymer-polymer coupling were to proceed solely by rearrangement, without dissociation at any stage, either of the ketals I or III would produce only block copolymer.

Oxidation of a mixture of equivalent weights of the two low-molecular-weight homopolymers at 25°C with a diethylamine–cuprous bromide catalyst yielded a copolymer that formed stable solutions in methylene chloride and could not be caused to crystallize by stirring with a 3:1 methanol/toluene mixture, a procedure that results in crystallization of DMP homopolymer or of the DMP portion of DMP–DPP block copolymers. The NMR spectrum was identical with that of the polymer obtained by simultaneous oxidation of the two monomers.

Exactly the same result was obtained when the homopolymers were oxidized at −25°C with a N,N,N',N'-tetraethylethylenediamine–cuprous chloride catalyst, conditions which have been reported to cause coupling of DMP homopolymers solely by rearrangement (14). The NMR spectrum of this polymer is shown in Figure 3, together with the spectra of a mixture of homopolymers and of a random copolymer formed by simultaneous oxidation of the monomers. Apparently, dissociation and redistribution occur often enough to determine the structure of the product in this system, even under conditions that favor coupling of polymer molecules by the rearrangement mechanism.

Spontaneous Redistribution of DMP and MPP. The stretching frequency of the phenolic hydroxyl group in DMP homopolymer occurs at 3601 cm^{-1} and that of MPP homopolymer at 3552 cm^{-1}, allowing the two different head groups to be distinguished. In the 1:1 DMP–MPP copolymers, 80 to 90% of the phenolic hydroxyls are of the MPP type. This can be explained, at least qualitatively, by the greater reactivity of

DMP head units; as oxidation proceeds, molecules with the less reactive MPP head would tend to accumulate. Surprisingly, however, the spectrum of a homopolymer blend, prepared by dissolving the two polymers in warm benzene and precipitating with methanol, showed the same preponderance of MPP phenolic groups although the degree of polymerization of the two polymers was such that more DMP than MPP end groups should have been present.

This is caused by a spontaneous redistribution between the two polymers:

$$H-\left[\underset{Me}{\overset{Me}{\bigcirc}}-O\right]_n-H \;+\; H-\left[\underset{Me}{\overset{Ph}{\bigcirc}}-O\right]_m-H \;\rightleftarrows$$

$$H-\left[\underset{Me}{\overset{Me}{\bigcirc}}-O\right]_n-\underset{Me}{\overset{Ph}{\bigcirc}}-OH \;+\; H-\left[\underset{Me}{\overset{Ph}{\bigcirc}}-O\right]_{m-1}-H \quad (14)$$

When the two polymers were separately dissolved in carbon disulfide, the solutions mixed, and the infrared spectrum immediately recorded, both hydroxyl peaks were observed in about the proper intensity. With each succeeding scan, the peak at 3601 cm^{-1} decreased and that at 3552 cm^{-1} increased; the change in intensity ratio with time is shown in Figure 4. The limiting ratio of 0.29 reached in this experiment probably does not represent the attainment of an equilibrium distribution. A more likely explanation is the exhaustion of a small amount of quinone or other impurity that is catalytically active at room temperature. When the solution is heated, or a small amount of tetramethyldiphenoquinone is added, the MPP peak continues to grow at the expense of the DMP peak; in the experiment of Figure 4, a ratio of 0.06 was ultimately reached. The redistribution process of reaction (10), if continued long enough, should lead to complete randomization, rather than merely to the exchange of head groups.

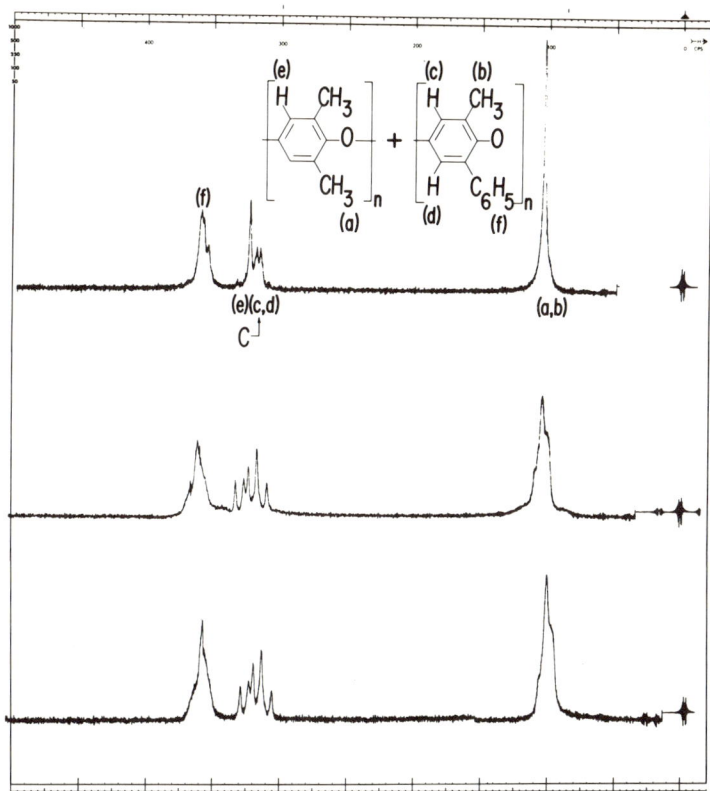

Figure 3. NMR spectra of 1:1 DMP–MPP copolymers; top: homopolymer blend; center: copolymer from oxidation of mixture of monomers at 25°; bottom: copolymer obtained by oxidizing mixture of homopolymers at −25°C

Some indication of randomization was seen when the polymers were heated under reflux in benzene for 30 days and then precipitated with methanol. The NMR spectrum showed a slightly broadened methyl signal and small peaks at δ 6.20 and 6.51 ppm, characteristics of random copolymers. The largest peaks by far were those corresponding to the homopolymers; redistribution apparently was not extensive enough to affect greatly the structure of the polymer chains.

Thermal Properties of DMP–MPP Copolymers

The thermal properties measured by differential scanning calorimetry (Table IV) provide no structural information in the DMP–MPP system. Neither of the two homopolymers undergoes thermally-induced crystallization. Furthermore, the homopolymers are compatible over the entire composition range; coprecipitated blends have a single glass transition at a temperature intermediate to the glass-transition tempera-

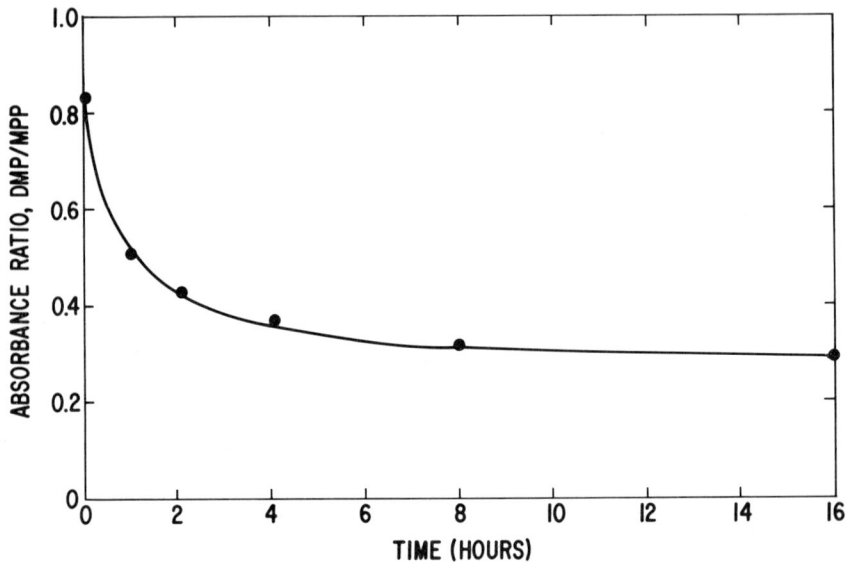

Figure 4. Change in relative intensity of DMP and MPP hydroxyl peaks in infrared spectrum of carbon disulfide solution of homopolymers

Table IV. Properties of 1:1 DMP–MPP Copolymers

Procedure	T(°C)	Intrinsic Viscosity (dl/g)	T_g	T_c	CH_2Cl_2	Type
MPP + DMP	25	0.61	192	—	Stable	Random
MPP + Growing DMP	25	0.72	193	—	Stable	Random
DMP + Growing MPP	25	0.58	194	—	Stable	Random
DMP + MPP Homopolymer	25	0.82	193	—	Stable	Random
MPP + DMP Homopolymer	25	0.61	193	—	Stable	Random
Both Homopolymers (Coprecipitated)	25	0.29	190	—	Ppt	Blend
Both Homopolymers (Cooxidized)	25	1.08	194	—	Stable	Random
Both Homopolymers (Cooxidized)	−25	0.90	194	—	Stable	Random

tures of the homopolymers. All of the 1:1 copolymers prepared in this work had a single glass transition, at 192°–194°C, almost identical to that of a coprecipitated blend. The glass transition of a 9:1 MPP–DMP copolymer, prepared by simultaneous oxidation of the two monomers, occurred at 176°C, the same as a blend of the same composition.

MPP–DPP Copolymers

The difference in reactivity of MPP and DPP in homopolymerization at 25°C is almost as great as that between DMP and DPP. It might therefore be expected that at this temperature, the behavior of MPP and DPP in copolymerization should resemble that of DMP and DPP—that is, simultaneous oxidation of both monomers or oxidation first of the less reactive DPP, followed by addition and oxidation of MPP, should yield random copolymer, while addition of MPP to growing DPP should form a block copolymer. At 60°C, however, MPP and DPP are of comparable reactivity, like DMP and MPP at 25°C, and perhaps only random copolymers could be obtained, no matter what procedure is followed. These expectations have been partially realized. The MPP–DPP copolymerization is more complex than either of the other two systems examined, with four distinguishable types of copolymer produced under different conditions.

Polymerization

Some typical polymerization procedures used in copolymerization of MPP and DPP are described here.

Procedure 1:—Simultaneous Oxidation of Monomers. A mixture of 0.288 gram of cuprous bromide, 0.36 ml of N,N,N',N'-tetramethylbutanediamine and 5 grams of anhydrous magnesium sulfate in 100 ml of benzene was vigorously stirred for five minutes at 25°C, with oxygen introduced near the bottom of the flask. A solution of 9.9 grams (0.04 mole) of DPP and 7.3 grams (0.04 mole) of MPP was added and oxidation continued for 90 minutes. The mixture was extracted with aqueous acetic acid and the polymer isolated by precipitation with methanol as previously described; the yield of copolymer was 14.4 grams (85%), with an intrinsic viscosity of 0.39 dl/g. Following the same procedure, but continuing the oxidation for 3.5 hours at 60°C, the copolymer was obtained in 84% yield, with an intrinsic viscosity of 0.34 dl/g.

Procedure 2: Oxidation of DPP with Redissolved MPP Homopolymer. A solution of 9.8 grams of DPP and 7.4 grams of a MPP homopolymer ($DP=190$) in 50 ml of benzene was added to the catalyst described under procedure 1. Oxidation was continued for five hours at 30°C; the copolymer was obtained in 83% yield, with an intrinsic viscosity of 0.45 dl/g.

Procedure 3: Oxidation of MPP with Redissolved DPP Homopolymer. A mixture of 7.3 grams of MPP and 9.8 grams of DPP homopolymer was oxidized for 15 minutes at 30°C, as described in the previous examples, yielding 86% of copolymer, with an intrinsic viscosity of 0.48 dl/g.

Procedure 4: Addition of MPP to Growing DPP Polymer at 60°C. The catalyst was prepared as in the previous examples, 9.9 grams of DPP was added, and oxidation continued at 60°C; the volume was maintained about constant by periodic addition of benzene. After four hours, 7.3 grams of DMP was added, causing a sharp decrease in solution viscosity. Oxidation was continued for four hours after addition of the DMP. The copolymer, obtained in 81% yield, had an intrinsic viscosity of 0.37 dl/g.

Procedure 5: Addition of DPP to Growing MPP at 60°C. The method of procedure 4 was followed with only the MPP initially present. The DPP was added after four hours. No decrease in solution viscosity was observed. Oxidation was continued for three hours, and the copolymer was obtained in 88% yield; it had an intrinsic viscosity of 0.35 dl/g.

Procedure 6: Oxidation of Mixture of Homopolymers. A solution of 7.3 grams of MPP homopolymer ($DP=195$) and 9.8 grams of DPP homopolymer ($DP=140$) was added to the catalyst described in procedure 1 and oxidized for one hour at 30°C. The polymer, obtained in 95% yield, had an intrinsic viscosity of 0.59 dl/g.

Characterization of Copolymers

Simultaneous Oxidation of Both Monomers. Random copolymers are produced by simultaneous oxidation of an equimolar mixture of the two phenols, either at 25° or 60°C. The NMR spectra of a blend of homopolymers and a 1:1 random copolymer are compared in Figure 5. Randomization clearly shows in the methyl region of the spectrum. Because only one type of phenoxy unit has a methyl substituent, the situation is the same as in the DMP–DPP system previously described. A methyl proton in a random copolymer has four possible magnetic environments, corresponding to the sequences MMM, PMP, MMP and PMM (M here represents a methylphenylphenoxy and P a diphenylphenoxy unit). Four methyl signals are observed; randomization appears, if anything, to be more complete than in DMP-DPP copolymers since the four signals are of comparable intensity in the MPP–DPP copolymer while the MMM signal in DMP-DPP copolymers is often appreciably stronger than the others. Further evidence of the random nature of the copolymers is seen in the region of the aromatic backbone protons. This region shows peaks in addition to those corresponding to the MMM and PPP sequences observed in the homopolymer.

The NMR and infrared spectra of copolymers produced by batch oxidation, with all of the monomers present from the beginning of the reaction, were identical with those of polymers prepared by gradual addition of a solution of the two monomers over an hour, showing that any tendency toward formation of blocks by preferential oxidation of the more reactive monomer was destroyed by the redistribution process.

Thermal Properties of Random MPP-DPP Copolymers and Homopolymer Blends. The thermal properties of some copolymers prepared by simultaneous oxidation of the two monomers, in varying proportions, are summarized in Table V. Solution-cast (CH_2Cl_2) films of copolymers containing a high proportion of DPP are partially crystalline, like DPP homopolymer, and undergo further crystallization on heating in the differential scanning calorimeter. A small amount (5 mole %) of MPP reduced by half both the initial and total crystallinity developed on heating, in addition to lowering the crystal melting temperature. Crystallization was not observed in copolymers containing more than 25 mole % MPP.

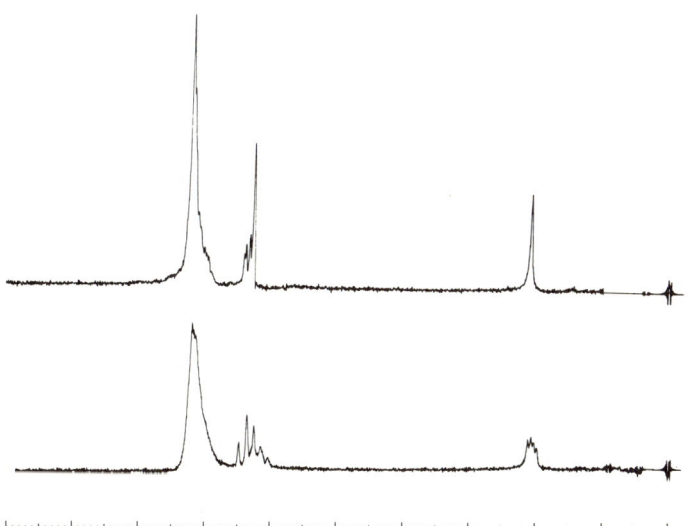

Figure 5. NMR spectra of 1:1 MPP–DPP copolymers; top: blend of homopolymers; bottom: random copolymer from simultaneous oxidation of monomers at 25°C

All of the random copolymers had a single glass transition, at a temperature approximately corresponding to the weighted average of the glass-transition temperatures of the two homopolymers. The homopolymers themselves showed only limited compatibility. Two glass transitions were observed in coprecipitated blends of equivalent weights of the two homopolymers. On the first heating, a transition at 225°-230°C was observed (DPP), and a transition at 160°-175°C could sometimes be distinguished. Crystallization took place at about 290°C. When the

Table V. Thermal Properties of Random MPP–DPP Copolymers

Composition (Mole % MPP)	T_g (°C)	T_m (°C)	Crystallinity [a]	
			Initial (%)	Total (%)
0	228	479	23	40
5		473	12	20
10	221	455	7	18
40	202	—	—	—
50	193	—	—	—
90	175	—	—	—
100	171	—	—	—

[a] Estimated from melt endotherm, assuming $\Delta Hf = 21$ cal/g

sample was then cooled and reheated, the MPP transition at 171°C was clear, and the DPP transition was not observed; this transition is usually not detected in DPP homopolymer after thermal crystallization, even though the crystallinity rarely exceeds 40% (17).

Sequential Oxidation at 25°C. Because of the slow polymerization rate of DPP at 25°C, most sequential polymerizations at this temperature were carried out with isolated homopolymers, separately prepared under conditions appropriate for the homopolymerization. Oxidation at 25°C of a 1:1 mixture of DPP and MPP homopolymer yielded a copolymer that did not crystallize, formed stable solutions in m-xylene, and had the NMR spectrum characteristic of a random copolymer.

The product of oxidation of a mixture of MPP and DPP homopolymer crystallized on heating. No clear glass transition was observed on the first heating, but on reheating, the transition occurred at 172°C. The NMR spectrum shows some randomization but resembles that of a mixture of homopolymers much more closely than that of a random copolymer (Figure 6). The product dissolves readily in m-xylene, but partially precipitates on standing overnight on a steam bath.

The soluble and insoluble fractions were examined separately. The insoluble fraction, which made up 35% of the total, had the NMR spectrum expected of a DPP-rich block copolymer, with a sharp methyl proton signal and only one strong signal, at δ 6.46 ppm (PPP), in the aromatic backbone region. The composition, from comparison of the integrated intensities of the methyl and backbone proton signals, was 82 mole % DPP and 18% MPP. The soluble fraction had the spectrum expected of a block copolymer with about 65% MPP units. Since a coprecipitated blend was separated almost quantitatively into the pure homopolymers with m-xylene under these conditions, the copolymer is characterized as a block copolymer.

Sequential Oxidation at 60°C. The NMR spectrum of the polymer obtained by adding DPP to growing MPP at 60°C showed extensive randomization but with a strong signal at δ 6.25 ppm, characteristic of PPP sequences. Solution cast films of this polymer were translucent. It crystallized on heating at about 300°C and had a single glass transition, at 194°C. It partially crystallized from hot m-xylene; the insoluble portion (32%) contained 80 mole % DPP and had a strong PPP signal at δ 6.25 ppm and four methyl proton signals of low intensity. This product clearly contains DPP blocks, either in a block copolymer or as DPP homopolymer. Its properties are consistent with those expected of a mixed block copolymer—that is, a copolymer containing blocks of DPP and blocks of randomly arranged MPP and DPP units.

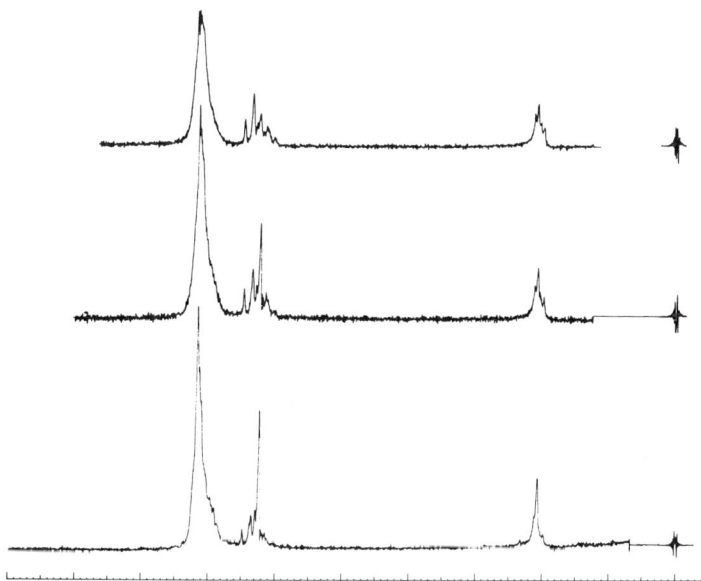

Figure 6. NMR spectra of 1:1 MPP–DPP copolymers; top: random copolymer from oxidation of DPP with MPP homopolymer at 25°C; center: mixed block copolymer from addition of DPP to growing MPP polymer at 60°C; bottom: block copolymer from oxidation of MPP with DPP homopolymer at 25°C

Addition of DPP to growing MPP at 60°C produced still another type of copolymer. Solvent-cast films of this material are transparent. The copolymer does not crystallize on heating, forms stable solutions in m-xylene, and has a single glass transition, at 190°C. The thermal behavior is similar to that of a random copolymer, but the NMR spectrum (Figure 7) is more nearly that expected of a block copolymer. The methyl proton peak is rather sharp with the chemical shift expected for

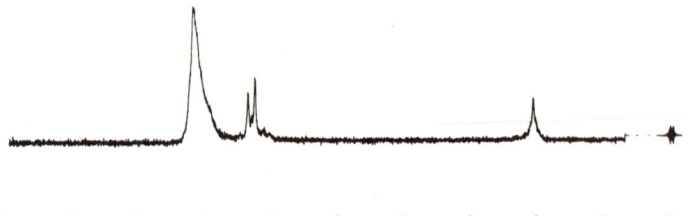

Figure 7. NMR spectrum of "short block" copolymer from addition of MPP to growing DPP at 60°C

the MMM sequence; the peaks in the aromatic region are those associated with MMM and PPP sequences. The polymer apparently is a block copolymer, but one in which the DPP blocks are too short to permit either thermal or solvent-induced crystallization.

Effect of Polymerization Conditions on Structure. The properties and structural classification of some polymers obtained from MPP and DPP under varying reaction conditions are summarized in Table VI.

Table VI. Properties and Characterization of 1:1 MPP–DPP Copolymers

Oxidation Procedure	T (°C)	Intrinsic Viscosity (dl/g)	T_g (°C) 1st Heat	T_g (°C) 2nd Heat	Tc (°C)	m-Xylene	Characterization
MPP + DPP	25	0.40	190	193	—	—	Random
MPP + DPP Homopolymer	25	0.26	196	172	294	Ppt	Block
DPP + MPP Homopolymer	25	0.45	192	193	—	—	Random
Both Homopolymers	25	0.59	—	190	300	Ppt	Mixed block [a]
Both Homopolymers	25	0.39	175, 221	171	290	Ppt	Homopolymers
MPP + DPP	60	0.38	192	191	—	—	Random
MPP + Growing DPP	25	0.39	175	186	292	Ppt	Mixed block [a]
MPP + Growing DPP	25	0.29	—	159	286	Ppt	Homopolymers
DPP + Growing MPP		0.40	196	193	—	—	Short block [b]
Coprecipitated Blend	—	0.19	165, 230	171	285	Ppt	

[a] Copolymer consisting of blocks of DPP with blocks of random MPP–DPP
[b] Block copolymer with DPP blocks not crystallizable

At 25°C, the products were largely those expected on the basis of previous work with the DMP–DPP system. Oxidation of the less reactive monomer, MPP, in the presence of DPP homopolymer yielded random copolymers; block copolymers were obtained when this order was reversed.

Reoxidation of a mixture of the two homopolymers sometimes produced block copolymers, but reproducibility was poor with this procedure. Other polymerizations under apparently identical conditions even with the same homopolymer samples yielded mixed block copolymers, or more frequently, a product that was essentially only a mixture of the two homopolymers.

The results at 60°C do not follow the pattern either of the DMP–DPP or DMP–MPP systems, which those results resemble in the relative reactivity of the components. Observation of the effect on solution viscosity of adding monomer to a solution of growing polymer provides no guide to the products. Addition of MPP to growing DPP causes a decrease in solution viscosity, indicating redistribution, but the product regularly observed was a block copolymer (although apparently with very short block length). Surprisingly, addition of the less-reactive DPP to growing MPP polymer caused no reduction in viscosity. The product was sometimes characterized as a mixed block polymer, but in other cases appeared to be a mixture of homopolymers. The reasons for the variety of products obtained from MPP and DPP, and for the sensitivity of this system to minor changes in reaction conditions, are not understood.

Discussion

The formation of random copolymer, even when the starting materials are preformed homopolymer blocks, as was observed with DMP and MPP, is reasonably explained by the monomer-polymer and polymer-polymer redistribution reactions of Reaction 3 and 9. Block copolymers are accounted for most easily by polymer-polymer coupling *via* the ketal arrangement mechanism (*see* Reaction 15, p. 256).

There are, however, other possible routes to block copolymers: successive addition of units of the reactive monomer to the polymer already present, Reaction 5; termination reactions between polymer molecules —side reactions of unknown nature lead to loss of reactive hydroxyl groups (*18*); possible reactions are ortho carbon-carbon coupling followed by dimerization, addition of amine or water to the ketal intermediate, etc. Block copolymers might even be formed by polymer-polymer redistribution assuming that such redistribution in polymers of greatly different reactivities (such as DMP and DPP), takes place almost exclusively in one type of polymer sequence—that is, that bond scission in a "mixed ketal" such as IV occurs always in the same direction—to produce the aryloxy radical corresponding to the more reactive monomer. None of these possible sources of block copolymer can be ruled out on the basis of available evidence. All could produce homopolymer in addition to block copolymer. All of the polymers produced in this work, except for those characterized as completely random copolymers, probably contained at least small amount of one or both homopolymers.

$$\text{[structure with Me, Ph groups]}_n + \text{[structure with Ph groups]}_m \xrightarrow{\text{Oxidize}}$$

IV

$$\text{IV} \xrightarrow{\text{Rearrange}} \text{H---[Ph/Ph ring---O]}_m\text{[Ph/Me ring---O]}_{n-1}\text{[Ph/Me ring=O]} \rightarrow P_m M_n \quad (15)$$

Literature Cited

1. Hay, A. S., Blanchard, H. S., Endres, G. F., Eustance, J. W., *J. Amer. Chem. Soc.* (1959), **81**, 6335.
2. Cooper, G. D., *Ann. N. Y. Acad. Sci.* (1969), **159**, 1.
3. Cooper, G. D., Bennett, J. G., Jr., *J. Org. Chem.* (1972), **37**, 441.
4. Cooper, G. D., Bennett, J. G., Jr., Katchman, A., *Advan. Chem. Ser.* (1971), **99**, 431.
5. Cooper, G. D., Katchman, A., *Advan. Chem. Ser.* (1969), **91**, 660.
6. Hay, A. S., *Advan. Polym. Sci.* (1967), **4**, 496.
7. Price, C. C., Nakaoka, K., *Macromolecules* (1971), **4**, 363.
8. Tsuchida, E., *Macromol. Chem.* (in press).
9. Bennett, J. G., Jr., Cooper, G. D., *Macromolecules* (1970), **3**, 101.

10. Cooper, G. D., Gilbert, A. R., Rinkbeiner, H., *Polymer Preprints* (1966), **7**, 166.
11. Factor, A., Heinshon, G. E., Vogt, L. H., Jr., *J. Polym. Sci., Part B* (1969), **7**, 205.
12. Hay, A. S., *Macromolecules* (1969), **2**, 107.
13. Cooper, G. D., Blanchard H. S., Endres, G. F., Finkbeiner, H., *J. Amer. Chem. Soc.* (1965), **87**, 3996.
14. Mijs, W. J., von Lohuisen, O. E., Bussink, J., Vollbracht, L., *Tetrahedron* (1967), **23**, 2253.
15. Endres, G. F., Kwiatek, J., *J. Polym. Sci.* (1962), **58**, 593.
16. White, D. M., *J. Org. Chem.* (1969), **34**, 297.
17. Wrasidlo, W., *Macromolecules* (1971), **4**, 442.
18. White, D. M., Klopfer, H. J., *J. Polym. Sci., Part A-1* (1972), **10**, 1565.

RECEIVED April 8, 1972.

17

Use of Gel Permeation Chromatography to Study the Synthesis of Bisphenol-A Carbonate Oligomers

A. B. ROBERTSON

Allied Chemical Corp., Morristown, N. J. 07208

J. A. COOK and J. T. GREGORY

PPG Industries, Inc., Barberton, Ohio 44203

A gel permeation chromatographic (GPC) method for characterizing bisphenol-A oligomer carbonates and the effects of synthesis procedure and reaction variables on the size distribution of the products are described. In comparison with end-group analysis the GPC method is the superior technique since it is relatively insensitive to low-molecular-weight contaminates in the oligomer, and because molecular weight by this method is independent of the type of end group present. The GPC method also gives information about molecular-weight distribution. The molecular characteristics of the products made by aqueous sodium hydroxide, interfacial, and homogeneous pyridine solution procedures are compared in this study. A simple fractionation procedure gave narrow-distribution, chloroformate-terminated prepolymers from the bisphenol-A oligomers prepared in homogeneous pyridine.

Bisphenol-A carbonate oligomers have been used in the syntheses of random and block copolycarbonates (1, 2). The physical properties of these polymers can be altered by tailoring sequence distribution and block size in the copolymer. To tailor sequence distribution and block size, it is necessary to know the molecular weight and molecular-weight distribution of bisphenol-A prepolymers present during synthesis.

Bisphenol-A oligomers can be characterized by end-group analysis and by other conventional methods for measuring molecular weight, but these techniques yield very little information about molecular-weight distribution and results are often affected by low-molecular-weight species such as solvent in the oligomer. Gel permeation chromatography (GPC), a technique first described by John Moore in 1964 (3), has been successfully used to characterize a number of low-molecular-weight oligomers (4, 5). Characterization by GPC yields information about both molecular weight and molecular-weight distribution, and, in addition, GPC results are relatively insensitive to low-molecular-weight contaminants in the oligomer. This paper describes a GPC method for characterizing bisphenol-A oligomers, and discusses the effect of synthesis procedure and reaction variables on the molecular characteristics of these materials.

Background

A knowledge of the chemistry of bisphenol-A oligomer carbonate synthesis is useful in the interpretation of GPC data. Bisphenol-A carbonate oligomers can be prepared by reaction of bisphenol-A with phosgene in the presence of an acid acceptor. The reaction can be conducted interfacially (6), in homogeneous pyridine solution (7), or by merely passing phosgene into an alkaline solution of bisphenol-A (8). The chemical reactions that occur in each of these systems are complicated and consist of multiple and competitive reactions.

The interfacial synthesis of bisphenol-A carbonate oligomers is conducted by passing phosgene into an agitated, two-phase mixture of a water-insoluble organic solvent and an alkaline solution of bisphenol-A. Reactions 1–4 shown on p. 260 illustrate some of the reactions believed to occur in the interfacial system.

Bisphenol-A is present as the disodium salt in the aqueous phase of the two-phase mixture. Phosgene enters the system and dissolves in the organic phase. It is believed that the reaction between phosgene and bisphenol-A occurs at the organic–aqueous interface to form the monochloroformate—Reaction 1—or bischloroformate—Reaction 2—ester of bisphenol-A. The chloroformate esters that form grow to oligomers by reaction with additional bisphenol-A or by self-condensation —Reactions 3 and 4, respectively.

The degree of polymerization of bisphenol-A oligomers prepared interfacially depends on stoichiometry, phosgenation rate, temperature, stirring speed, and other variables. In general, the degree of polymerization may be increased by increasing sodium hydroxide concentration, by raising temperature (in the range of 0° to 39°C.), by decreasing phosgenation rate, or by increasing stirring speed. Such variations apparently affect the relative rates Reactions 1–4.

$$HO-\underset{CH_3}{\underset{|}{\overset{CH_3}{\overset{|}{C}}}}-\phi-OH + COCl_2 \xrightarrow{\underset{NaOH/H_2O}{CH_2Cl_2}}$$

$$NaO-\phi-\underset{CH_3}{\underset{|}{\overset{CH_3}{\overset{|}{C}}}}-\phi-OCCl + NaCl \tag{1}$$

$$NaO-\phi-\underset{CH_3}{\underset{|}{\overset{CH_3}{\overset{|}{C}}}}-\phi-OCCl + COCl_2 \xrightarrow{\underset{NaOH/H_2O}{CH_2Cl_2}}$$

$$ClCO-\phi-\underset{CH_3}{\underset{|}{\overset{CH_3}{\overset{|}{C}}}}-\phi-OCCl + NaCl \tag{2}$$

$$ClCO-\phi-\underset{CH_3}{\underset{|}{\overset{CH_3}{\overset{|}{C}}}}-\phi-OCCl + NaO-\phi-\underset{CH_3}{\underset{|}{\overset{CH_3}{\overset{|}{C}}}}-\phi-ONa \tag{3}$$

$$\xrightarrow{\underset{NaOH/H_2O}{CH_2Cl_2}} ClCO-\phi-\underset{CH_3}{\underset{|}{\overset{CH_3}{\overset{|}{C}}}}-\phi-OCO-\phi-\underset{CH_3}{\underset{|}{\overset{CH_3}{\overset{|}{C}}}}-\phi-O-Na$$

(X) $$Na-O-\phi-\underset{CH_3}{\underset{|}{\overset{CH_3}{\overset{|}{C}}}}-\phi-OCCl + \xrightarrow{\underset{NaOH/H_2O}{CH_2Cl_2}}$$

$$Na\left[O-\phi-\underset{CH_3}{\underset{|}{\overset{CH_3}{\overset{|}{C}}}}-\phi-OC\right]_x Cl + (x-1) NaCl \tag{4}$$

Teritary amines such as pyridine may also be used as acid acceptors in the preparation of bisphenol-A carbonate oligomers. Tertiary-amine-catalyzed reactions are usually conducted in homogeneous solution. For instance, bisphenol-A oligomer bischloroformates may be prepared by dripping a solution of bisphenol-A in pyridine into a pool of phosgene. A complex probably forms between pyridine and the carbonyl chloride group, and that complex is more reactive with aromatic hydroxy compounds than the corresponding uncomplexed chlorocarbonic acid. Some reactions thought to occur in the pyridine system are shown in Reactions 5 through 7.

$$\text{HO}-\text{C}_6\text{H}_4-\text{C}(\text{CH}_3)_2-\text{C}_6\text{H}_4-\text{OH} + \text{COCl}_2 \xrightarrow[\text{Benzene}]{\text{Pyridine}}$$

$$\text{ClCO-O}-\text{C}_6\text{H}_4-\text{C}(\text{CH}_3)_2-\text{C}_6\text{H}_4-\text{OH} + \text{HCl} \quad (5)$$

$$\text{ClCO-O}-\text{C}_6\text{H}_4-\text{C}(\text{CH}_3)_2-\text{C}_6\text{H}_4-\text{OH} + \text{COCl}_2 \xrightarrow[\text{Benzene}]{\text{Pyridine}}$$

$$\text{ClCO-O}-\text{C}_6\text{H}_4-\text{C}(\text{CH}_3)_2-\text{C}_6\text{H}_4-\text{OCCl} + \text{HCl} \quad (6)$$

$$x \text{ ClCO-O}-\text{C}_6\text{H}_4-\text{C}(\text{CH}_3)_2-\text{C}_6\text{H}_4-\text{OH} + \text{COCl}_2 \xrightarrow[\text{Benzene}]{\text{Pyridine}}$$

$$\text{Cl}-[\text{CO-O}-\text{C}_6\text{H}_4-\text{C}(\text{CH}_3)_2-\text{C}_6\text{H}_4-\text{O}]_x-\text{H} + (x-1)\text{ HCl} \quad (7)$$

The degree of polymerization of bischloroformate oligomers prepared in the pyridine system is controlled primarily by the molar ratio of phosgene to bisphenol-A, as long as the ratio is greater than 1. Increasing the ratio of phosgene to bisphenol-A tends to lower the degree of polymerization of the product. The cosolvent (benzene, THF, etc.) has a marked effect upon the degree of polymerization achieved since the rates of the various reactions are affected. In some cases, the low-molecular-weight oligomers become insoluble and precipitate.

Bisphenol-A oligomer carbonates can also be prepared by passing phosgene into an aqueous solution of the sodium salt of bisphenol-A. The reactions that occur are probably similar to those that occur in the interfacial system. The major difference from the interfacial system is that oligomers that form are insoluble and precipitate as a white solid. Precipitation of the oligomer occurs at a relatively low degree of polymerization. Products made by this procedure have a relatively high percentage of hydroxyl end groups.

Experimental

Product Characterization by GPC. A Waters gel permeation chromatograph, model 200, was used in this work. The instrument was operated at room temperature, and tetrahydrofuran was used as the elution solvent. Samples were analyzed on a four-column system; column permeabilities were 2.5×10^5, 1.5×10^4, 10^3, and 10^2 A (Waters designation). Oligomer samples (0.5% by weight in tetrahydrofuran) were injected for 120 seconds. A solvent flow rate of 1 ml/minute was used.

The GPC column system was calibrated with respect to molecular weight by noting the elution positions of several, well-characterized, bisphenol-A carbonate standards. The standards were either pure bisphenol-A oligomer bischloroformates or narrow distribution bisphenol-A carbonate oligomers. Narrow-distribution oligomers were obtained by fractionating broad-distribution oligomers using solvent evaporation (9), and their absolute molecular-weights were determined by vapor-pressure osmometry. A calibration curve was constructed by plotting measured number-average molecular weight against observed GPC elution volume. The calibration data are shown in Figure 1.

It was necessary to include a refractive index correction in the molecular-weight calculations because experimental data on samples of the same concentration by weight established that monomer bisphenol-A bischloroformate gave a smaller detector response than did higher oligomers. The effect fell off rapidly, however, as there was no difference between dimer and trimer bisphenol-A bischloroformate. By multiplying measured heights by 1.4, the curve area for bisphenol-A bischloroformate could be equated to the area for higher oligomers. The validity of using this correction factor was established by comparing calculated molecular weights of synthetic mixtures of bisphenol-A bischloroformate and well-characterized standards to molecular weights determined by GPC. The GPC values agreed with calculated values within 5%.

Interfacial Syntheses of Bisphenol-A Carbonate Oligomers. Oligomers that were prepared interfacially were made in a 5000-ml, three-necked-baffled flask equipped with a water condenser and a Dry Ice condenser. Cooling was provided by an ice bath around the reactor.

The usual reaction mixture consisted of methylene chloride (1500 ml), distilled water (1000 ml), bisphenol-A (456.6 grams), and sodium hydroxide (269 grams). These were added to the cooled (25°C) flask and stirred at about 300 rpm. The stirrer speed was set at the desired level (200-500 rpm) and phosgene added to the mixture at a rate of 4.06 grams per minute. The flask was cooled with ice during phosgenation when temperatures below 39°C were desired, or was allowed to warm as the reaction proceeded to methylene chloride reflux temperature (39°C). The polymerization was terminated by separating the methylene chloride layer and washing it first with 1% sulfuric acid, then with water until the water extract was free of inorganic chloride. The washed product was left in methylene chloride solution and dried over magnesium sulfate. Solid product was isolated by evaporation to dryness on a rotary film evaporator.

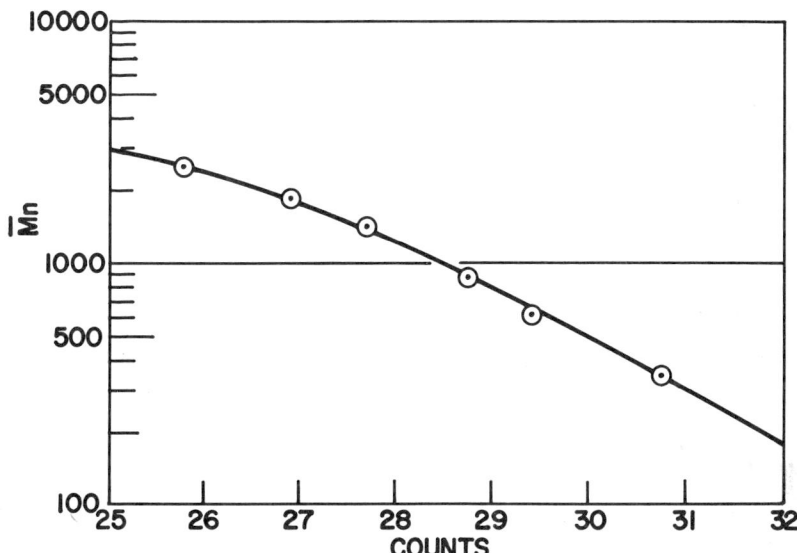

Figure 1. Plot of GPC calibration data showing the relationship of the number-average molecular weight vs. the elution volume (counts) of the particular oligomer

Syntheses of Bisphenol-A Carbonate Oligomers in Aqueous Caustic. Oligomers made by the aqueous caustic method were made in an apparatus identical with that used in the interfacial method; experimental conditions were similar. Methylene chloride, though, was not added to the reaction mixture. The product precipitated on formation and was isolated by filtration.

Syntheses of Bisphenol-A Carbonate Oligomers in Homogeneous Pyridine. Oligomers made by the pyridine method were prepared in the same apparatus except that drying tubes were added where necessary. It was also necessary to use carefully dried solvents since the undesirable side reaction with water converts chlorocarbonic acid to hydrochloric acid and carbon dioxide, thus upsetting the reaction stoichiometry.

The cooling bath was charged with ice and salt water so the reaction temperature could be held as low as −5°C. Gaseous phosgene (112 grams in the case of 11% excess phosgene) was condensed in the cooled flask. A solution consisting of bisphenol-A (114.1 grams), pyridine (79 grams), and dry benzene (100 ml) was slowly added to the phosgene pool over 77 minutes. The temperature was held between 0° and 5°C during addition of the bisphenol-A solution. Almost immediately after starting addition of the bisphenol solution, pyridine hydrochloride appeared as a white solid in the reaction mixture. When all of the bisphenol-A solution was added, 300 ml of dry benzene were added, the cooling bath removed, and the reaction allowed to stand overnight. The next day, the reaction mixture was heated to 65°C and flushed with dry nitrogen to blow off residual phosgene. Pyridine hydrochloride was separated from the product by filtration, and a solid product was obtained by evaporation to dryness in a rotary evaporator.

Tetrahydrofuran was used instead of benzene in the synthesis of some bisphenol-A oligomers by the pyridine method. The reaction was run in an almost identical manner except for the change in solvent.

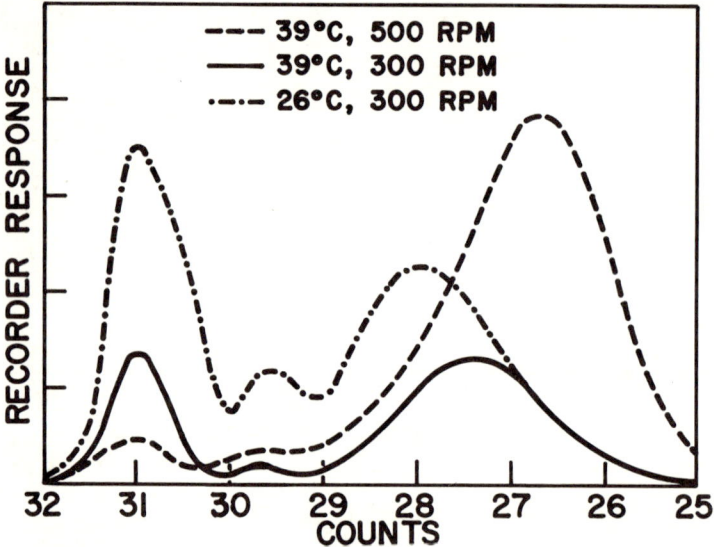

Figure 2. GPC curves for bisphenol-A oligomer bischloroformates prepared by the interfacial technique at the reaction temperatures and stirrer speeds shown

Analysis of Bisphenol-A Carbonate Oligomers for Phenolic Hydroxyl. Analysis of oligomers for phenolic hydroxyl end groups was conducted by quantitative infrared spectroscopy. Hydroxyl absorbance at 3580 cm^{-1} was measured for a number of synthetic mixtures of bisphenol-A and bisphenol-A homopolycarbonate. A calibration curve for hydroxyl absorbance *vs.* weight percent hydroxyl end groups was constructed. Hydroxyl content of bisphenol-A oligomers was calculated from the calibration data.

Analysis of Bisphenol-A Carbonate Oligomers for Chloroformate Chlorine. Chloroformate chlorine end groups were determined by a modified Volhard procedure in which the bischloroformate first reacts with aqueous pyridine to convert chloroformate to chloride ion. The chloride ion is extracted into water and titrated by the Volhard procedure.

Determination of Molecular Weights. Number-average molecular weights were determined by vapor-pressure osmometry using a Mechrolab osmometer, model 302. Samples were run in chlorobenzene solution (10). The weight-average molecular weights were calculated from the number-average molecular weights using the peak widths of the GPC curves, without correcting for axial broadening, and the calibration from low-molecular-weight polyether standards to determine the ratio M_w/M_n. The results for weight-average molecular weight are within 10% of the true value (11).

Hydroxyl Content by NMR Analysis. Initially, the presence of hydroxyl-terminated oligomers was determined by NMR analysis. However, this method was not sensitive to low hydroxyl content and the infrared method was developed.

The NMR analysis was based on the fact that the aromatic protons of BPA or hydroxyl-terminated bisphenol-A bischloroformate oligomers have an A_2B_2 pattern centered at 7.16 δ, while the aromatic protons of carbonate joined or chloroformate end-grouped oligomers have a slightly degenerated singlet at 7.5 δ. In an NMR of a mixture of the two types of oligomers, the up-field portion of the A_2B_2 pattern of the hydroxyl terminated oligomer (7.0 ppm δ) is plainly visible. The presence of this doublet in the NMR of the oligomers was taken as evidence for hydroxyl-terminated oligomer.

Discussion

Bisphenol-A Carbonate Oligomers Prepared Interfacially. Figure 2, shows the GPC curves for three bisphenol-A oligomer bischloroformates. These oligomers were synthesized by the interfacial technique. Molecular-weight values calculated from the GPC data and the corresponding average degrees of polymerization are listed in Table I.

Listed in Table I along with molecular-weight values are the temperatures and stirring speeds at which the interfacial syntheses were

Table I. Molecular-Weight Data on Interfacially Prepared Bisphenol-A Carbonate Oligomers

	Reaction Conditions					Average
		Stirring				Degree of
Oligomer	Temp	Speed	M_n	M_w	M_w/M_n	Polymerization
I	39°C	500 rpm	1354	1916	1.42	4.92
II	39°C	300 rpm	1123	1687	1.50	4.03
III	26°C	300 rpm	680	1184	1.74	2.28

conducted. GPC analysis of the products rapidly determines how variations in reaction conditions affect the molecular weight and molecular-weight distribution of these oligomers.

The results in Table I show that oligomer molecular weight is lowered by decreasing either stirring speed or temperature. Both variables tend to alter the rates of the competitive reactions described for this system. Lowering stirring speed decreases the interfacial surface area and thus decreases the rate of reaction of chloroformate-terminated oligomers with phenolate-terminated oligomers. Lowering the temperature probably decreases the rate of chloroformate hydrolysis relative to the reaction of phosgene with phenol. Thus, chloroformates, once formed, are not as rapidly reconverted to phenolates that can undergo further chain building reactions.

In the range of these experiments, the molecular-weight distribution narrows as oligomer degree of polymerization increases. When the reaction is allowed to proceed to high molecular weight, bisphenol-A homopolycarbonate is obtained; this product has a molecular-weight distribution of 2.3, comparable with the most probable value of 2 predicted for polycondensation reactions.

The various peaks in Figure 2 occur often in GPC analysis of bisphenol-A oligomer bischloroformates. The peaks that elute at 30.9 and 29.6 counts, respectively, were identified as monomer and dimer bisphenol-A bischloroformate. Trimer bisphenol-A bischloroformate elutes at 28.8 counts, but it is not resolved in Figure 2. The broad portion of the elution curve that peaks around 27 counts consists of all oligomers that are three and higher in degree of polymerization.

Bisphenol-A Carbonate Oligomers Prepared in Aqueous Caustic. Bisphenol-A carbonate oligomers that are relatively low in molecular weight and which have a relatively high percentage of hydroxyl end groups can be prepared by passing phosgene into an aqueous solution of the sodium salt of bisphenol-A. The product separates as an insoluble white solid at a low degree of polymerization. The GPC curves of products made in the aqueous system are shown in Figure 3, and the molecular-weight data on the products are given in Table II.

The results in Table II indicate that both the molecular weight and the molecular-weight distribution of these oligomers increase as the base concentration in the media is increased. The GPC curve of the oligomer made with a 3:1 ratio of sodium hydroxide to BPA (Figure 3) shows that the proportion of high-molecular-weight oligomers has increased. The high-molecular-weight end of the distribution curve drops off less rapidly than the curve for the product made with a 2:1 ratio of sodium hydroxide to BPA. Thus, to minimize molecular weight and molecular-

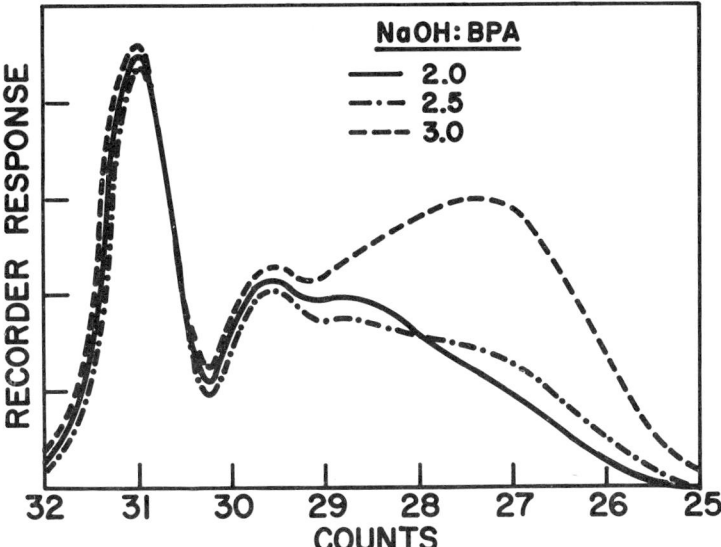

Figure 3. GPC curves for the bisphenol-A oligomer bischloroformates prepared by the aqueous alkaline technique at the mole ratios indicated

Table II. Molecular-Weight Data on Bisphenol-A Oligomer Carbonates Prepared in Aqueous Caustic

Oligomer	Reactant Ratio			M_n	M_w	M_w/M_n	Average Degree of Polymerization
	NaOH	BPA	Phosgene				
IV	2	1	1.5	529	791	1.50	1.70
V	2.5	1	1.5	545	861	1.58	1.75
VI	3.0	1	1.5	656	1117	1.71	2.17
[a]VII	—	—	—	348	385	1.11	1.0
[b]VIII	—	—	—	891	1387	1.56	3.12

[a] Boiling hexane-soluble fraction of a bisphenol-A oligomer prepared in aqueous caustic
[b] Hexane-insoluble fraction of a bisphenol-A oligomer prepared in aqueous caustic

weight distribution of such oligomers, the reaction should be conducted at a low base concentration.

When product made with a 2:1 ratio of BPA to sodium hydroxide was extracted with boiling hexane, a 45% yield of monomer bisphenol-A bischloroformate could be isolated from the hexane-soluble fraction. The GPC curves of the product isolated by the boiling-hexane extraction are shown in Figure 4. The GPC curve indicates that a small amount of dimer and trimer bischloroformate was present in the hexane-soluble fraction. These higher oligomers were not removed by three recrystallizations from boiling hexane. The aqueous preparation of monomer bisphenol-A bischloroformate is a novel and facile way to prepare this compound, although product purity is low and further work needs to be done to develop a suitable purification procedure. Calculations based on end-group analysis indicate the hexane-soluble extract contains 92% monomer bisphenol-A bischloroformate.

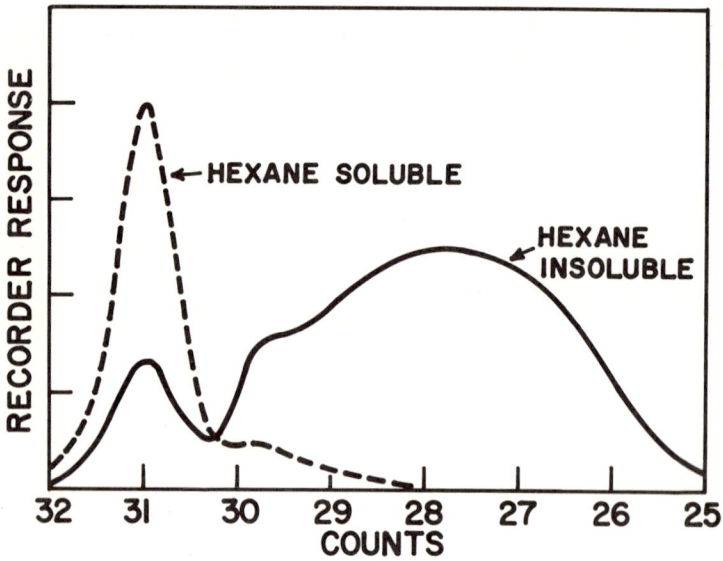

Figure 4. GPC curves of the products separated by the boiling-hexane extraction of bisphenol-A oligomer bischloroformates prepared by the aqueous alkaline technique at mole ratio (NaOH: BPA) of 2

The boiling-hexane-insoluble fraction had an average degree of polymerization of 3.1. Infrared analysis indicates the hexane-insoluble fraction contained about 60% hydroxyl end groups.

Bisphenol-A Carbonate Oligomers Made in Pyridine. A third method for preparing bisphenol-A oligomer bischloroformates is for bisphenol-A to react with phosgene in a pyridine medium. Usually, an inert solvent that will dissolve the bisphenol-A oligomer is used in conjunction with pyridine in such preparations. Some GPC results on products made by the pyridine method are shown in Table III. GPC curves for these products are shown in Figures 5 and 6. The GPC data show the expected effect of stoichiometry—that is, that molecular weight decreases as the amount of excess phosgene in the system increases.

Table III. GPC Data on Bisphenol-A Carbonate Oligomers Prepared in Pyridine

Oligomer	Solvent	Mole % Excess Phosgene	GPC Values		
			M_n	M_w	M_w/M_n
IX	Benzene	11	622	1102	1.77
X	Benzene	50	562	898	1.60
XI	Tetrahydrofuran	12.5	807	1189	1.47
XII	Tetrahydrofuran	25	575	886	1.54

Cosolvent also affects molecular weight; Table III gives results with benzene and tetrahydrofuran. At about the same excess phosgene levels —11% for benzene and 12.5% for tetrahydrofuran—product made in the latter solvent is higher in molecular weight; see Figure 5. In fact, product made with 25% excess phosgene using tetrahydrofuran as cosolvent has about the same molecular weight as product made with 50% excess phosgene using benzene as cosolvent; see Figure 6.

Benzene is a poor solvent for bisphenol-A oligomers, compared with tetrahydrofuran. At a rather low degree of polymerization, oligomers precipitate from benzene, thus suppressing the chain-building reaction. The GPC curves for benzene-soluble and benzene-insoluble fractions are shown in Figure 7. The benzene-insoluble product contains significantly

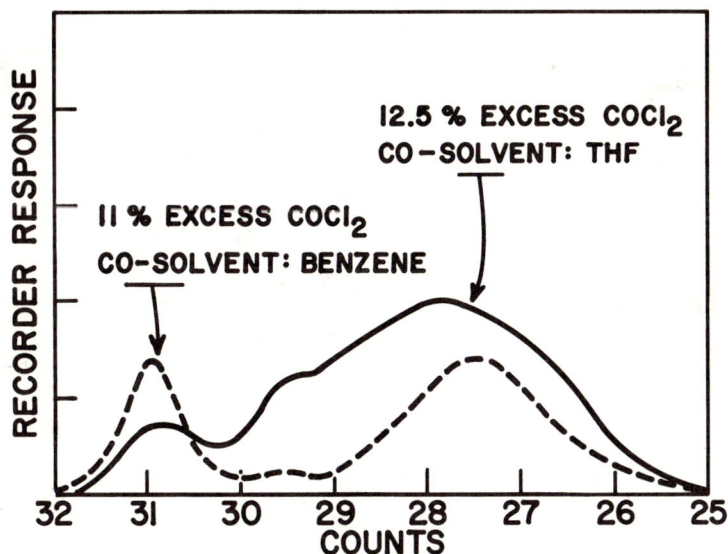

Figure 5. GPC curves of bisphenol-A oligomer bischloroformates prepared in benzene and THF using the same phosgene excess

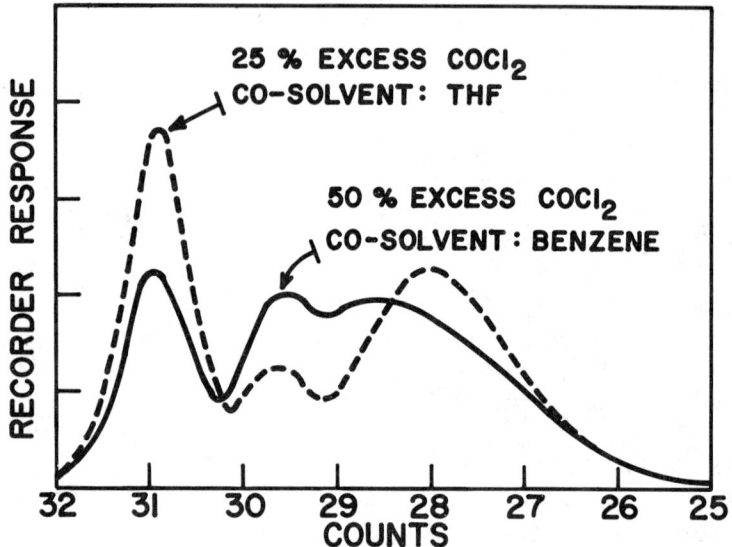

Figure 6. GPC curves of bisphenol-A oligomer bischloroformates prepared in benzene and THF. When the phosgene excess is increased from 11 to 50% with benzene as cosolvent, the molecular-weight distribution is relatively unchanged, compared with Figure 5. However, with THF as cosolvent, a change in phosgene excess from 12.5 to 25% causes a marked change in molecular-weight distribution

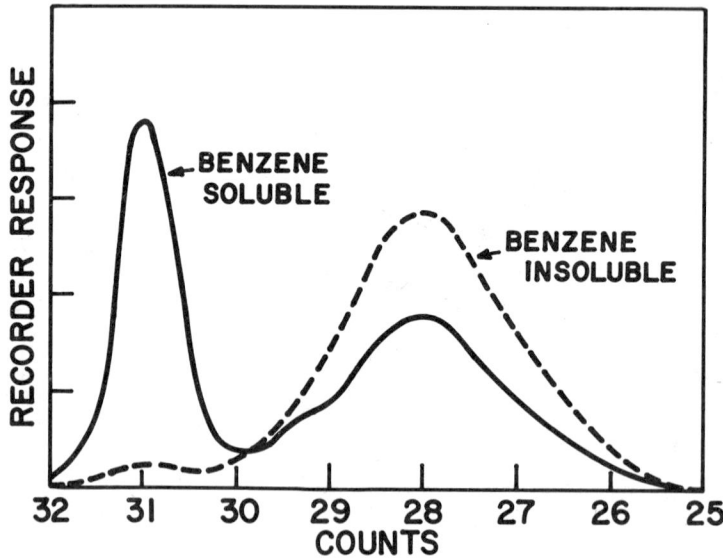

Figure 7. GPC curves for the benzene-soluble and insoluble bisphenol-A oligomer bischloroformates prepared using benzene as cosolvent

less monomer bischloroformate, and has an average degree of polymerization of 3.03. The benzene-soluble product has an average degree of polymerization of 1.56. The higher-molecular-weight oligomers prepared in tetrahydrofuran also have a more narrow molecular-weight distribution. The main difference in the distribution curves is the decreased amount of monomer and dimer bischloroformate in oligomers prepared in tetrahydrofuran.

Narrow-distribution bisphenol-A oligomer bischloroformates can be isolated by fractional precipitation of pyridine-prepared oligomers. Oligomer XII was separated into three fractions by adding hexane to a

Table IV. GPC Results on Fractionated Bisphenol-A Carbonate Oligomers

Oligomer	M_n	M_w	M_w/M_n	Degree of Polymerization
XIII [a]	1548	1776	1.15	5.71
XIV [b]	1010	1213	1.20	3.59
XV [c]	455	539	1.19	1.40

[a] Precipitated on addition of 1.5 volumes of hexane to 1 volume of 4% solution of product in CH_2Cl_2
[b] Precipitated on addition of an additional 1.5 volumes of hexane to the filtrate
[c] Remained soluble in 3:1 mixture of hexane to methylene chloride

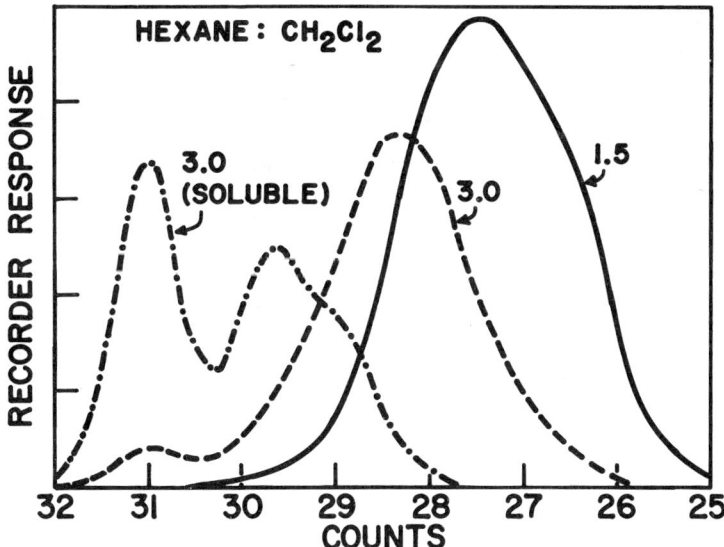

Figure 8. GPC curves of the bisphenol-A oligomer bischloroformates prepared using a 25% excess of phosgene in THF as cosolvent, and fractionated by precipitating from methylene chloride solution by addition of n-hexane. The volume ratios of solvent and nonsolvent are noted

methylene chloride solution of oligomer. GPC results on the three fractions are shown in Table IV, and the GPC curves in Figure 8.

Comparison of GPC and End-Group Molecular-Weight Values. Molecular-weight values determined by end-group analysis and by GPC are compared in Table V.

Table V. Comparison of GPC and End-Group Molecular-Weight Values for Bisphenol-A Carbonate Oligomers

Oligomer	Per cent Chloroformate Chlorine	Per cent Hydroxyl	End-Group Value	GPC Value
XVI	3.92	0.41	1483	1438
XVII	14.6	0.18	473	489
XVIII	3.20	0.48	1690	1418
XIX	12.4	0.21	554	529
XV	14.5	0.30	464	454
XIV	6.86	0.19	980	1010
XIII	3.46	0.34	1700	1549

Molecular weights from end-group analysis were calculated from the total of per cent hydroxyl and chloroformate chlorine. This equation was used:

$$2.09 \text{ (weight \% hydroxyl)} + \text{weight \% chlorine} = 7100\,\overline{M}_n$$

This relationship is also conveniently used to calculate the number of hydroxyl end groups in an oligomer of known chloroformate content and known GPC number-average molecular weight.

GPC and end-group analyses in general agree quite well. A notable exception is oligomer XVIII, which has a 19% higher end-group molecular weight than GPC value. The reason for this discrepancy is not known. In defense of the GPC value, it may be said that oligomer XVI and oligomer XVIII were expected to have the same molecular weight because they were prepared by the same synthesis procedure. GPC values come much closer to satisfying this expectation than do end-group values. End-group values are quite sensitive to error in chloroformate or hydroxyl analysis; for instance, an error of 0.5 weight % in chloroformate or of 0.2 weight % hydroxyl analysis would bring the end-group molecular weight of oligomer XVIII into agreement with the GPC molecular weight. GPC values, by contrast, are relatively insensitive to analyst errors because the analysis only involves passing a weighed sample through a column.

Acknowledgment

Vapor-pressure osmometry values were determined by J. Humes, and the infrared method for percent hydroxyl was developed by G. Reid. The help of these persons is sincerely appreciated.

Literature Cited

1. Bissinger, W. E., Strain, F., Stevens, H. C., Dial, W. R., Chisholm, R. S., U.S. Patent 3,215,668 (1965).
2. Merrill, S. H., *J. Polym. Sci.* (1961) **55**, 343.
3. Moore, J. C., *J. Polym. Sci.* (1964) **58**, A-2, 835.
4. Sotobayashi, V. H., Lie, S. L., Springer, J., Veherreiter, K., *Makromol. Chem.* (1968), **111**, 172.
5. Heitz, W., "Separation of Oligomers by Use of Gel Chromatography," Abstracts, Third Prague Microsymposium, IUPAC, September (1968).
6. Einhorn, A., *Ann.* (1898) **300**, 135.
7. Wittbecker, E. L., U. S. Patent 2,731,445 (1965).
8. Schnell, H., "Chemistry and Physics of Polycarbonates," Interscience, New York, (1964).
9. Cantow, M. J., "Polymer Fractionation," Academic Press, New York (1967).
10. Bonnar, R. U., Dimbat, M., Stross, F. H., "Number-Average Molecular Weight," Interscience, New York (1958).
11. Determan, H., "Gel Chromatography," Springer-Verlag, New York (1968).

RECEIVED April 13, 1972.

18

Kinetics and Mechanism of Urethane Formation in DMF

The Reaction of 4,4'-Diphenylmethane Diisocyanate and Alcohols Catalyzed by Dibutyltin Dilaurate

G. BORKENT and J. J. VAN AARTSEN

Akzo Research Laboratories, Corporate Research Department, Arnhem, The Netherlands

The catalysts of reactions between 4,4'-diphenylmethane diisocyanate (MDI) and alcohols in N,N-dimethylformamide (DMF) by dibutyltin dilaurate has been investigated. The reaction rate of the catalyzed urethane formation in DMF is proportional to the square root of dibutyltin dilaurate concentration. This result differs from that of similar studies on apolar solvents. The catalysis in DMF can be explained very well by a mechanism in which a small amount of the dibutyltin dilaurate dissociates into a catalytic active species.

In continuation of a study of the uncatalyzed reactions between MDI (4,4'-diphenylmethane diisocyanate) and alcohols in DMF (N,N-dimethylformamide) (1), the effect of dibutyltin dilaurate on the same reactions has been studied. The results were compared with those found in studies on the mechanism of catalysis of urethane formation in apolar solvents (2-6).

For the catalysis of isocyanate-alcohol reactions in apolar solvents, several mechanisms have been proposed. However, the results of the kinetic measurements in DMF could not be explained with these mechanisms. So we concluded that, in the polar solvent DMF, the mechanism of the catalyzed urethane formation differs from the published mechanisms in apolar solvents. The behavior in DMF can be explained from a mechanism in which dibutyltin dilaurate dissociates into a catalytic active species.

From measurements at different temperatures, the activation parameters $\Delta S^{\neq}$ and $\Delta H^{\neq}$ for the uncatalyzed and the catalyzed urethane formation were calculated.

Results and Discussion

The reaction rates of alcohols and MDI in DMF in the presence of different amounts of dibutyltin dilaurate were measured by a UV-spectroscopic method following the formation of urethane at 300 nm (Figure 1). In each experiment, a 20- to 100-fold excess of alcohol was used. The reactions are pseudo-first-order as the alcohol and dibutylin dilaurate concentrations are constant in one experiment.

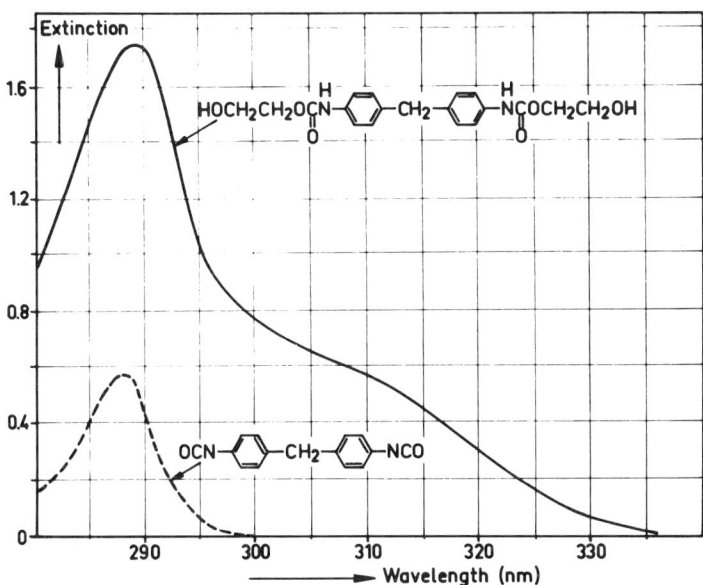

Figure 1. UV spectra of MDI and a urethane from MDI and glycol, both 6.2×10^{-3} eq/liter

Reactions of MDI with alcohols proceed *via* a competitive, consecutive second-order reaction through an intermediate urethano-isocyanate:

$$ROH + OCN-\text{C}_6\text{H}_4-CH_2-\text{C}_6\text{H}_4-NCO \xrightarrow{k'}$$

$$OCN-\text{C}_6\text{H}_4-CH_2-\text{C}_6\text{H}_4-NHCOOR$$

$$ROH + OCN-\text{C}_6\text{H}_4-CH_2-\text{C}_6\text{H}_4-NHCOOR \xrightarrow{k''}$$

$$ROOCNH-\text{C}_6\text{H}_4-CH_2-\text{C}_6\text{H}_4-NHCOOR$$

Earlier work (1) shows that, in DMF, $k'/k'' = 2.0 \pm 0.1$. This means that we obtain simple kinetics because the reactivities of the NCO groups in MDI and the intermediate urethano-isocyanate are equal.

From the pseudo-first-order rate constants k_1, the second-order rate constants k_2 are obtained by dividing k_1 by the alcohol concentration. It was found that the reaction rate constant k_1 is proportional to the alcohol concentration (at the same catalyst concentration). Table I gives the k_2 values for the reaction between methanol and MDI catalyzed by dibutyltin dilaurate at 25.1°C. A plot of the k_2 values vs. the dibutyltin dilaurate concentration (Figure 2) apparently deviates from a straight line, indicating that the mechanism of the catalyzed urethane formation in DMF differs from the mechanisms observed in apolar solvents (2-6). Most workers have assumed that in apolar solvents the mechanism involves formation of a complex between alcohol and dibutyltin dilaurate or the formation of a ternary complex between alcohol, isocyanate, and catalyst. In these cases, the relation between k_2 and catalyst concentration differs from the relation observed in DMF.

The nonlinear relation between k_2 and catalyst concentration can be understood by assuming a mechanism with fast dissociation of the catalyst into a catalytically active species.

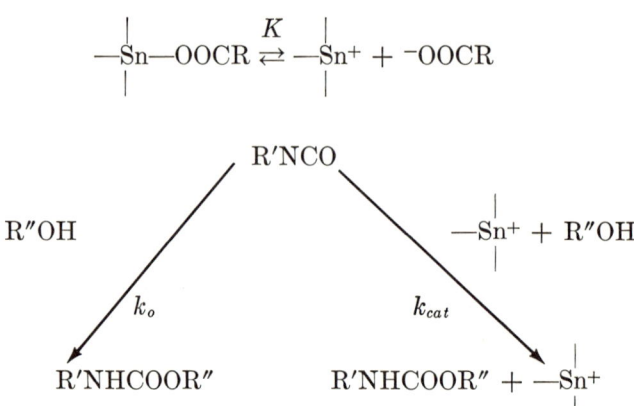

where: k_o = rate of the uncatalyzed reaction; k_{cat} = rate of the catalyzed reaction; and K = dissociation constant of the catalyst

When it is assumed that only a very small fraction of dibutyltindilaurate is dissociated, the rate of formation of urethane (U) is:

$$\frac{dU}{dt} = k_o \times [\text{R'NCO}] \times [\text{R''OH}] + k_{cat} \times [\text{R'NCO}] \times [\text{R''OH}] \times [-\text{Sn}^+]$$
$$= \{k_o + k_{cat} \times K^{0.5} \times [\text{Sn}_o]^{0.5}\} \times [\text{R''OH}] \times (U_\infty - U_t)$$
$$= k_1 \times (U_\infty - U_t)$$

Table I. The Reaction between MDI and Methanol in DMF at 25.1°C; Catalyst: Dibutyltin Dilaurate

[Catalyst] $\times$ 10^6 mole/liter	[Catalyst]$^{0.5}$ $\times$ 10^3	$k_2 \times 10^3$ [a] $l\ mole^{-1}\ sec^{-1}$
0.83	0.91	5.4
1.64	1.28	6.4
3.30	1.82	6.6
6.60	2.57	7.1
9.9	3.14	8.2
16.4	4.05	8.8
20.2	4.50	10.1
38.4	6.20	11.2
62.5	7.90	13.5
81.0	9.0	14.5
110.0	10.5	15.2
130.0	11.4	17.1
161.0	12.7	19.0

[a] The k_2 values are calculated from rate measurements made at least at three alcohol concentrations

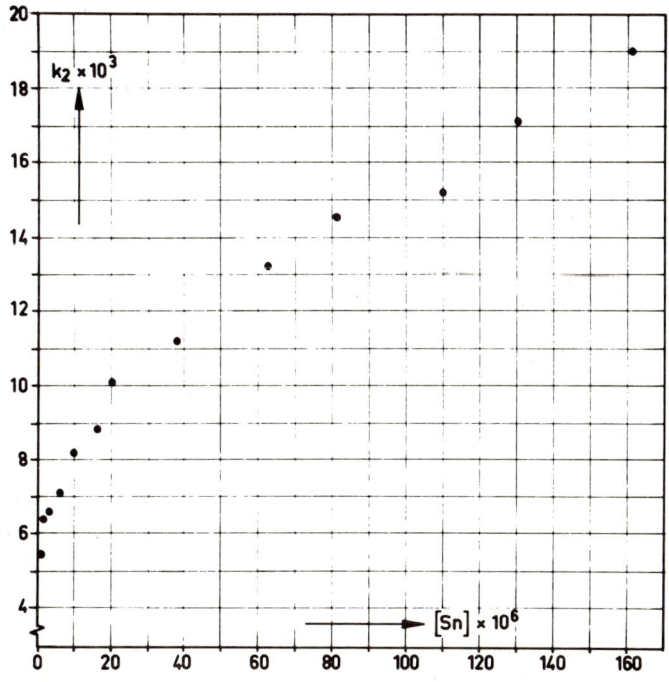

Figure 2. Rate of urethane formation vs. [Sn_o]

Integration and introduction of Lambert-Beer's law give:

$$\ln(A_\infty - A_t) = -k_1 \times t + \text{constant};$$

$[Sn_o]$ = concentration of dibutyltin dilaurate, U_t = concentration of urethane at time t, U_∞ = concentration of urethane at the end of the reaction, A_t = absorbance at time t; and A_∞ = absorbance at the end of the reaction.

In each experiment, a pseudo-first-order reaction will be observed with rate constants k_1. The second-order rate constants k_2 are obtained by dividing k_1 by $[R''OH]$ ($k_2 = k_o + k_{cat} \times K^{0.5} \times [Sn_o]^{0.5}$).

It follows that a plot of the k_2 values against $[Sn_o]^{0.5}$ should give a straight line with slope $k_{cat} \times K^{0.5}$ and intercept k_o. Figure 3 gives the result of the reaction between MDI and methanol.

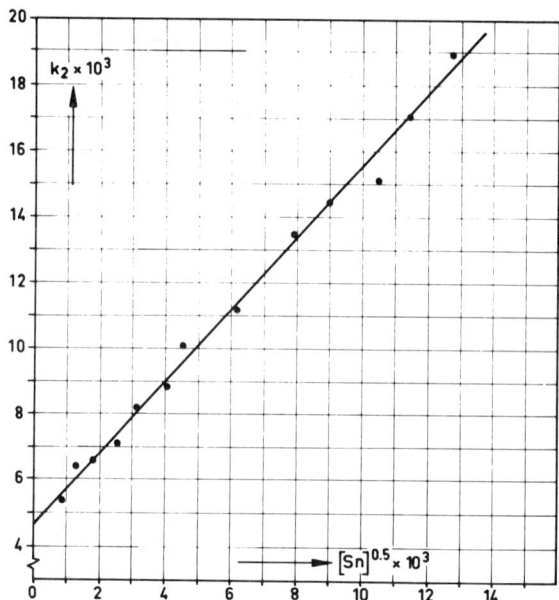

Figure 3. *Rate of urethane formation vs.* $[Sn_o]^{0.5}$

Table II. Reaction of Alcohols with MDI in DMF Catalyzed by Dibutyltin Dilaurate at 25.1°C

Alcohol	$k_o \times 10^3$, l $mole^{-1}$ sec^{-1}	$k_{cat} \times K^{0.5}$, $l^{1.5}$ $mole^{-1.5}$ sec^{-1}
CH_3OH	4.68 ± 0.24	1.08 ± 0.03
C_2H_5OH	2.73 ± 0.31	0.93 ± 0.04
n-C_4H_9OH	2.67 ± 0.29	0.90 ± 0.04
iso-C_4H_9OH	1.93 ± 0.19	0.73 ± 0.03
$CH_3OCH_2CH_2OH$	1.08 ± 0.11	0.53 ± 0.02

The linear relation (with coefficient of correlation $r = 0.99$) agrees with the proposed mechanism. The reaction of a number of primary alcohols with MDI in DMF catalyzed by dibutyltin dilaurate are given in Table II, which also gives the calculated k_o and $k_{cat} \times K^{0.5}$ values.

A comparison of the calculated k_o values (rates of uncatalyzed reactions) with the experimentally determined k_o values from (1) appears in Table III.

A good agreement is found between the k_o values (except for ethanol). This supports the proposed mechanism.

For methanol and 2-methoxyethanol, measurements have been carried out at 25.1° and 60.1°C. Table IV gives the k_o and $k_{cat} \times K^{0.5}$ values together with the calculated values for the entropy and enthalpy of activation of the catalyzed and uncatalyzed reaction. For the uncatalyzed reaction, the $\Delta S^{\neq}$ and $\Delta H^{\neq}$ values are about the same as found in previous measurements (1).

Table III. Comparison Between k_o Value from Table I and Ref. 1

Alcohol	$k_o \times 10^3$ (Table I)	$k_o \times 10^3$ (1)
CH_3OH	4.68 ± 0.24	4.77 ± 0.25
C_2H_5OH	2.73 ± 0.31	3.91 ± 0.17
n-C_4H_9OH	2.67 ± 0.29	3.20 ± 0.09
iso-C_4H_9OH	1.93 ± 0.19	1.95 ± 0.10
$CH_3OCH_2CH_2OH$	1.08 ± 0.11	1.36 ± 0.04

Table IV. Activation Parameters for Urethane Formation in DMF

Alcohol	Temp °C	$k_o \times 10^3$	ΔH_{uncat} kcal/mole	ΔS_{uncat} e.u.	$k_{cat} \times K^{0.5}$	ΔH_{total} kcal/mole	ΔS_{total} e.u.
CH_3OH	25.1	4.68	5.9	−49	1.08	9.8	−25
	60.1	15.0			6.94		
$CH_3OCH_2CH_2OH$	25.1	1.08	5.4	−54	0.54	11.8	−20
	60.1	3.12			4.90		

Calculation of activation parameters of the catalyzed reaction is somewhat complicated, because we have a temperature dependence for k_{cat} and the dissociation constant K of the catalyst. In general, the temperature dependence of a dissociation reaction is given (7) by:

$$K = e^{-\Delta F^o/RT} = e^{-\Delta H^o/RT + \Delta S^o/R}$$

where: ΔF^o = change in standard free energy for the equilibrium reaction; ΔH^o = change in enthalpy; and ΔS^o = change in entropy.

The temperature dependence of k_{cat} is given (8) by:

$$k_{cat} = \frac{RT}{N \cdot h} \times e^{-(\Delta H^{\neq} + \frac{1}{2}\Delta H o)/RT + (\Delta S^{\neq} + \frac{1}{2}\Delta S o)/R}$$

The two formulas show that the temperature dependence of $k_{cat} \times K^{0.5}$ is given by:

$$k_{cat} \times K^{0.5} = \frac{RT}{N \cdot h} \times e^{-(\Delta H^{\neq} + \frac{1}{2}\Delta H o)/RT + (\Delta S^{\neq} + \frac{1}{2}\Delta So)/R}$$

So the calculated activation parameters for the catalyzed reaction (ΔH_{total} and ΔS_{total}) are given by:

$$\Delta H_{total} = \Delta H^{\neq} + \tfrac{1}{2}\Delta H^{o}$$
$$\Delta S_{total} = \Delta S^{\neq} + \tfrac{1}{2}\Delta S^{o}$$

For dissociation reactions in DMF, ΔH^o and ΔS^o values are not known. In general, the ΔH^o values are between -6 and 3 kcal/mole and the ΔS^o values are always negative (9) (from 0 to -50 eu). When ΔS^o is negative, it follows that $\Delta S^{\neq}_{cat}$ is much less negative than $\Delta S^{\neq}_{uncat}$. This is important for the mechanism of the catalytic reaction. Probably the transition state of the catalyzed urethane formation in DMF is much less rigid than the transition state in the uncatalyzed urethane formation.

Experimental

The reactions were done in the thermostated cell compartment of a Shimadzu QV-50 spectrophotometer. The wavelength was fixed at 300 nm. The sample and reference cells contained a solution of alcohol and dibutyltin dilaurate in dry DMF and were thermostated before use in the cell compartment. At zero time, a small quantity of solid MDI was rapidly dissolved in the sample cell and the absorbance recorded as a function of time. The end value of the absorbance (A_{∞}) was determined after eight to 10 half-life periods.

Literature Cited

1. Borkent, G., van Aartsen J. J., *Rec. Trav. Chim. Pays Bas* (1972), **91**, 1079.
2. Entelis, S. G., Nesterov, O. V., *Russ. Chem. Revs.* (1966), **35**, 917.
3. Lipatova, T. E., Bakalo, L. A., Niselsky, Yu. N., Sirotinskaya, A. L., *J. Sci.* (1970), **A4**, 1743.
4. Entelis, S. G., Nesterov, O. V., Tiger, R. P., *J. Cell. Plast.* (1967), **3**, 360.
5. Frisch, K. C., Reegen, S. L., Thir, B., *J. Polym. Sci., Part C*, (1967), **16**, 2191.
6. Reegen, S. L., Frisch, K. C., *J. Polym. Sci. A-1* (1970), **8**, 2883.
7. Leffler, J. E., Grunwald E., "Rates and Equilibria of Organic Reactions," John Wiley, New York, (1963).
8. Frost, A. A., Pearson, R. G., "Kinetics and Mechanism," John Wiley, New York (1961).
9. Bolton, P. D., Hepler, L. G., *Quart. Revs.* (London) (1971), **25**, 521.

RECEIVED April 1, 1972.

INDEX

A

Absorbance	278
Absorption isotherms, vinyl chloride polyethylene	137
Absorption, vinyl chloride	141
Acrylonitrile–styrene copolymerization	117
Activation energy	69
Activation parameters	279
Activation of polymers, mechanical	68
Active chain ends	2
Addition	127
polymerization, ionic ring-opening	177
Aging test, oven	130
Agitated films	51
Agitated reactors	37
Agitators for large reactors	40
bottom-entering	41
Alcohols and MDI in DMF, reaction rates of	275
Allyl glycidyl ether, cured copolymers of propylene oxide and	220
Ammonium persulfate	160
Amorphus, poly(propylene oxide)	208
Analysis, end-group	272
Analysis, NMR	265
Anhydrides, copolymerization of epoxide and cyclic	225
Anionic polymerization	1
influence of solvents in	7
Applications of graft copolymers	151
Aqueous caustic, bisphenol-A carbonate oligomers prepared in	266
Aromatic backbone protons	240
Aryloxy radicals	242
Atactic polymer	21

B

Backbone polymer	142
Benzene	219, 269
Benzeneboronic anhydride	186, 198
Benzoyl peroxide	156
initiator	180
Benzyl methacrylate	2
Bischloroformate oligomers	261
Bisphenol-A carbonate oligomers	258
interfacial synthesis	259
made in aqueous caustic	266
made in pyridine	268
Bisphenol-A prepolymers	258

Blends	126
polyvinyl chloride	130
Block copolymers	255
Bonds, pendant double	109
Bottom-entering agitators	41
Branching	108
and crosslinking in styrene–butadiene polymerizations	102
Bulk polymerization	218
Butadiene polymerizations, branching and crosslinking in styrene–	102
n-Butylacrylate	116
glycol dimethacrylate growth stages in	111
n-Butylmethacrylate	119

C

Cable insulation, graft copolymers for	152
Carbanion, free	1
Carbon–sodium bond, transition between	14
Carbon tetrabromide	121
Carboxyl content, terminal	190
Catalyst	
for epoxide polymerizations, hexacyanometalate salt complexes as	208
for epoxide polymerizations, zinc compounds as	212
lithium bromide	179
mode of action of	226
zinc hexacyanocobaltate-glyme-zinc chloride	218
Catalytic activities	216
Catalyzed urethane formation	274
Caustic, bisphenol-A carbonate oligomers prepared in aqueous	266
Cellulose	78
Chain	115
cleavage	115
ends, active	2
entanglements	117
extenders	188
composition of PET made with	192, 193
initiation	2
length, primary mean	114
scission	69, 111
termination	1
transfer	171
Characterization of copolymers	250
Characterization by GPC	259

Chemical clevage of crosslinks ... 116
Chlorinated isobutenes, infrared
  spectra of 168
Chlorinated polyvinyl chloride,
  polyether modifiers for 125
Chlorination of polyisobutene,
  radiation-induced 161
Chlorination, rate of 164
p-Chlor-styrene 116
Chromatography, gel
  permeation 191, 258
Cleaning, reactor 48
Cleavage of crosslinks, chemical .. 116
Combined polymerization and
  swelling 115
Compensation effects 29
Compensation enthalpy 23
Compensation temperature 23, 32
Complexation, mechanochemical .85, 87
  products 90
Complexes, mechanoexcited 69
Composition of PET made with
  chain extenders 192, 193
Condensing agent 80
Conformational effects 35
Cooling 42
Copolymer(s)
  alloys, impact strength of PVC-
    VC/PE graft 146
  applications of graft 151
  block 255
  for cable insulation, graft 152
  characterization of 250
  DMP-DPP 234
  DMP-MPP 238
  thermal properties of 247
  melt viscosity of PVC-VC/GE
    graft 149
  MPP-DPP 249, 254
  preparation and properties of
    poly(arylene oxide) 230
  properties of graft 142
  of propylene oxide and allyl
    glycidyl ether 220
  PVC graft 145
  in a PVC matrix, dispersion of
    graft 144
  random 251, 255
  MPP-DPP 252
  structure of 1:1 DMP-DPP 238
  VC/EPR graft 153
  VC/PE graft 142
Copolymerization
  of acrylonitrile–styrene 117
  of epoxide and cyclic anhydrides 225
  of tetrachloroethylene and ethyl-
    ene 156
Cost of polyester resin 177
Coupling 243
  redistribution in polymer 240
Crack formation 71
Crosslinked glassy polymers that
  proliferate 115
Crosslinked polytetrahydrofuran .. 130

Crosslinking 106, 108, 153, 180
  agents 123
  by dipolar interaction of nitrile
    groups 119
  range, optimum 112
  in styrene–butadiene polymeriza-
    tions 102
Crosslinks, chemical cleavage of .. 116
Crosslinks, formation of 107
Cured copolymers of propylene
  oxide and allyl glycidyl
  ether 220
Cured polyester, one-step synthesis
  of 176
Cured polyesters, properties of ... 180
Cyclic anhydrides, copolymerization of epoxide and 225

D

DDP copolymers, DMP- 234
Decarboxylation 85
Decay, thermal 121
Degradation 167
  of polyisobutene chain 166
Degree of polymerization 259
  weight-average 105
Design of large polymerization
  reactors 37
Destruction, interspherulite 71
Destruction of polyamides, mech-
  anochemical 76
Diad
  formation, rate constants of ... 23
  fractions 27
  isotactic 27
  syndiotactic 21
Diamines 80
Dibutyltin dilaurate 275
Dicyclohexyl terephthalate 189
Dielectric constant 96
Diethylene glycol 205
Diffusion 7
Diisopropyl xanthogen disulfide .. 107
Dilatometric techniques 171
Dimethacrylate Schiff base 117
Diphenylbenzene phosphite 189
Diphenyl carbonate 186, 202
Diphenyl malonate 186, 200
4,4′-Diphenylmethane diisocya-
  nate 274
Diphenyl oxalate 186, 200
Diphenyl sulfite 189
Diphenyl terephthalate ..186, 191, 199, 200
Dipolar interaction of nitrile
  groups, crosslinking 119
Disaggregation, mechanical 74
Dispersion 147
  of graft copolymer in a PVC
    matrix 144
Dissociation constant 13
Distribution, molecular-weight ... 259
Divinylbenzene 111, 116 123

INDEX

DME, kinetics and mechanism of urethane formation in 274
DMF, reaction rates of alcohols and MDI in 275
DMP-DPP copolymers 234
  structure of 1:1 238
DMP-MPP copolymers 238
  thermal properties of 247
tera-Dodecyl mercaptan 106
DPP copolymers, structure of 1:1 DMP- 238

E

Elastomer 220
Electrical conductivity 92
Electron microphotographs 111
Elementary reactions in radical and anionic polymerizations 1
Emulsion polymerization, vinyl ester 170
Emulsion polymerization of vinyl hexanoate 170
End-group analysis 272
Enthalpy, compensation 23
Epoxide and cyclic anhydrides, copolymerization of 225
Epoxide polymerizations 208, 212
EPR graft copolymers, VC/- ..147, 153
Error slope. 31
ESR signals 119
Ester emulsion polymerization vinyl 170
Ethyl acrylate 119
Ethylene, copolymerization of tetrachloroethylene and 156
Ethylene glycol 63

F

Falling films 51
Filling ratio 81, 87
Film(s) 51
  agitated 51
  apparatus, thin 51
  evaporator, thin 52
  falling 51
  formation 52
  reaction in an agitated thin ... 51
First-order Markov process 22
Flakes, popcorn polymer 121
Formation of crosslinks 107
Formation of gel 107
Formation of a popcorn polymer .. 112
Free carbanion 1
Free radical(s) 102, 119
  chain reaction 243
  polymerization 177
  monomer constitution and stereocontrol in 21
Fumaric acid 181

G

Gel, formation of 107

Gel permeation chromatography 191, 258
Glassy polymer into a popcorn polymer, transformation of .. 111
Glassy polymers that proliferate, crosslinked 115
Glycol dimethacrylate 116, 117, 123
  growth stages in n-butylacrylate– 111
Glyme 216
  complexes, zinc hexacyanometalate 214
GPC, characterization by 259
Graft copolymer(s)
  alloys, impact strength of PVC-VC/PE 146
  for cable insulation 152
  melt viscosity of PVC-VC/PE .. 149
  properties of the 142
  PVC 145
  VC/EPR 147, 153
  VC/PE 142
  vinyl chloride 135
Graft polymerization processes .136, 138
Grafting of vinyl chloride onto polyethylene 136
Growth of globular particles in styrene–divinylbenzene 111
Growth stages in n-butylacrylate-glycol dimethacrylate 111

H

Half-life, 100°C 121
Heat distortion temperature 132
Heat of monomers, polymerization 44
Heterotactic triad 27
Hexacyanometalate salt complexes 208
  preparation and properties of .. 213
tert-Hexadecyl mercaptan 108
Hexaphenyl ortho-terephthalate .. 186
High-impact resins from polytetrahydrofuran 127
High molecular-weight poly(propylene oxide) 217
Homopolymer properties 232
Homopolymerization 231, 249
Homopolymers oxidation 245
Hydrolytic stability 197
Hydroxyl, phenolic 264

I

Impact strength of PVC-VC/PE graft copolymer alloys 146
Incompatible mixtures 131
Induction period 217
Influence of solvents on polymerization 1
  anionic 7
  radical 2
Infrared specta of four chlorinated isobutenes 168

Initiation, chain	2
Initiator, benzoyl peroxide	180
Initiator, 1,5-pentanediol	222
Insulation, graft copolymers for cable	152
Interfacial synthesis of bisphenol-A carbonate oligomers	259
Interspherulite destruction	71
Intrinsic viscosity	183, 219
Ion pairs	1
Ionic ring-opening addition polymerization	177
Isobutenes, infrared spectra of chlorinated	168
Isotactic diads	27
Isotactic triads	28
Isotherms, vinyl chloride/polyethylene absorption	137
Izod impact strengths, notched	125

K

Kinetic behavior of styrene and vinyl acetate	171
Kinetics and mechanism of urethane formation in DME	274
Kinetics, polymerization	170

L

Ligand synthesis	88
Liquid monomer–polymer system	113
Lithium bromide catalyst	179
Low-density polyethylene	144
Luwa high-viscosity machine	56

M

Magnetic properties	97
Maleic anhydride	177
Markov process, first-order	22
MDI in DMF, reaction rates of alcohols and	275
Mechanical activation of polymers	68
Mechanical disaggregation	74
Mechanism of urethane formation in DME, kinetics and	274
Mechanochemical complexation	85, 87
products of	90
Mechanochemical destruction of polyamides	76
Mechanochemical polycondensation	76
polymers produced by	79
and polycomplexations	68
Mechanochemical synthesis	76
Mechanochemically synthesized polychelates	91
Mechanochemistry	76
Mechanoexcited complexes	69
Melt viscosity	193
of PVC-VC/PE graft copolymer	149
Metallic salt	89
Methanol	279

Methyl acrylate	119, 123
Methyl methacrylate	2, 116
polymerization	26
Milling	86
duration	81
temperature	128
time	131
vibratory	77
Mixing	38
Mixtures, incompatible	131
MPP-DPP copolymers	249, 254
random	252
Mode of action of the catalyst	226
Model, termination	4
Modifier	107, 127
polyether	125
Molecular weight	259, 265
distributions	1, 191, 259
partial conversion	103
of poly(propylene ether) diols	223
Monoallyl maleate	123
Monomer constitution	33
and stereocontrol in free radical polymerizations	21
Monomer–polymer system, liquid	113
Monomers, polymerization heat of	44

N

Nitrile groups, crosslinking by dipolar interaction of	119
NMR analysis	265
NMR spectra	239
Notched Izod impact strengths	126

O

Oligomers	
bischloroformate	261
bisphenol-A carbonate	258, 268
interfacial synthesis of	259
prepared in aqueous caustic, bisphenol-A carbonate	266
One-step synthesis of cured polyester	176
Opacity	148
Optimum crosslinking range	112
Orthoesters	204
Osmometry	163
Oven aging test	130
Oxidation, homopolymers	245
Oxidation, simultaneous	249
Oxidative coupling of phenols	240
Oxygen-free storage	120-121

P

Paraffins, perchlorinated	156
Partial conversion molecular weight	103
PE graft copolymer, melt viscosity of PVC-VC/	149
PE graft copolymers, VC–	142
Pendant double bonds	109
1,5-Pentanediol as initiator	222

INDEX

Peptide groups 85
Perchlorinated paraffins 156
Permeation chromatography, gel .. 258
PET made with chain extenders 192, 193
PET polycondensation 184, 188
Phenolic hydroxyl 264
Phenols 199
  oxidative coupling of 240
  reactivity of 231
Phenylenediamine 80
Phosgene 262
Phthalic anhydride 177
Plastic deformation 70
Plasticity, Williams 225
Polyamides, mechanochemical destruction of 76
Poly(arylene oxide) copolymers, preparation and properties of 230
Polychelate(s) 92, 99
  mechanochemically synthesized 91
Polycomplexations, mechanochemical polycondensations and ... 68
Polycondensation 186, 241
  of poly(ethylene terephthalate) 183
  mechanochemical 76
  PET 184, 188
  and polycomplexations, mechanochemical 68
  of polyester 62, 66
  polymers produced by mechanochemical 79
Polyepichlorohydrin, VC 148
Polyether modifiers for polyvinyl chloride and chlorinated polyvinyl chloride 125
Polyethers, rubbery 125
Polyesterification 176
Polyesters 97, 208
  one-step synthesis of cured 176
  polycondensation of 62, 66
  properties of cured 180
  resin, cost of 177
  unsaturated 176
Polyethylene
  absorption isotherms, vinyl chloride/ 137
  grafting of vinyl chloride onto .. 136
  low-density 144
  oxide 133
  stressed 70
  terphthalate poly(ε-caprolactam) cellulose 77
  terephthalate, polycondensation of 183
Polyisobutene chain, degradation of 166
Polyisobutene, radiation-induced chlorination of 161
Polymer(s)
  atactic 21

Polymer(s) (*Continued*)
  backbone 142
  coupling, redistribution of 240
  formation of popcorn 112
  mechanical activation of 68
  popcorn 110
  into a popcorn polymer, transformation of a preformed glassy 111
  produced by mechanochemical polycondensation 79
  that proliferate, crosslinked glassy 115
  solutions, solvent stripping of .. 57
  spherulitic 71
  structure 237
  system, liquid monomer– 113
  thermostable 91
  transformation of a preformed glassy polymer into a popcorn 111
  viscous 51, 61
Polymerization(s)
  addition 177
  anionic 1
  branching and crosslinking in styrene–butadiene 102
  bulk 218
  curves 172
  degree of 259
  elementary reactions in radical and anionic 1
  epoxide 212
  free-radical 177
  graft 136, 138
  heat of monomers 44
  hexacyanometalate salt complexes as catalysts for epoxide 208
  influence of solvents 1
    in anionic 7
    in radical 2
  ionic ring-opening addition 177
  kinetics 170
  methyl methacrylate 26
  monomer constitution and stereocontrol in free radical 21
  of propylene oxide 210
  radical 1
  reactors, design of large 37
  and swelling, combined 115
  temperature influence on stereocontrol 25
  vinyl ester emulsion 170
  vinyl hexanoate 173
  emulsion 170
  weight-average degree of 105
  zinc compounds as catalysts for epoxide 212
Poly(2-methyl-6-phenyl-1,4-phenylene oxide) 234
Polyols, preparation of poly(propylene ether) 221

Polypropylene 161
  ether diols 224, 225
    molecular weights of 223
  ether polyol 208
    preparation of 221
  ether triol 211
  maleate 211
  oxide, high-molecular-weight .. 217
Polystyrene 56
Polytetrahydrofuran 125
  crosslinked 130
  high-impact resins from 127
Polyvinyl alcohol 78
Polyvinyl chloride 162
  blends 130
  polyether modifiers for 125
Popcorn polymers 110, 121
  formation of 111, 112
Preparation
  of hexacyanometalate salt complexes 213
  of poly(arylene oxide) copolymers 230
  of poly(propylene ether) polyols 221
Prepolymers, bisphenol-A 258
Primary mean chain length 114
Products of mechanochemical complexation 90
Propagation step 2
Propagation reaction 164
Properties
  of cured polyesters 180
  of graft copolymers 142
  of hexacyanometalate salt complexes 213
  homopolymer 232
  magnetic 97
  thermal 237
Propylene glycol 176
Propylene oxide cured copolymers 220
Propylene oxide, polymerization of 210
Protons, aromatic backbone 240
PVC
  graft copolymer mixtures 145
  matrix, dispersion of graft copolymer in a 144
  suspension 46
  -VC/PE graft copolymer, melt viscosity of 149
Pyridine 261
  bisphenol-A carbonate oligomers made in 268

Q

Quinone ketal 243

R

Radiation-induced chlorination of polyisobutene 161

Radical(s)
  and anionic polymerizations ... 1
  aryloxy 242
  formation 111
  free 119
  polymerization, influence of solvents on 2
  polymerizations 1
Random copolymer 251, 255
  MPP-DPP 252
Randomization 246
Rate
  of chlorination 164
  constants of diad formation 23
  constants of transition between carbon-sodium bond 14
  of formation of urethane 276
Reaction(s)
  in an agitated thin film 51
  free-radical chain 243
  propagation 164
  in radical and anionic polymerizations, elementary 1
  rates of alchohols and MDI in DMF 275
  termination 165
Reactivity of phenols 231
Reactor(s)
  agitated 37
  agitators for large 40
  cleaning 48
  design of large polymerization . 37
  operation, safety features in ... 40
  shell 46
Redistribution 243
  of polymer coupling 240
  spontaneous 245
Resin, cost of polyester 177
Resin from polytetrahydrofuran, high-impact 127
Ring-opening addition polymerization, ionic 177
Rubbery polyethers 125
Rupture 71

S

Safety features in reactor operation 49
Salt complexes as catalysts for epoxide polymerization, hexacyanometalate 208
  preparation and properties of .. 213
Salt, metallic 89
Schiff base, dimethacrylate 117
Scission 73
Sebacic acid dichloride 78
Seeded runs, unseeded and 170
Segment diffusion 1
Sequential oxidation 252
Shear modulus 148
Simultaneous oxidation 249
Slope, error 31

INDEX

Sodium bond, rate constants of the transitions between carbon– . 14
Solutions, solvent stripping of polymer 57
Solvent(s)
  influence on polymerization ... 1, 2, 7
  removal 61
  stripping 51
  of polymer solutions 57
Spectra of four chlorinated isobutenes, infrared 168
Spectra, NMR 239
Spherulitic polymer 71
Spontaneous redistribution 245
Stability, thermal 193
Stereocontrol of polymerization, temperature influence on the . 25
Storage, oxygen-free 120, 121
Strain-destruction relationship 71
Straining in a microscopic range .. 111
Stressed polyethylene 70
Stressed systems 75
Stripping, solvent 51, 57
Structure of 1:1 DMP-DPP copolymers 238
Structure, polymer 237
Styrene 2, 42, 116, 123, 177
  –butadiene polymerization, branching and crosslinking in 102
  copolymerization 117
  –divinylbenzene, growth of globular particles in 111
Suspension PVC 46
Swelling, combined polymerization and 115
Syndiotactic diads 21
Syndiotactic triads 28
Synthesis
  of bisphenol-A carbonate oligomers 259
  of cured polyester, one-step ... 176
  ligand 88
  mechanochemical 76

T

Telomers 156
  waxlike 159
Temperature 23, 32
  heat distortion 132
  influence on the stereocontrol of polymerization 25
  milling 128
Tensile properties 148
Tensile strength 132, 221
Terminal carboxyl content 190
Termination model 4
Termination reactions 165
Tetrachloroethylene 159
  and ethylene, copolymerization of 156
Tetrahydrofuran 219, 269

Tetraphenyl orthocarbonate 186, 191, 205
Thermal
  decay 121
  properties 237, 251
  of DMP-MPP copolymers ... 247
  resistance 197
  stability 91, 193
Thermostable polymer 91
Thin-film apparatus 51
  evaporator 52
Time-conversion curves 164
Time, milling 131
Transeterification 186
Transformation of a preformed glassy polymer into a popcorn polymer 111
Transparency 147, 148
Triads, heterotactic 27
Triads, isotactic 28
  syndiotactic 28

U

Unsaturated polyesters 176
Unseeded and seeded runs 170
Urethane formation, catalyzed ... 274
Urethane rate of formation 276
Urethano-isocyanate 275
Uses of poly(propylene ether) polyols 225

V

VC/PE graft copolymer, melt viscosity of PVC- 149
VC/EPR graft copolymers . 142, 147, 153
VC/polyepichlorohydrin 148, 154
  co-ethylene oxide 154
Vibratory milling 77
Vinyl acetate 116, 123
Vinyl chloride
  absorption 141
  graft copolymers 135
  /isobutene 162
  /polyethylene absorption isotherms 137
  onto polyethylene, grafting of .. 136
Vinyl ester emulsion polymerization 170
Vinyl hexanoate, emulsion polymerization 170, 173
Viscosity
  instrinsic 183, 219
  machine, Luwa high- 56
  melt 193
  of PVC-VC/PE graft copolymer, melt 149
  of solvent 3
Viscous polymers 51, 61
Void contents 121

W

Waxlike telomers 159

Weight-average degree of polymerization 105

Williams plasticity 225

Z

Zinc compounds as catalysts for
    epoxide polymerizations 212
Zinc hexacyanocobaltate 214
    -glyme-zinc chloride catalyst ... 218
Zinc hexacyanoferrate 214
Zinc hexacyanometalate-glyme
    complexes 214

The text of this book is set in 10 point Caledonia with two points of leading. The chapter numerals are set in 30 point Garamond; the chapter titles are set in 18 point Garamond Bold.

The book is printed offset on Danforth 550 Machine Blue White text, 50-pound. The cover is Joanna Book Binding blue linen.

*Jacket design by Norman Favin.
Editing and production by Mary Westerfeld.*

The book was composed by Modern Linotypers, Inc., Baltimore, Md., printed and bound by The Maple Press Co., York, Pa.

QD
1
A355
#128

AUG 30 1974